T0235766

The Foundations of Spacetime Physics

This book provides an up-to-date overview of the foundations of spacetime physics. It features original essays written by world-class experts in the physics and philosophy of spacetime.

The foundational questions regarding the origin and nature of spacetime are branching into new and exciting directions. These questions are not restricted to the quantum gravity program but also arise in the context of a well-established theory like general relativity. Against the background of these quick and diverse developments, this volume features a broad range of perspectives on spacetime. Part I focuses on the nature of spacetime in non-quantum theories, such as Newtonian mechanics and relativity. Part II explores some intriguing conceptual implications of developing a quantum theory of spacetime.

The Foundations of Spacetime Physics is an essential resource for scholars and advanced students working in philosophy of physics, philosophy of science, and scientific metaphysics.

Antonio Vassallo is Assistant Professor of Philosophy at Warsaw University of Technology, Poland, where he coordinates the philosophy of physics research group. His recent publications include "Does General Relativity Highlight Necessary Connections in Nature?" (*Synthese*, 2021) and "Dependence Relations in General Relativity" (*European Journal for Philosophy of Science*, 2020).

Routledge Studies in the Philosophy of Mathematics and Physics

For more information about this series, please visit: https://www.routledge.com/Routledge-Studies-in-the-Philosophy-of-Mathematics-and-Physics/book-series/PMP

The Foundations of Spacetime Physics

Philosophical Perspectives

Edited by
Antonio Vassallo

Routledge
Taylor & Francis Group

LONDON AND NEW YORK

First published 2023
by Routledge
605 Third Avenue, New York, NY 10158

and by Routledge
4 Park Square, Milton Park, Abingdon, Oxon, OX14 4RN

*Routledge is an imprint of the Taylor & Francis Group, an
informa business*

Library of Congress Cataloging-in-Publication Data

A catalog record for this book has been requested

ISBN: 978-1-032-10720-2 (hbk)

ISBN: 978-1-032-11244-2 (pbk)

ISBN: 978-1-003-21901-9 (ebk)

DOI: 10.4324/9781003219019

Typeset in Sabon
by Apex CoVantage, LLC

The Foundations of Spacetime Physics

Philosophical Perspectives

Edited by
Antonio Vassallo

Routledge
Taylor & Francis Group

LONDON AND NEW YORK

First published 2023
by Routledge
605 Third Avenue, New York, NY 10158

and by Routledge
4 Park Square, Milton Park, Abingdon, Oxon, OX14 4RN

Routledge is an imprint of the Taylor & Francis Group, an informa business

Library of Congress Cataloging-in-Publication Data

A catalog record for this book has been requested

ISBN: 978-1-032-10720-2 (hbk)

ISBN: 978-1-032-11244-2 (pbk)

ISBN: 978-1-003-21901-9 (ebk)

DOI: 10.4324/9781003219019

Typeset in Sabon

by Apex CoVantage, LLC

Contents

List of contributors

Sam Baron (Australian Catholic University)

Tomasz Bigaj (University of Warsaw)

Bryan Cheng (University of Oxford)

Karen Crowther (University of Oslo)

Sebastian De Haro (University of Amsterdam)

Juliusz Doboszewski (University of Bonn)

Tim Koslowski (University of Würzburg)

Vincent Lam (University of Grenoble Alpes and University of Bern)

Baptiste Le Bihan (University of Geneva)

Laurie Letertre (University of Grenoble Alpes and Czech Academy of Science)

Cristian Mariani (University of Grenoble Alpes and University of Barcelona)

Vera Matarese (University of Bern)

Pedro Naranjo (Warsaw University of Technology and University of Warsaw)

Daniele Oriti (Ludwig Maximilian University of Munich)

J. Brian Pitts (University of Lincoln, University of Cambridge, University of South Carolina)

James Read (University of Oxford)

Antonio Vassallo (Warsaw University of Technology)

Introduction

Antonio Vassallo

Now, more than ever, the foundational questions regarding the origin and nature of spacetime occupy a privileged place in the research agenda of many physicists and philosophers. This lively interest has been fueled by the astounding recent progress in gravitational physics, especially under the light of the general relativistic unification of spacetime and gravitational effects. For example, the detection of gravitational waves from a black hole merger (Abbott et al., 2016) forcefully poses the question as to how to interpret this empirical result: Is it a novel, *sui generis* physical phenomenon, or is this an indication that spacetime is a substance akin to other physical fields such as the electromagnetic one? Or, perhaps, what we call "gravitational wave" has no fundamental spatiotemporal connotation and can be reduced to ordinary material facts? Another example is represented by the tremendous advancements in black hole imaging techniques, which have shown with surprisingly high resolution the "shadow" of such mysterious objects (The Event Horizon Telescope Collaboration et al., 2019). Also in this case, a plethora of interpretive questions is prompted: Are black holes some sort of "physical" singularities in the fabric of spacetime? Do they signal the existence of a more fundamental layer of reality, where gravitational phenomena assume markedly quantum connotations? These examples clearly indicate that the field of research on the foundations of spacetime physics is currently flourishing and constantly branching into new and exciting directions. Against the background of these rapid developments in the field, the present book seeks to provide a window into the state-of-the-art in the research by collecting a series of essays on topical issues in the physics and philosophy of spacetime.

The contents of the book

The book is divided into two main parts. The first one focuses on the philosophical discussion about the nature of spacetime in non-quantum spacetime theories. One of the main themes explored is whether spacetime is a substance in the sense of having independent

DOI: 10.4324/9781003219019-1

existence, or it is metaphysically dependent—in one way or another—on material structures. This distinction is at the root of the centuries-old *substantivalism vs. relationalism* debate, which can be traced back at least to Descartes, Newton, and Leibniz.

Newton famously maintained—*contra* Descartes—that the motions of material bodies cannot be consistently described just in terms of their mutual displacements. In particular, Newton's Bucket thought experiment showed that relative motions are neither necessary nor sufficient to account for inertial effects such as—case in point—the concave shape that the water surface takes inside a rotating vessel. For Newton, there exists an objective, absolute state of motion. This kind of motion is referred to distinguished underlying spatial and temporal structures, which can be used as an absolute standard because they are immutable— i.e., they are not affected by any physical happening. Leibniz was suspicious of these Newtonian structures because, if "true" material motions are referred to such a background, then it is possible to have ontologically different yet observationally indistinguishable states of affairs, and this would be in conflict with two staples of Leibnizian philosophy: The Principle of Sufficient Reason (PSR) and the Principle of the Identity of Indiscernibles (PII). This is the gist of his arguments from static and kinematic shifts against Newtonian space and time.

In their contribution, **Bryan Cheng and James Read** consider the static shift argument in the context of Newtonian gravity, further diving it into a *spatial* and a *temporal* component. The core of the argument is well-known: Consider a counterfactual situation where the entire material content of the universe is rigidly shifted spatially, say, three meters to the left with respect to its actual position—or, in the temporal case, say, three hours into the past. The actual and the counterfactual situations differ regarding the locations occupied in the Newtonian background structure. At the same time, however, both situations would look and feel exactly the same to the observers inhabiting these universes, so that they have no means to tell which of the two cases they are presented with. Leaving aside any consideration on PSR and PII, such a situation looks awkward in the eyes of the modern philosopher of physics since it seems to point at the fact that there can be features of the world that are "transparent" to physics.

The most straightforward way to escape the conclusion of Leibniz' shift arguments is, of course, to give up any commitment to Newtonian spatial and temporal substances and embrace a relationalist ontology that backs up the observable states of affairs only. This is not to say that the substantivalist has no defense. Tim Maudlin, for example, articulated a response to the static shift argument that rests on epistemic grounds. In a nutshell, Maudlin defuses the charge of underdetermination brought about by the indistinguishability of the two situations— the actual and the counterfactual—by simply pointing out that we *know* that we inhabit the actual world. So, for example, I know that the laptop I am typing these words with is located *here* and not three meters to the left of here. Cheng and Read's chapter offers a detailed analysis of Maudlin's epistemological argument and Shamik Dasgupta's criticism against it. In doing so, the authors take advantage of some conceptual tools developed in the context of the philosophy of language (e.g., David Kaplan's theory of indexicals). The discussion focuses not only on the Newtonian case but also on how the debate plays out in electromagnetism and general relativity. In this latter theory, the substantivalist is not committed anymore to the existence of an immutable spatiotemporal background. Indeed, general relativity is the epitome of a *background independent* theory, which means that spatiotemporal structures are "dynamical actors" on a par with material fields. Clearly, no static shift argument can be proposed in this context since general relativistic spacetime is not merely a "container" for matter. However, substantivalists must still face a thorny issue involving underdetermination. This issue is exploited in the so-called *hole argument*.

Einstein himself first articulated the hole argument in his struggle to grasp the notion of general covariance. However, the argument's modern (and more philosophically cogent) formulation is due to John Earman, John Norton, and John Stachel. The argument can be summarized as follows. Take two diffeomorphism-related solutions (or, *models*) of Einstein's field equations: On the surface, these two models represent different "placements" of the metrical and material fields over the base manifold that represents the "scaffolding" of spacetime. However, all the observable magnitudes pertaining to each model are diffeomorphically invariant, which means that the two models are empirically equivalent. This empirical equivalence represents the same sort of underdetermination already encountered in the static shift argument. But there is more. Imagine that the two models can be foliated employing Cauchy 3-surfaces and assume, for the sake of argument, that the two foliations coincide up to a particular surface Σ_0, after which a certain diffeomorphism acts on one of the two. If we insist that the two models *differ* on some factual respect after Σ_0, then we are forced to accept that there is no way in which the dynamical development of a model before Σ_0 uniquely fixes which of the two states of affairs will come afterward. This uncertainty represents a failure of determinism. Initially, it was thought that this argument impacted spacetime substantivalism *tout-court*. However, the modern consensus is that the hole argument goes against a very particular—and naive—type of substantivalism, i. e., the one that identifies spacetime with the base manifold. There are, in fact, other types of substantivalism that can escape the clutch of the hole argument.

In his chapter, **Tomasz Bigaj** provides an in-depth discussion of this latter class of substantivalist positions, usually referred to as *sophisticated*.

Roughly, the leitmotif of sophisticated substantivalism denies that two diffeomorphically related models represent distinct situations. So, for example, Maudlin's response, in this case, is that two models represent the very same situation if and only if the fields redistribution represented by the diffeomorphic transformation leaves the starting metrical properties of each spacetime point untouched, otherwise such a transformation does not generate a metaphysically possible scenario. If this "modal amputation" sounds too radical, the response can be made more moderate by arguing that it is sufficient that, for any point *P* in the starting model, the diffeomorphism assigns *P*'s metrical properties to some other point *Q* in the second model. In this way, it becomes trivially true that the two models represent the same state of affairs. Bigaj aims to defend sophisticated substantivalism in the face of several counter-arguments. To do so, he first provides a self-contained technical characterization of diffeomorphisms, arguing that the usual picture of these transformations in terms of the "dragging" of the fields along the manifold is misleading. He also uses this technical machinery to question the validity of the claim made in the literature that the hole argument can be mounted using any arbitrary permutation of manifold points. He then proceeds to flesh out the heavy modal undertones borne by sophisticated substantivalism, discussing, in particular, the notion of modality *de re*. This extended conceptual setup makes it possible to frame both the objections to sophisticated substantivalism and Bigaj's responses in a clear and organic manner.

The challenge posed by the hole argument is just the tip of the iceberg of the debate about the ontology of spacetime in general relativity. Another controversial aspect of the discussion concerns whether spacetime carries physically genuine energy-momentum. *Prima facie*, this seems to be the case if we think that general relativistic spacetime "influences" in a clear physical sense the motion of material bodies. However, such an impression comes under fire when realizing that gravitational energy-momentum in general relativity is represented by a non-generally covariant, pseudo-tensorial object. This lack of a generally covariant expression means that different reference frames would disagree on the gravitational energy-momentum content of the very same spacetime region. Clarifying this aspect is of utmost importance for the substantivalist because, if it turns out that spacetime does indeed carry energy-momentum, that will constitute robust evidence that spacetime is a physical substance, not just some sort of vocabulary shortcut to refer to relations among material bodies.

J. Brian Pitts devotes his chapter to this particular topic, making an unorthodox but ingenious proposal to make sense of gravitational energy-momentum in general relativity. Pitts stresses how general relativity can be approached from two different physical and conceptual perspectives. The first is the textbook understanding of the theory in terms of a series of tensor fields spread across the manifold. This attitude naturally suggests that spacetime—as described by the manifold plus the metrical field—is inherently different from other fields (such as the electromagnetic one) since it is quite literally the base on which the other fields live. Pitts dubs this attitude *GR exceptionalism* and notes how the skeptic responses to the reality of gravitational energy-momentum are rooted in this exceptionalist tradition. The second attitude, championed by Richard Feynman, is instead to regard spacetime as yet another field and apply the perturbative methods of particle physics to it. This approach, called *particle physics egalitarianism*, treats spacetime as a spin-2 field. Pitts argues that tackling the problem of gravitational energy from this egalitarian viewpoint helps clarify the issue substantially.

A remarkable shift in methodological perspective brought about by adopting the particle physics-inspired approach is the realization that the formalism of general relativity is not bound to using standard tensorial objects. In fact, generalized geometric objects can be used, which obey transformation rules that go beyond the tensorial one. Given this setup, Pitts' proposal is conceptually simple. First, he decomposes spacetime into a metrical part consisting of the manifold M plus a matrix $\eta = (-1, 1, 1, 1)$, and a gravitational one represented by a potential $\gamma_{\mu\nu}$, which is an "hybrid" between a tensor and a connection. Then, he shows that the part of this expression that actually carries energy-momentum (also described by a pseudo-tensorial object) is the gravitational one $\gamma_{\mu\nu}$. Pitts argues that this split is not arbitrary and does no violence to the background independent spirit of general relativity. Moreover, he considers and contrasts several arguments against taking pseudo-tensorial objects physically seriously. It is important to note that Pitts' proposal to characterize gravitational energy in general relativity does not necessarily support substantivalism, given that the "spacetime" part $\langle M, \eta \rangle$ does not carry energy-momentum. Hence, Pitts' framework is ontologically neutral because both substantivalists and relationalists can exploit it to make sense of gravitational energy in general relativity.

After such an extensive discussion of the substantivalist doctrine, the need for an equally detailed presentation of relationalism becomes pressing. **Antonio Vassallo, Pedro Naranjo, & Tim Koslowski** take up precisely this task. The authors' goal is to present the latest developments in the quest for constructing a fully relational theoretical framework for physics and, at the same time, outline the conceptual evolution of modern relationalism. The authors acknowledge the complaint commonly framed in the literature that relationalism is a somewhat elusive doctrine, whereas substantivalism has very clear conceptual connotations. To this complaint, they reply conceding that relationalism may *have been* evasive since, historically, it developed much more slowly than substantivalism (which was mostly set out already in Newton's

Scholium to his *Principia*). At the same time, however, they submit that *modern* relationalism is a very well-developed stance, which can now be considered more of a nuanced thesis rather than an elusive one. They focus on the relational program championed by Julian Barbour and Bruno Bertotti to drive this point home.

The Barbour-Bertotti program seeks to develop a fully relational framework for physics that is faithful to the core insights—philosophical *and* technical—put forward by Leibniz, Mach, and Poincaré. The central tenet of this version of relationalism is that the physical description of a closed system—in particular, its dynamical evolution—should be given only in terms of the degrees of freedom intrinsic to the system itself, without reference to anything external: This is the essence of what Barbour and Bertotti called the "Mach-Poincaré principle." Such a principle is implemented by requiring that (i) the only physically meaningful spatial information of a system is encoded in its dimensionless and scale-invariant configuration, and (ii) the "temporal" development of the system should be accounted for in terms of changes in its relational configuration only.

At this point, one may ask what makes this relational program *modern*, as opposed to the ancient doctrine originally put forward by, e.g., Descartes and Leibniz. For Vassallo, Naranjo, and Koslowski, the answer lies in the heavy empiricist undertones that the motivations for going relational have acquired. Indeed, the starting point of the whole Barbour-Bertotti program was the realization that any physical measurement in physics invariably amounts to the comparison of the magnitude to be measured with a reference taken as measurement unit. In the case of spatial length, for example, to say that a rope is two meters long is to say that it is twice the length of a rod that, by stipulation, is taken to be one meter. Hence, what matters to physics is the *dimensionless ratios* of magnitudes, rather than their "absolute" values—which are empirically inaccessible since no comparison is involved. It is precisely this comparativist attitude towards physical magnitudes—in particular length, size, and duration—that motivates the relational aversion to structures whose physical description is given in absolute terms. Note how this discussion is not framed in terms of PII or PSR (even though it is compatible with these "old" principles).

With this conceptual characterization in place, Vassallo, Naranjo, and Koslowski provide a brief discussion of the technical developments that led from Barbour and Bertotti's original best-matching theory, through shape dynamics, up to the latest refinement of the framework, called *pure shape dynamics*. The final part of the chapter is devoted to exploring the possible metaphysical readings of pure shape dynamics, highlighting the surprisingly wide range of interpretations that the framework supports—thus giving a concrete picture of the conceptual nuances involved in modern relational physics.

The first part of the book ends with a chapter written by **Juliusz Doboszewski**, which presents an inquiry into the philosophical issues involved in a particular class of models of general relativity, i.e., Kerr spacetimes. These spacetimes represent vacuum solutions to Einstein's field equations that exhibit a peculiar feature, i.e., the presence of a ring-shaped curvature singularity. Kerr solutions are taken to represent the spacetime geometry outside an uncharged rotating black hole.

Doboszewski starts by considering the current empirical evidence for Kerr black holes and concludes that, although there is still no decisive evidence for their existence, there are significant indications that these objects may inhabit the center of galaxies like ours. Having established that Kerr spacetimes represent physically realistic models of general relativity, the author considers some philosophically interesting features of these models. The most intriguing among such features is the presence of closed timelike curves that can be exploited to construct a "machine" for time travel. Doboszewski sketches the main conditions under which time machines are physically possible and finds that several issues bar Kerr spacetimes from straightforwardly meeting such requirements. The most delicate issue at stake is whether Cauchy horizons can occur in the interior of Kerr black holes: If it turns out that this is not the case, then the possibilities to have a time machine in this setup will decrease dramatically. Doboszewski notes that the question about the occurrence of Cauchy horizons is linked to the so-called *Strong Cosmic Censorship* hypothesis, which, roughly, states that stable Cauchy horizons cannot occur inside physically realistic black holes. The fate of Kerr time machines is then bound to the validity (or lack thereof) of such a conjecture. With this question in mind, Doboszewski reviews some recent results in the literature, which seek to place some constraints on the occurrence of Cauchy horizons in spacetimes similar to the Kerr one. His interim conclusion is that, although we do not have enough conceptual reasons to exclude the possibility of time travel in Kerr spacetimes as yet, an increasing number of physical results within general relativity and its extensions seem to point in that negative direction consistently.

The second part of the book focuses on the new frontiers in spacetime physics and the novel challenges (conceptual and technical) that these developments bring about. All the chapters in this part revolve around the quantum gravity program, which seeks to construct a theory that merges general relativity with quantum theory. Although the problem of quantum gravity is anything but novel in itself, it has originated a field of research that is in constant development, and which stimulates and deepens the debate about the nature of spacetime. Indeed, the main suggestion that the quantum gravity program makes—independently of the particular theoretical approach considered—is that we should radically revise our classical notion of spacetime. Some approaches go as far as to suggest that the physical world is fundamentally non-spatiotemporal.

Such a puzzling picture prompts the question as to how we should conceive classical spacetime as non-fundamental, or "emergent" from a non-spatiotemporal fundamental ontology. Much of the debate about the emergence of classical spacetime in quantum gravity revolves around the best characterization of what spatiotemporal features should be recovered from this "emergence" process. **Sam Baron and Baptiste Le Bihan** argue that this is not the right way to address the question if we want to reach some consensus.

Baron and Le Bihan individuate two approaches to understanding spacetime: a *narrow* one, which adopts the notion of spacetime specific to a given theory, and a *broad* one, which conceives of spacetime in theory-independent terms. These different approaches generate a disagreement about which of the two is the "correct" one. But even when settling on a specific approach, the question about the nature of spacetime remains open. For example, even if we adopt the notion of spacetime proper to general relativity, we still have to face the substantivalism/relationalism debate mentioned earlier. The authors argue that there is little to no hope of progressing in the discussion as long as we insist on these attitudes. Their way out of the impasse is embracing a completely different and novel attitude, which they dub *spacetime quietism*: Just be silent about the existence and nature of spacetime.

Baron and Le Bihan argue that spacetime quietism is the best option if we want to progress in addressing the problem of spacetime emergence in quantum gravity. The motivation for this choice is simple: Establishing the existence and nature of spacetime is not needed in addressing the emergence problem. To illustrate this point, they consider the problem of empirical incoherence in quantum gravity. Roughly, the problem is the following: Given that the empirical evidence that corroborates a theory is cast in terms of "local beables," i.e., material objects localized in space and time (e.g., measurement apparatuses), if such a theory denies the existence of local beables it undermines its own empirical evidence and, thus, our reasons for believing it to be true. Clearly, any approach to quantum gravity that cannot address this problem is not a viable candidate to deliver the "right" theory. The question that Baron and Le Bihan now ask is: Is it necessary to say anything about the existence and nature of spacetime to decide whether an approach to quantum gravity is empirically incoherent? They answer this question in the negative because, they argue, to assess whether a quantum gravity theory is empirically incoherent, we do not need to ascertain if and how the theory addresses the emergence of spacetime. Instead, we should check if the theory allows for the recovery in the appropriate limit of notions such as "observable" and "observer." More precisely, they argue that these latter notions should be understood in functionalist terms. Hence, quietism about spacetime shifts the accent from questions about spacetime to questions about the realization of the functional roles that can back up a robust notion of empirical corroboration.

In her chapter, **Vera Matarese** delineates two different attitudes to investigating the nature of space that are interestingly resemblant to the broad and narrow approaches introduced earlier. On the one hand, according to Matarese, metaphysicians consider different conceptually possible characterizations of space and assess their metaphysical coherence. On the other hand, philosophers of physics tend to tie the picture of space to a particular physical theory. However, differently from Baron and Le Bihan, Matarese's goal is not to advocate a dismissal of these attitudes altogether. Instead, she seeks to bridge the two camps by offering a model of discrete space that is derived from physics and, at the same time, effectively addresses some concerns about the metaphysical tenability of discrete space. In particular, the author considers Hermann Weyl's famous argument against the view that space is made up of mereological atoms. Weyl considers a simple model of discrete space consisting of a square regularly divided into a finite amount of smaller squares, or "tiles." He notices that there are as many tiles along the diagonal as there are along the sides of the square. Weyl then restricts the attention to half of the square, which is an isosceles rectangular triangle. He points out that if we count the length of its sides as the number of tiles that constitute them, then the hypotenuse—which coincides with the diagonal of the starting square—has the same length of the other two sides, which violates Pythagoras' theorem. Given that Weyl's argument is valid for any arbitrarily fine-grained tiling of the square, the conclusion is that discrete space so conceived cannot approximate Euclidean space. But if we anchor any ontic commitment to discrete space to the requirement that it approximates Euclidean space, then Weyl's argument forces us to abandon any such commitments.

Matarese counters Weyl's challenge by proposing a model of discrete space based on loop quantum gravity. This latter approach to quantum gravity suggests that, at the fundamental level, space is composed of finite "chunks" with different shapes, lengths, and areas, held together by an "adjacency" relation. Matarese takes this picture at face value, arguing that space is the mereological fusion of these many substances. This model defuses Weyl's challenge in that it does not assign the same length to all atoms so that it can recover a smooth, Euclidean space with an appropriate limiting procedure. Matarese is careful in pointing out that loop quantum gravity, being a *quantum* theory, is not concerned with a single configuration of chunks but with a superposition of them (called *spin-network* states). Hence, her model does not automatically follow from this theoretical framework. In order to sidestep this problem, she argues for a realist interpretation of spin networks reminiscent of the primitive ontology approach to quantum physics, which avoids the indeterminacy suggested by quantum superpositions by postulating the existence of structures that remain stable and determinate throughout the dynamical development.

The lack of ontologically definite structures implied by quantum superpositions is just a facet of the conundrum that physicists and philosophers have to face when tackling the question of what the world is like if quantum theory is true. This conceptual tangle becomes even more complicated when the quantum aspects of spacetime are also considered. **Vincent Lam, Laurie Letertre, and Cristian Mariani** embark on a mission to investigate the interplay between the conceptual issues arising in quantum gravity and quantum foundations. The authors start by discussing entanglement and non-locality. They point out that these two features of quantum physics are "interpretation neutral" in the sense that there exists some consensus in the literature about the fact that the world—in one way or another—must exhibit these features if quantum theory is true.

Quantum entanglement and the related non-local correlations are highly counter-intuitive features since they imply non-separability, i.e., the fact that the whole entangled system cannot be univocally decomposed into parts (subsystems). The authors point out that these features can be accounted for either in structuralist or holistic terms. On the first reading, quantum systems are a web of entanglement relations among subsystems. The key aspect of this picture is that the entanglement relation connects the subsystems so that these latter have no independent existence. A similar moral is shared by the holistic approach, which maintains that the whole system is ontologically prior to its entangled subsystems—thus subverting the usual mereological picture where the parts of a composite system are prior to the whole. The critical point highlighted by Lam, Letertre, and Mariani is that, on both structuralist and holistic conceptions, the characterization of entangled systems is not tied to spatiotemporal notions—e.g., entanglement relations are independent of spatial distances.

The authors go on to discuss another peculiar notion in quantum foundations, i.e., causal non-separability. To do so, they briefly consider the so-called "process matrix formalism." The core concept of this formalism is that of a process matrix, which is a formal tool to describe how the inputs and outputs of a certain number of local quantum operations are combined to form a global quantum process. By construction, a process matrix gives information about the causal order among local quantum operations. A counter-intuitive feature of this framework is that it allows for quantum processes that are incompatible with any fixed causal structure among the local operations that constitute them. This feature is somewhat the "causal" analog to non-separability and, hence, is dubbed *causal non-separability*. This type of non-separability signals a tension with classical spacetime, which possesses a definite and objective causal structure. The authors suggest interpreting causal non-separability realistically as a concrete case of ontic indefiniteness: There is no fact of the matter about a quantum process having a fixed causal structure. The central thesis of Lam, Letertre, and Mariani's chapter is that these metaphysical tools can help address the questions about the nature of spacetime in quantum gravity. Indeed, they argue, given that structuralism, holism, and ontic indefiniteness as discussed earlier are independent of any preexisting spatiotemporal notion, they can be exploited for giving a metaphysical characterization of a non-spatiotemporal fundamental ontology.

So far, much has been said about the possible *consequences* of a quantum treatment of spacetime, but the discussion has not touched upon the possible *motivations* for pursuing this strategy. **Karen Crowther & Sebastian De Haro** consider an obvious motivation for the quantum gravity program, i.e., the desire to dispose of the singularities that plague both general relativity and quantum field theory. The authors' goal is to clarify whether these singularities really need to be resolved, how compelling this motivation is for seeking a quantum gravity theory, and how this strategy can shape the structure of such a theory.

The chapter starts with an overview of the most common types of singularities occurring in general relativity and quantum field theory. The presence of singularities in general relativity signals a pathological behavior of spacetime structures. Thus, *geodesic incompleteness* implies that there can be worldlines of freely falling objects that end abruptly—i.e., within a finite amount of proper time—meaning that these objects can appear out of nowhere or disappear without an underlying physical explanation. Instead, we talk of a *curvature singularity* when a region of spacetime exhibits divergent curvature values, which implies that the tidal forces experienced by an object approaching this region will grow without an upper bound.

As regards quantum field theory, it is very well-known that the perturbative calculations on which the framework is based involve summation over an infinite number of intermediate energy states and, thus, lead to divergent integrals (*ultraviolet divergencies*). These singularities can be avoided by introducing a finite energy scale cutoff through renormalization techniques. This, however, raises the problem about the nature of the cutoff itself: Should we take it physically seriously (i.e., as signaling an actual energy "threshold") or should we consider it just as a convenient mathematical approximation (which would render quantum field theory an effective theory, rather than a fundamental one)? But, even if renormalization is taken into account, there can be processes in which the strength of the interaction diverges at very high energy scales. This is the case for the so-called *Landau pole*, which may point at the fact that the framework of quantum field theory becomes inconsistent at very high energies—which would lend support to the thesis that this framework is an effective one.

Crowther and De Haro analyze several arguments that assess the severity of all these kinds of singularities. The analysis distills four different attitudes present in the literature towards singularities. The first is to recognize that the singularities represent a conceptual problem but not one that needs a theory of quantum gravity in order to be solved. Accordingly, this attitude seeks to resolve the singularities already at the level of the theories they figure in. The second attitude is the opposite: Singularities represent pathological aspects of general relativity and quantum field theory, which need a more advanced theoretical framework to be resolved. The third attitude is to take singularities as "genuine" features of the theories they figure in, thus attributing them a predictive and explanatory value in these theories' context. The fourth attitude suggests that all these singularities can be ignored because both general relativity and quantum field theory are not "final" theories, i.e., they will be replaced by more accurate theories in the future. In the end, Crowther and De Haro's discussion clarifies that, although the existence of singularities is a powerful motivation for pursuing the quantum gravity program, still such a motivation can be resisted on solid physical and philosophical grounds.

The final chapter of the book, written by **Daniele Oriti**, provides a technical characterization of a possible physical scenario where spatiotemporal structures emerge from an underlying non-spatiotemporal regime. The author sets the stage with a thorough discussion of the notion of emergence, clarifying that he is interested in phrasing this notion in physical terms. He recognizes two main types of emergence: *Inter-theoretic* and *ontological*. The first type of emergence relates models pertaining to different frameworks—e.g., a Newtonian model emerging from a general relativistic one in the appropriate limit or approximation. Although this relation is mathematical, it constitutes the backbone for inferring an emergence relation between the natural phenomena described by those models. Ontological emergence is instead a relation holding between physical entities themselves and, as such, it has an inherent metaphysical flavor—as opposed to a merely formal, mathematical one. Hence, inter-theoretic emergence can be considered the basis for inferences that lead to ontological emergence. Oriti points out that inter-theoretic emergence alone is not enough to ground claims about ontological emergence: A pivotal role in this regard is played by the particular ontic commitments associated with models. This remark is particularly significant in the case of a non-spatiotemporal model of a quantum gravity theory from which a general relativistic model can be recovered. Indeed, whether this type of inter-theoretic emergence can be taken to point towards an ontological type of emergence depends on whether one is willing to commit oneself to the existence of non-spatiotemporal entities.

With this clarification in place, Oriti proceeds to sketch a layered structure representing spacetime emergence in quantum gravity. Although such a structure is general enough to be compatible with many different approaches to quantum gravity, Oriti anchors the discussion to the case of topological quantum field theories. These models feature quantized tetrahedra whose discrete geometry is encoded in algebraic data: Each of the triangles on the boundary of each tetrahedron is associated with a particular element of a Lie group. Similar to the standard quantum field case, where quanta of the field can be created (and annihilated), an arbitrary number of tetrahedra can be created by "gluing" together existing ones. In a certain sense, these tetrahedra represent the fundamental "atoms" according to this class of theories.

However, the framework carries little to no geometric connotations at this fundamental level: A generic quantum state corresponds to arbitrary numbers of disconnected quantum tetrahedra, which do not represent classical geometric shapes. This, for Oriti, represents the fundamental non-spatiotemporal discrete ontology of the world. The starting layer of emergence is reached in this context when considering the collective dynamical behavior of the atoms. In the appropriate hydrodynamics approximation, this collective description eventually leads to a condensate phase where some weak geometric properties start to appear. It is then possible to recast the condensate hydrodynamic equations into equations for geometric observables, thus getting a model that can be considered a quantized version of general relativity. The further step is to adopt the appropriate limit in which the quantum fluctuations of the geometric observables are effectively dampened, thus reaching the usual general relativistic description of spatiotemporal structures. In the end, Oriti acknowledges that this framework presents many open issues, both technical and philosophical—not least the problem of providing a sharp metaphysical characterization of the fundamental atoms. However, he submits that this framework represents a concrete and fairly well-developed example of spatiotemporal emergence, which shows the depth of complexity that such a process involves.

Acknowledgements

This book originates from a series of online talks dubbed "Warsaw Spacetime Colloquium" and organized by the philosophy of physics group at Warsaw University of Technology since October 2020. I am grateful to the speakers—many of whom ended up contributing to this volume—and to the highly engaged audience for providing solid evidence that the foundations of spacetime physics represent a thriving field of research. I am also indebted to the experts who devoted time and effort to review the chapters' manuscripts, all in the name of science. Special thanks go to Andrea Oldofredi for taking up the editor's duties for the chapter that I co-authored with Pedro Naranjo and Tim Koslowski. Finally, I wish to express my gratitude to Elaine

Landry, Dean Rickles, and Andrew Weckenmann for their trust and support throughout the production process of this volume.

Antonio Vassallo
Marki, Poland, February 2022

Bibliography

Abbott, B., R. Abbott, T. Abbott, M. Abernathy, F. Acernese, K. Ackley, . . ., and Virgo Collaboration (2016). Observation of gravitational waves from a binary black hole merger. *Physical Review Letters* 116 (6), 061102.

The Event Horizon Telescope Collaboration, K. Akiyama, A. Alberdi, W. Alef, K. Asada, R. Azulay, . . . and L. Ziurys (2019). First M87 Event Horizon Telescope results. I. The shadow of the supermassive black hole. *The Astrophysical Journal Letters* 875 (1), L1.

Part I

Classical spacetime physics

1 Shifts and reference

Bryan Cheng and James Read

1.1 Introduction

Few (if any) topics in the philosophy of spacetime have received more robust study over the past thirty years than the hole argument. Recall that the target of this argument, as presented in its modern form by Earman and Norton (1987), is "manifold substantivalism"—the view that the spacetime manifold of general relativity (GR) is physically real. The argument proceeds as follows. If $\mathcal{M} := \langle M, g_{ab}, \Phi \rangle$ is a model of GR (here, M is a four-dimensional differentiable manifold, g_{ab} is a generic Lorentzian metric field obeying the Einstein equation (presented explicitly later), and Φ is a placeholder for matter fields), then so too is the model $d_*\mathcal{M} := \langle M, d_*g_{ab}, d_*\Phi \rangle$ related to $\mathcal{M}$ by a diffeomorphism d (stars indicate push-forwards); *prima facie*, these models represent different distributions of metrical and material fields over spacetime, yet nevertheless they are empirically equivalent: having fixed a representational context, no observer "embedded" in the worlds represented by either $\mathcal{M}$ or $d_*\mathcal{M}$ would be able to tell which model truly represents the physical goings-on.[1] Thus, there arises in GR an underdetermination problem *vis-à-vis* solutions of the theory such as $\mathcal{M}$ and $d_*\mathcal{M}$, related by diffeomorphism. As Earman and Norton note, this can also be converted into a problem of indeterminism in GR: suppose that $\mathcal{M}$ and $d_*\mathcal{M}$ are identical up to some spacelike Cauchy surface Σ, but differ in some designated region (the "hole") thereafter. Just given data on (the physical correlate of) Σ, no observer would be able to tell whether their world will evolve as per $\mathcal{M}$, or as per $d_*\mathcal{M}$—indeed, there is no fact about which of these will occur, given the dynamics of GR.[2]

Earman and Norton's solution to the hole argument is to reject the physical reality of the manifold M—in which case, one can no longer articulate the difference between the worlds represented by $\mathcal{M}$ and $d_*\mathcal{M}$, since there no longer exist manifold points to which the positions of (the physical correlates of) g_{ab} and Φ may be referred.[3] Maudlin's own response to the hole argument, first presented in Maudlin (1988), differs radically, for Maudlin sees the problem not as an argument for

DOI: 10.4324/9781003219019-3

relationism about the manifold, but rather as an argument for a particular form of *substantivalism*. On Maudlin's "metric essentialist" view, we are to reject the possibility of all but one of the worlds represented by the elements of the class of models of GR related by a hole diffeomorphism—because the manifold of GR is taken to have its metrical properties *essentially*. If only one of these worlds is possible, then the hole argument is blocked, in both its underdetermination and indeterminism forms.[4]

All of this is well known. What is also known is that Maudlin embraces a *different* solution to what is often regarded as a parallel problem in the context of Newtonian gravitation theory—*viz.*, the static shift. Recall that, if $\mathcal{N} := \langle M, t_{ab}, h^{ab}, \nabla, \varphi, \rho \rangle$ is a model of Newtonian gravitation theory set in Galilean spacetime (NGT) (here, t_{ab} and h^{ab} are fixed temporal and spatial (degenerate) metric fields on M of respective signatures $(1, 0, 0, 0)$ and $(0, 1, 1, 1)$; ∇ is a derivative operator on M compatible with t_{ab} and h^{ab}; and φ and ρ are real scalar fields on M which represent, respectively, the gravitational potential and matter density—for more details, see Malament, 2012, ch. 4), then so too is $d_*\mathcal{N} := \langle M, t_{ab}, h^{ab}, \nabla, d_*\varphi, d_*\rho \rangle$, in which d is a diffeomorphism implementing the displacement of the matter content of the universe by a certain vector from its position in $\mathcal{N}$ (again, here we have fixed a representational context).[5] For example, if $\mathcal{N}$ represents the centre of mass of the universe as being located *here*, then $d_*\mathcal{N}$ represents the centre of mass of the universe as being located (e.g.) *five metres to the left of here*. Again, the static shift is often taken to lead to a problem of underdetermination: no observer "embedded" in the worlds represented by either $\mathcal{N}$ or $d_*\mathcal{N}$ would be able to tell which model truly represents their world.

In response to the static shift, Maudlin writes the following:

> If Clarke is right, the material universe *could have been* located elsewhere in absolute space—that is, located some other place than it is, keeping all the relative positions the same. But we do not need to make any observation to know that this did not *actually* happen: by hypothesis, the other placement of matter is counterfactual.
> (Maudlin, 2012, p. 46)

Maudlin is here stating that the force of the static shift can be blunted by appeal to a particular *epistemological* argument: I *know* that I am *here*, not five metres to the left of here. Thus, the static shift does not lead to a genuine problem of underdetermination.

This response to the static shift on Maudlin's part is strikingly different from his response to the hole argument. The first central aim of this chapter, addressed in §1.2, is to consider this argument in detail, as

Maudlin presents it. That achieved, we then seek in §1.3 to identify the reasons for which Maudlin offers a different response to the hole argument than to the static shift. In §1.4, we address Dasgupta's critiques of this argument, which are based around a notion of "inexpressible ignorance," and dissect said critiques using resources from the philosophy of language. Finally, in §1.5, we explore how these responses play out in a structurally similar case—namely, the interpretation of gauge transformations in electromagnetism; we find that the situation is interestingly different in that context. This chapter has an appendix, in which we provide a short history of the response which Maudlin offers to the static shift.

1.2   Spatial and temporal static shifts

It is worth dwelling on the exact nature and form of Maudlin's epistemological argument regarding the static shift in NGT. Here is how he puts the point in an earlier paper:

> Various positional states of the universe as a whole are possible: It could be created so my desk is *here*, or three meters north of here, or 888 meters from here in the direction from Earth to Betelgeuse, and so on. Which is the *actual* state of the world? Now the answer is easy: In its actual state, my desk is here, not three meters north or anywhere else.
>
> (Maudlin, 1993, p. 190)

An alternative way to put Maudlin's point is the following: What *exactly* does the spatial static shift suggest that we are ignorant of? I *know—qua* observer "embedded" in the world represented by the relevant Newtonian model—that I am *here*, not three metres to the north of here or anywhere else. Indeed, as Maudlin points out, the specification of any shifted scenario will always be such so as to determine antecedently whether such a scenario is actual or not. As he writes:

> [O]ne finds that the static shift does not result in an indistinguishable state of affairs, nor does it imply that there are any real but empirically undeterminable spatiotemporal facts about the world. The world described by the shift may be *qualitatively* indistinguishable from the actual world in the sense that no purely qualitative predicate is true of the one which is false of the other. But we have more than purely qualitative vocabulary to describe the actual world; we have, for example, the indexicals, without which the Leibniz shift cannot be described.
>
> (Maudlin, 1993, p. 190)

Maudlin's point is an interesting and important one. In particular, it suggests that, in the case of models of NGT related by a Leibnizian static shift, no substantive epistemological problem would arise *even if we assumed that such models represent physically distinct (but empirically indistinguishable) scenarios.* By extension—and as we discuss later—in the case of models of GR related by a hole diffeomorphism in the underdetermination version of the problem, a substantive epistemological problem should also be avoidable. (The question of whether Maudlin's argument suffices to overcome any epistemological challenge presented by the hole argument in its indeterminism form is more delicate, but will also be addressed later.)

So far, we have considered just *spatial* static shifts—but, of course, in NGT *temporal* static shifts are also possible—these involve a global, time-independent repositioning of the world's matter content *in time.* For instance, a temporal shift might move the world's entire material content three seconds to the future of where it actually happens to be in absolute time. By analogy with the spatial static shift, one might be led to infer that Maudlin's point would, again, be that such a shift also generates no genuine epistemological problem, for similar reasons to those discussed previously. (For instance: is the present time *now*, or three seconds to the future of now?) However, the reasons Maudlin adduces for thinking that temporally shifted scenarios generate no epistemological problem appear to be crucially distinct from those adduced in the case of the spatial static shift:

> A universe created 15 billion years ago is observationally distinguishable from one just like it (i.e., having a qualitatively identical total history) which began within the last four minutes. Things would look *awfully* different if the big bang had occurred in the last half hour.
>
> (Maudlin, 1993, p. 190)

Maudlin, it seems, is not merely claiming that the temporal shift generates no *epistemological* problem for any Newtonian observer. Rather, he is claiming that such worlds are straightforwardly *empirically* distinct: intuitively, things would *not* look and feel and taste and sound and smell the same for any observer in the two shifted scenarios. ("Things would look *awfully* different if the big bang had occurred in the last half hour.")

One might read Maudlin as having made a straightforward mistake here: *in enacting the temporal shift, he has forgotten to shift the observer.* In enacting the spatial static shift, recall, Maudlin's point—at least as we interpret it—was not that everything would "look *awfully* different" if everything was moved three metres due north. It was that, *in spite* of things looking awfully *similar*—the worlds being, in a simple and intuitive sense, empirically indistinguishable—no genuine *epistemological* problem is generated (even if we regard such shifted scenarios as being genuinely physically distinct). Indeed, it is straightforward to recognise that if *everything else* were shifted three metres due North, but *I* remained fixed, things would tend to be noticeably different. (My desk, for instance, would be three metres further away from me.) But to understand static shifts in this fashion is to misconstrue their nature.[6]

In fairness to Maudlin, he immediately goes on to note:

> Of course, if the big bang had occurred four minutes ago then in another 15 billion years there might be someone who looks just like me writing a sentence that looks just like this. But that person would have no difficulty determining that he is not alive *now*, just as I have no difficulty knowing that I will not be alive *then*. And though he would produce the same characters and phonemes as I, the indexicals in his language would guarantee that his utterances would not mean the same thing as mine.
>
> (Maudlin, 1993, p. 190)

Understood in *this* way, the analogy with the spatial static shift—suitably construed, so as to involve a shift of the relevant observer—is straightforward. That is, Maudlin's point (again) is, or at least seems to be, that neither observer, in the temporal shift or spatial static shift case, would face any genuine epistemological problem ("Is this time *now*? Am I located *here*?"), in spite of the observational indistinguishability of their respective situations.[7]

Ultimately, all of this is to say that one must be careful in how one formulates the epistemological argument—for while this argument (if successful) *does* afford a means of identifying which of a class of statically shifted worlds is one's own, it does *not* render those worlds *empirically distinguishable* for an observer embedded in the world, as Maudlin might be read as suggesting in the case of the temporal static shift. There is some evidence that other authors do commit this mistake—for example, in a footnote to his recent paper on the hole argument, Weatherall writes in the context of the hole argument in GR (in its underdetermination form) that

> if one has an observer at a given point p, the situation where the metric at p is g_{ab} and the metric at that point is $\tilde{g}_{ab}$ will in general be distinguishable—for instance, in one case, one might be happily working at one's desk; and in the other, plummeting into a black hole.
>
> (Weatherall, 2018, p. 336, fn. 20)

Whether (as with Maudlin) such a reading is truly fair to Weatherall requires more detailed discussion—for which we refer the reader to Pooley and Read (2021). The only point which we wish to register here is that there is a natural sense in which Weatherall *could* be read as making the same mistake that one might attribute to Maudlin, in the case of the temporal static shift.

1.3   Static shifts and holes

In the previous section, we explored Maudlin's epistemological argument, given in the case of the (spatial) static shift in NGT. Our purpose in this section is to explicate (one potential—and we think plausible—reconstruction of) why Maudlin offers a *different* response to the static shift in NGT than to the hole argument in GR.

The central dynamical equation of general relativity is the Einstein equation,

$$G_{ab} = 8\pi T_{ab}, \tag{1.1}$$

where G_{ab} is the Einstein tensor (ultimately a complicated expression in the metric field g_{ab} and its derivatives, assuming metric compatibility), and T_{ab} is the stress-energy tensor associated with the matter fields Φ in the solution of the theory under consideration. (Ultimately, T_{ab} is some complicated expression in terms of the Φ and the metric field g_{ab} and their derivatives—see Lehmkuhl, 2011.) Famously, this equation is *diffeomorphism invariant*: arbitrary diffeomorphisms take solutions to solutions. (For a precise contemporary discussion of diffeomorphism invariance and its relation to general covariance, see Pooley, 2017.)

The analogous dynamical equation of NGT is the Newton-Poisson equation,

$$h^{ab}\nabla_a\nabla_b\varphi = 4\pi\rho. \tag{1.2}$$

Unlike (1.1), (1.2) is not diffeomorphism invariant—acting on (1.2) with a diffeomorphism associated with an affine transformation, for example, one finds that solutionhood is preserved only when the condition

$$d_*h^{ab} = h^{ab} \tag{1.3}$$

is satisfied (note that, in deriving this result, we do not transform the fixed fields—cf. Pooley, 2017); that is, solutionhood is only retained when an affine transformation corresponds to a global *Galilean transformation*. (In addition to rigid Galilean transformations, (1.2) is also invariant under a richer class of time-dependent transformations; this will be of relevance later.) Invariance of (1.2) under Galilean transformations is related to the *kinematic shift* problem in Newtonian gravitation: while kinematically shifted solutions are physically distinct in Newtonian mechanics set in Newtonian spacetime, due to this space-time setting having extra structure (namely, a fixed vector field σ^a, representing facts about the persistence over time of points of absolute space) with respect to which these kinematic shifts are referred, such is *not* the case for Newtonian mechanics set in Galilean spacetime (i.e., what we are calling "NGT"), in which this extra structure has been excised.[8]

In this chapter, our central concern is with a *different* class of transformations under which (1.2) is invariant—namely, the static shifts introduced in §1.1. There is a sense in which static shifts in NGT are analogous to hole diffeomorphisms: in both cases, solutions related by the diffeomorphism under consideration are isomorphic. However, it is also important to stress a clear difference between the two transformations: in GR, hole diffeomorphisms (in the indeterminism version of the problem) act non-trivially only to the future of the spacelike hypersurface Σ; by contrast, static shifts in NGT are rigid: they act in the same way on all manifold points. There is no NGT analogue of a shift (i) entirely to the future of Σ, such that (ii) the shifted models are isomorphic.

This point is crucial in accounting for why Maudlin addresses the static shift via an epistemological argument (of the kind we have witnessed), yet addresses the hole argument via metric essentialism. We will very shortly give this account; however, before doing so, it will be illuminating to first present a natural candidate answer to this question—which, for reasons we will explain, is in fact specious. This candidate answer runs as follows: while the epistemological argument is sufficient to address the static shift, and *also* the underdetermination problem in the case of the hole argument, it is insufficient to address the indeterminism problem in the case of the hole argument. The reason is that the worlds under consideration in the latter case are identical up to (the physical correlate of) Σ, but differ thereafter. In this case, any claim that the observer *just knows* that she is *here* is insufficient to determine the exact manner in which one's world will evolve to the future.

This candidate answer is specious for the following reason. As already discussed, in the case of the hole argument, one has a class of models related by hole diffeomorphisms; supposing that one selects one such model to represent the actual world, one knows immediately that all the other worlds represent merely counterfactual possibilities. Crucially, such models are *complete*: they are models of the entire physical world, including regions to the future of (the physical correlate of) Σ. Suppose that the model is specified concretely via a coordinatization of the manifold and an expression of the metric and matter fields in the given coordinate system. Then, one has a name for every point (the quadruple of its coordinates) and the values of the fields at every point. So, one knows precisely and uniquely how one's world will evolve in the future—contrary to the line of argument given earlier.

So why, then, *does* Maudlin embrace a different response to the hole argument as compared with the static shift? The true answer to this question has nothing to do with epistemology—for, as we have just seen, one *can* know how one's world will evolve to the future of (the physical correlate of) Σ. Rather, the issue is this: the hole argument generates a problem of (radical!) indeterminism for GR: given (the history associated with) hole-diffeomorphic solutions of GR up to Σ, the theory (it seems) simply does not adjudicate on which of the relevant class of possible worlds will be realised. This problem of indeterminism—which, recall, does not arise in the Newtonian case—is not resolved merely by noting that once we have stipulated which model represents our actual world, we no longer have an epistemic problem concerning our world's future development (cf. Pooley and Read, 2021, fn 8).[9]

Since the epistemological argument is, therefore, insufficient to resolve the indeterminism problem in the case of the hole argument, Maudlin must recourse to a different tactic. Enter metric essentialism. And since, having introduced metric essentialism, it turns out that this can *also* answer the underdetermination problem in the case of the hole argument, Maudlin (we take it) embraces this solution *tout court*.[10] Ultimately, then, Maudlin's reasons for endorsing different solutions to the static shift versus to the hole argument seem to us to be ones of modesty: while the epistemological argument suffices in the former case since there no indeterminism problem arises, in the case of the hole argument, on the other hand, the indeterminism problem calls for a different solution—*viz.*, metric essentialism (for Maudlin, at least!)—which one then notices can be applied to *both* versions of the hole argument, and which Maudlin therefore embraces *tout court* in that context.

There are two final observations to be made on these matters. Against Stachel's point that there is no analogue of the indeterminism version of the hole argument in the case of NGT (see Stachel, 1993, p. 152), Saunders points out that one can use the invariance of (1.2) under arbitrary non-rotating accelerative transformations in order to generate a Newtonian version of the hole argument: let the acceleration associated with this transformation be trivial up to Σ, and non-trivial thereafter (Saunders, 2003, §1).[11] In this case, for the reasons already explained, metric essentialism does not suffice to resolve the problem, for only the material content of the universe is shifted. In addition, the epistemological argument also does not suffice to resolve the challenge, for just as with e.g., the kinematic shift, the models related by the transformation are not isomorphic; the worlds represented by them differ more than merely haecceitistically (more on this later).

The second observation is this. Consider NGT set in Galilean space-time (as opposed to NGT set in Newtonian spacetime, which was the case considered by Maudlin). In this case, the notion of the absolute velocity of objects in spacetime remains meaningful (in the sense that the idealised one-dimensional timelike paths through the spacetime representing the trajectories of material bodies still have tangent vectors), but there is no sense in which objects have *this* absolute velocity rather than *that* absolute velocity (this is part of the import of "sophistication" about symmetries: see Dewar (2019), Jacobs (2021), Martens and Read (2020). In this case, one might argue that, just as one is able to identify indexically one's position in the case of the static shift in NGT set in Newtonian spacetime, so too is one able to identify indexically one's *velocity* in the case of the kinematic shift in NGT set in Galilean space-time. (Question: What is my absolute velocity? Answer: *This* one; my absolute velocity could not have been otherwise.) Assuming that there is no disanalogy between position and velocity *vis-à-vis* our ability to make reference to these quantities (on the issue of reference, see the following section), we concur with this verdict.[12,13]

1.4 Responses to the epistemological argument

In this section, we address two critical reactions to this epistemological argument, both due to Dasgupta (2011, 2015a, 2015b). The first regards the notion of "inexpressible ignorance" (§1.4.1); the second regards the idea of 'God's favourite point' (§1.4.2). We argue that the second of these responses to the epistemological argument is straightforwardly unsuccessful; the first of these responses is also unsuccessful, albeit for more nuanced reasons.

1.4.1 *Inexpressible ignorance*

Dasgupta's response to Maudlin's epistemological argument spans three papers: "The bare necessities" (Dasgupta, 2011), "Substantivalism vs relationalism about space in classical physics" (Dasgupta, 2015b) and "Inexpressible ignorance" (Dasgupta, 2015a). Over the course of these papers, Dasgupta develops gradually his response to Maudlin based upon the notion of inexpressible ignorance. In this subsection, we track this evolution: §1.4.1.1 deals with his two earlier papers, which follow Maudlin closely; §1.4.1.2 analyzes Dasgupta (2015a) and its apparent divergence from Maudlin; finally, §1.4.1.3 suggests an alternative means of bolstering this particular response to Maudlin offered by Dasgupta (albeit one which we also ultimately find to be problematic).

1.4.1.1 "The bare necessities" and "Substantivalism vs relationalism about space": Dasgupta's initial reactions

In the earliest of the three previously mentioned papers, "The bare necessities," Dasgupta presents the following reconstruction of Maudlin's epistemological argument:

> In the case of velocity, we can ask the question "Am I in a state of absolute rest or a state of uniform motion?"... Maudlin's observation is that in the case of location there is no analogous question: given the resources we have by which to ask the question of where we are in Newtonian space, the only questions we can ask are questions we can readily settle. The point might be put like this. Call a question about our absolute location or velocity *open* if it cannot be reliably answered by verifying facts about the relative positions or relative velocities of material bodies. Then Maudlin's point is that we have the conceptual resources to ask an open question about or [sic] velocity through Newtonian space but not about our location in Newtonian space.
>
> (Dasgupta, 2011, pp. 145–146)

While Maudlin focuses on the fact that we can give *answers* to questions regarding our positions in Newtonian absolute space, and so according to him we are not ignorant about such matters, Dasgupta is concerned with the form of the *question*. He claims that we are unable to formulate an open question regarding the static shift, and instead suggests reversing Maudlin's reasoning, to argue that not only are we ignorant of such matters, but, moreover, "we cannot even express what it is that we cannot detect" (Dasgupta, 2011, p. 146).

We find Dasgupta's interpretation of Maudlin through the lens of open questions to be problematic: Maudlin does not appeal to "facts about the relative positions or relative velocities of material bodies"; he uses indexicals to refer directly to absolute spacetime points, hence answering the question. Even granting that his characterization of Maudlin is correct, given Dasgupta's definition that "a question about our absolute location or velocity" is "open if it cannot be reliably answered by verifying facts about the relative positions or relative velocities of material bodies," is it not in fact a *positive quality* that we have a closed question? Maudlin would likely see a closed question as simply a question that can be answered successfully, hence removing ignorance from the picture. As Perry contends, Dasgupta must "provide some reason to believe there's inexpressible, in-principle ignorance *in this particular case*" (Perry, 2017, p. 233) in order to motivate his reversal of Maudlin's argument, without which the closed question about our position in space appears satisfactory as a resolution to the epistemological problem presented by static shift scenarios.

In "Substantivalism vs relationalism about space," Dasgupta follows a similar line of reasoning to his first paper, though in a more refined manner. There, he reconstructs Maudlin's argument as follows:[14]

1   We cannot formulate an unanswerable question about where we are in Galilean spacetime.
2   Therefore, we can know where we are in Galilean spacetime.

(Dasgupta, 2015b, p. 619)

Dasgupta disputes (2), preferring to argue that,

> (1) is true because of our expressive limitations. On this view, (1) is true not because we can know where we are in Galilean space-time, but because we lack the capacity to refer to regions of space-time in a way that would allow us to formulate an unanswerable question.

(Dasgupta, 2015b, p. 619)

Before investigating Dasgupta's motivations for reversing Maudlin's reasoning, we note that there is a particular change of terminology from Dasgupta's previous paper which mandates further scrutiny: his forgoing of the terminology of "open questions" in favour of the terminology of "unanswerable questions." If one assimilates the latter to the former, then our earlier criticisms of the concept of open questions still stand. Let us grant, though, that there is a difference in the two notions. Recalling Dasgupta's definition of an open question in the context of the static shifts, we would remark that even if "a question about our absolute location or velocity … cannot be reliably answered by verifying facts about the relative positions or relative velocities of material bodies" (Dasgupta, 2011, p. 146), and so is open on Dasgupta's definition, there might nevertheless remain *in principle* possibilities for answering such a question. This, perhaps, motivates Dasgupta's move from "open questions" to "unanswerable questions." In the ensuing, we will follow Dasgutpa's lead in focusing on the latter.[15]

We consider now two different responses to Maudlin's epistemological argument offered by Dasgupta, which make use of the notion of an "unanswerable question."[16] The first of these motivations is found in "The bare necessities":

> when one remembers that there are uncountably many shifted worlds that would all look and feel and taste exactly the same as the actual world, there is a clear feeling that our location in space is therefore in some sense beyond our epistemic grasp. And (speaking for

myself) this feeling is not dissipated by being told that one cannot formulate an unanswerable question about one's position in space.

(Dasgupta, 2011, p. 146)

This is representative of what we imagine to be a widespread initial reaction to Maudlin, but it is inadequate on deeper consideration. According to Maudlin, we *can* refer to spacetime points using indexicals, such that they are indeed within our epistemic grasp, and therefore cannot lead to ignorance.[17]

The second motivation is found in "Substantivalism vs relationalism about space," and is (in our view) even less convincing:

> the argument that we are ignorant of location was exactly the same as the argument that we are ignorant of velocity. In both cases, there are infinitely many boosted or shifted worlds that look and smell and taste and feel exactly alike, and so are indiscernible in that sense. Why then should the situation be any different in the case of location than velocity? Why, just because I cannot formulate an unanswerable question about my location, should it follow that I am not ignorant at all?
>
> (Dasgupta, 2015b, p. 619)

Maudlin's entire argument is that the structure of the static shift argument is *in fact different* from the kinematic shift, due to the difference in our referential capabilities towards velocities and positions; Dasgupta simply appears to be begging the question here.

Overall, then, these papers do not present a compelling motivation for a shift towards Dasgupta's perspective. In fact, the situation reduces to two different attitudes towards interpreting the case at hand, aligning with the authors' desired result in the case of the static shift. To Dasgupta, one *cannot* formulate an unanswerable question; to Maudlin, one is *able* to give an answer to the question. As Perry notes, the burden should remain with Dasgupta to articulate a reason why it is problematic not to have an unanswerable question, or what has gone wrong with Maudlin's question (or answer) in the first place. Nevertheless, Dasgupta's approach suggests a subtle issue with Maudlin's account: the latter does not consider whether the *nature* of the question used to interrogate our absolute position will affect his argument, and simply (and somewhat uncritically) presents the answer making appeal to indexicals. Given this opportunity for criticism, an alternative approach to critiquing Maudlin will be presented in §1.4.1.3. To formulate this approach, however, we must first analyze Dasgupta's third paper on the subject.

1.4.1.2 *"Inexpressible ignorance": divergence from Maudlin*

We now move on to consider "Inexpressible ignorance," which, as the title suggests, is the most direct exposition of Dasgupta's notion of inexpressible ignorance. We contend, however, that this paper constitutes a greater departure from Maudlin's original argument than the two articles considered up to this point. To understand this divergence, it is crucial to refer to a passage in "Substantivalism vs relationalism about space" in which Dasgupta concludes that the conclusion of Maudlin's argument does not follow:

> To the contrary, I would appear to have two cognitive failings: a failure to know, and a failure to be able to ask a certain kind of question. Maudlin's view has the bizarre consequence that this double failure amounts to a success!
>
> To defend his view, Maudlin might appeal to a general principle to the effect that one is ignorant about some topic if and only if one can formulate a question about it that one does not know the answer to. Call this the principle that all ignorance is expressible. . . . But I think that the principle is false.
>
> (Dasgupta, 2015b, pp. 619–620)

Curiously, Dasgupta portrays in "Inexpressible ignorance" this "principle that all ignorance is expressible" as having originated with Maudlin. The following is his reconstruction of Maudlin in that paper:

> Grant that our (supposed) ignorance about position in Newtonian space is inexpressible in this sense. That is the first part of Maudlin's view, and I agree with it. Turn now to his second claim, that *ignorance is always expressible*. . . .
>
> Maudlin gave no argument for this second claim. And it is, on the face of it, most implausible. There are (remember) infinitely many shifted worlds that differ with regard to where we are in Newtonian space, all of which look *exactly* the same. In the case of velocity, this kind of proliferation of indiscernible possibilities suggested that I am ignorant. Why is the situation any different in the case of position? Why, just because I cannot express my ignorance, should it follow that I am not ignorant at all? Indeed, in the case of position, I appear to have *two* cognitive failings: a failure to know, and a failure to express that ignorance. Maudlin's view has the bizarre consequence that this *double* failure amounts to no failure at all!
>
> (Dasgupta, 2015a, p. 446)

It seems to us that Dasgupta's response here is unsuccessful, for he has not characterised accurately Maudlin's argument. It is peculiar that, having presented a particular solution to the problem at hand (a solution which is, as Dasgupta notes, hardly defensible in its generality), Dasgupta simply attributes this view to Maudlin. Given that Maudlin (1993) never mentions the notion of inexpressibility, there are two different possible readings of his work on these issues. We argue in this subsection that the latter of these two readings is more plausible—and that, on this reading, Dasgupta's response to Maudlin in this paper fails, if certain philosophical commitments underlying Maudlin's response can be defended successfully.

Reading 1: Statically shifted worlds are distinct, and there is inexpressible ignorance in the case of static shifts, but nevertheless these shifts do not generate an epistemological challenge.[18]

We have already seen that Maudlin regards statically shifted worlds as being distinct. On this first reading of Maudlin, undergirding this difference between such shifted worlds is that there are facts—facts of which we are inexpressibly ignorant—regarding the natures of spacetime points. Thus, on this reading, Maudlin accepts the first of Dasgupta's two premises in his reconstruction of Maudlin's argument in this quotation: that there is inexpressible ignorance. However, Maudlin (on this reading) denies Dasgupta's second premise in his reconstruction of Maudlin's argument: that ignorance is always expressible. Rather, he maintains (on this reading) that there *can* be cases of inexpressible ignorance, but denies that these always need lead to epistemological problems such as identifying which of a class of statically shifted worlds is one's own. On this reading, Maudlin maintains that there *is* inexpressible ignorance, but nevertheless deploys the indexical argument to state that he *knows* which of these worlds is one's own. The point is that this knowledge regarding these identity of one's world does not necessarily presuppose knowledge of the facts about spacetime points of which one is inexpressibly ignorant.

Ultimately, we do not think that this reading of Maudlin is stable. However, in order to articulate our reasons for thinking this, we should first introduce our second possible reading of Maudlin: this will require a more detailed discussion.

Reading 2: Statically shifted worlds are distinct, but there is no inexpressible ignorance in the case of static shifts, so these shifts do not generate an epistemological challenge.

On this reading of Maudlin, he denies the first of Dasgupta's two earlier premises: in spite of the statically shifted worlds being distinct, there is *no* inexpressible ignorance here regarding spacetime points. Rather, on this reading, Maudlin's point is that *there is no fact about which we are ignorant, and nothing is inexpressible.*[19] This reading is supported by (i) Maudlin's writing that "[t]o even formulate the appropriate question in the static case one must indexically pick out a spatiotemporal location" (Maudlin, 1993, p. 190), (ii) his subsequently stating that "[f]or the substantivalist, terms such as 'here' or 'now' can be used to drive linguistic pegs into the fabric of absolute space and time. Without such pegs, the static Leibniz shift cannot even be formulated" (Maudlin, 1993, p. 191) and (iii) his arguing that precisely this "linguistic wherewithal needed to establish the coordinates also provides us the means of answering the question" of one's absolute position (Maudlin, 1993, p. 191). On this reading, there is no ignorance in the first place, because the tools we use to pose the question of our position in Newtonian space—indexicals—are precisely those that can be used to answer the question.

Now, of course, in response to this, there is room for Dasgupta simply to deny that the use of indexicals can provide all answers to questions regarding one's position in Newtonian space. Indeed, Dasgupta presents an account of inexpressible ignorance according to which this phenomenon can arise with respect to some entity when one is (a) not acquainted with that object (*nota bene*: Dasgupta does not proffer an analysis of acquaintance—see Dasgupta, 2011, p. 464), or (b) does not know the "full essence" of that object (Dasgupta, 2011, p. 464).[20] Dasgupta then argues that, if we want there to be inexpressible ignorance in the case of static shifts, we had better buy into this account of inexpressible ignorance (which, he claims, is superior to all extant alternatives—see Dasgupta §§2–3), but then we see that (a) and (b) do indeed fail, so (*pace* this current reading of Maudlin) there is inexpressible ignorance here.

In response to this, one can take the bull by the horns: even accepting this analysis of the origins of inexpressible ignorance, it is not clear that there is indeed inexpressible ignorance in the case of the static shift. There are two ways to do this: deny (b), or deny (a). On denying (b): one might simply object to the heavy-duty metaphysics of essences.[21] However, given Maudlin's own commitment to spacetime points having essential properties—as made clear in his metric essentialist response to the hole argument—it is not obvious that this response is available to him. On denying (a), there is room to do this, although it will depend upon the particular analysis of acquaintance which one endorses.

Perhaps a more effective response to Dasgupta than either denying (b) or denying (a), however, is to question the appropriateness of Dasgupta's approach to inexpressible ignorance to begin with. Suppose, in particular, that one adopts the "liberalism about reference and singular thought" defended by Hawthorne and Manley (2012), according to which acquaintance is *not* necessary in order to make reference to a given entity. Witness, for example, the following passage:

> Suppose John and David are talking on the telephone, and David says "It's raining" in a way that makes it clear he is talking about rain at his own location. Suppose further that John's grasp on David's location is very "thin": he knows only that it is the place

David is currently in, but not under various paradigmatic guises—he cannot see it or point it out on a map, and he does not know any proper name for it. Moreover, because the acceptability of "knowing where" reports require that John grasp the location under the guise salient in the context, he may not count as knowing where David is, or as knowing which location is supplied by context. But liberals about singular thought have no reason to deny that John can think about David's location, and thus no reason to deny that he is capable of grasping the proposition David expressed, understanding what he said, and so on. In this respect, it is just as though David had used an indexical or a newly introduced name to refer to his location.

(Hawthorne and Manley, 2012, p. 137)

The point is that, given liberalism about singular thought, indexicals and other contextually specified information are perfectly sufficient to make reference to spatial locations. Thus, Dasgupta's conception of inexpressible ignorance (as presented at this point in his paper) seems, one might say, to rest on a neo-Russellian notion of reference-by-acquaintance which one might reasonably reject (cf. Dasgupta, 2011, p. 464). (We discuss later whether there is any sense in which the notion of inexpressible ignorance might be taken to be consistent with liberalism about reference and singular thought.)

1.4.1.3   *Kaplan's account of indexicals: an alternative attack*

Despite all this, there is a possibility of salvaging Dasgupta's account; even if one accepts this liberalism about reference and singular thought, it appears possible to critique Maudlin's invocation of indexicals in his argument, which relies on several unspoken assumptions. Maudlin *does* appear to assume said liberalism, so as to accommodate our lack of causal acquaintance with spacetime points. In addition, however, he makes implicit use of an analysis of indexicals and demonstratives which aligns closely with the influential approach due to Kaplan, most famously presented in Kaplan (1989).

There are three components of Kaplan's approach to indexicals which are particularly useful for Maudlin. First, Kaplan argues that indexicals are directly referential—which, roughly, means that the content of an indexical is just the object to which it refers. This property makes indexicals suitable for the function of driving "linguistic pegs into the fabric of absolute space and time" (Maudlin, 1993, p. 191). Second, Kaplan's Corollary 2 states that "[i]gnorance of the referent does not defeat the directly referential character of indexicals" (Kaplan, 1989). This ensures that the account is compatible with liberalism about reference and singular thought, since Kaplan aims explicitly to reject "direct acquaintance theories of direct reference," one example of which is Russell's theory (Kaplan, 1989). Third, there is the consequence of rigid designation, deriving from Kripke's use of the term (Kripke, 1980), which essentially means that once the context of an indexical is fixed, it holds true across all possible worlds—particularly useful for the consideration of static shifts and Maudlin's counterfactual formulation.

There are, however, well-known problems with Kaplanian direct reference accounts of indexicals—some of the most notable among which are analogues to Frege's puzzles regarding proper names, in which two statements that refer to the same object have different truth conditions. One such problem case is that of the the messy shopper (due to Perry, 1979): in this case, a shopper sees in a mirror what appears to be another shopper making a mess; unbeknownst to him, *he* is that very shopper. In this case, the shopper simultaneously believes both the statement "He is making a mess," and the statement "I am *not* making a mess." On direct reference theories, these express contradictory propositions, since "he" refers back to the shopper himself—meaning that the shopper, according to direct reference theories, holds contradictory beliefs. Those who do not wish to accept such a conclusion may accordingly reject Kaplan's account of indexicals; this, in turn, might seem to impair the effectiveness of Maudlin's argument.

Alternatively, the rival "descriptivist" theory of indexicals attributes "purely qualitative descriptive content" (Braun, 2017, §4.2) to indexicals. A possible consequence of this view is that the semantic category to which indexicals belong changes. A passage from Hawthorne and Manley aligns with this view:

> Take "here" and "now." On the standard Kaplanian conception, these terms as used at a context are referential devices whose semantic value are a place and a time respectively, and hence are rigid designators. [Footnote suppressed.] But this simple picture arguably takes liberties with the semantic category of those expressions: they are plausibly better construed as modifiers that generate property expressions out of property expressions. (That is, they are of the same semantic type as "in Texas.") Thus, in "John is smoking here," "here" does not simply supply a location, as though John is being said to smoke a location. Instead, the result of, say, modifying "smokes" by "here" while located at l will be a complex expression that expresses a property that applies to a thing iff that thing smokes at l (similarly, *mutatis mutandis*, for "now"). In this way, "here" and "now" have a spatially or temporally constraining effect on a predicate. But that does not make it correct to say that "here" and "now" simply refer to spatial or temporal locations.
>
> (Hawthorne and Manley, 2012, p. 245)

If Maudlin accepts a descriptivist account of indexicals, then his epistemological response to the static shift seems to falter (that said, whether it actually does so is not completely clear, as we discuss later), given his having stated explicitly that "all absolute places are qualitatively identical" (Maudlin, 1993, p. 191). According to this point of view, spatial indexicals alone would be insufficient to defuse the epistemological challenge posed by the static shift. Thus, our second reading of Maudlin has three key inputs: (i) Hawthorne-Manley liberalism about reference and singular thought, (ii) a Kaplanian account of indexicals, and (iii) an assumption that Kaplanian indexicals can fully characterise a spacetime point. Hawthorne and Manley, in fact, repudiate Kaplan's account of indexicals (see the earlier quotation)—in which case, the possibility of Dasgupta's notion (though perhaps not account) of inexpressible ignorance appears to re-arise in this context. While Maudlin in fact (contrary to Dasgupta's reading) forecloses such inexpressible ignorance at the very outset, by tacitly making use of Kaplan's account of indexicals, if one is to reject this account (and, indeed, we have seen that there are reasonable grounds to do so), then it seems that Maudlin must find other resources to refer to spatiotemporal locations.

There is, however, (at least) one possible way to save Maudlin's account, even when working within the framework of a descriptivist conception of indexicals. While Maudlin does indeed need to be able to latch onto particular spacetime points, he need not, in fact, do so via the machinery of indexicals (so this line of thought goes). The reason for this is due to the fact that some names may be introduced via the help of definite descriptions, yet thereafter function semantically as names (e.g., "Maudlin") or demonstratives (e.g., while pointing, "that spacetime point"). These names contribute the referent to the determination of truth conditions, and allow for rigid designation; moreover, such naming works without necessarily any direct acquaintance with the object being so designated. Given this, Maudlin may still be able to refer to particular spacetime points. Indeed, this point is arguably present in the quote from Hawthorne and Manley given earlier: in that passage, the authors offer a semantic analysis of "John is smoking here" distinct from that offered by Kaplan, but their analysis nevertheless introduces a name l for the place in question, and that name may (in line with the previous) function semantically thereafter to contribute the named place to the content (or truth conditions) of statements. This, in turn, seems to be sufficient for Maudlin's response to the static shift to go through, even when working within the framework of a descriptivist account of indexicals.[22]

Returning to our first reading of Maudlin, according to which there is inexpressible ignorance in cases of the static shift, this discussion raises questions about whether the account can, in fact, go through. The reason for this is that one must disambiguate the account of indexicals at play. It seems that something like liberalism about reference and singular thought is a necessary condition for identifying which of a class of statically shifted worlds is one's own, and thereby resolving the epistemological challenge posed by such shifts. Taking this liberalism for granted, then, if Kaplan's account of indexicals is also endorsed, there is in fact no inexpressible ignorance at all, contrary to the claims of our first reading of Maudlin.[23] If, on the other hand, the descriptivist account of indexicals is endorsed, then it is not obvious that one can, in fact, identify indexically which of the class of shifted worlds is one's own—as we have seen in the case of our second reading of Maudlin. However, one might—even in the context of the descriptivist account of indexicals—be able to make (rigid) reference to spacetime points in one's world via definite descriptions and the act of naming (as discussed earlier), so granting this would again seem to imply that there is no inexpressible ignorance at all. Given all this, we therefore find our second reading of Maudlin to be the more plausible of the two.

Finally, to reprise our readings of Dasgupta in §1.4.1.1, it is possible that he is taking issue with questions (such as Maudlin's) that are necessarily answerable by virtue of the resources used in phrasing them being the only possible answers. If we reject direct reference in favour of a descriptivist account of indexicals, then we cannot use those indexicals to refer to specific spacetime locations. As a result, the closed question that Maudlin has constructed no longer possesses any meaning involving spacetime; the fact that the same resources are used in both question and answer renders a quasi-analytic connection between them, yet they cannot refer to any particular region of spacetime. In this situation, Dasgupta's instinct that we're missing something from the picture—that there's more inexpressible ignorance behind the closed question— might be claimed to yield fruit, allowing his concerns of inexpressible ignorance to resurface in a stronger form. However: even in the case in which Dasgupta does indeed appeal to a descriptivist account of indexicals, it bears stressing that it is far from obvious that his argument succeeds. The reason is that—as we have already seem in our discussion of fixing reference via definite descriptions and the act of naming—there are different means of answering the question of where I am located in absolute space beyond mere appeal to indexicals. Thus, it is ultimately not clear that Dasgupta's argument against Maudlin succeeds, on either the Kaplanian or descriptivist conception of indexicals.

In response to our reading of Dasgupta (as embracing, ultimately unsuccessfully, a descriptivist account of indexicals in order to foreclose the success of Maudlin's epistemological argument), one might point to the fact that Dasgupta (2015a, pp. 453–454) considers a "modest externalism" akin to Hawthorne-Manley liberalism about reference and singular thought, according to which "I can know where my desk is in absolute space after all" (Dasgupta, 2015a, p. 454). Note, however, that Dasgupta regards this modest externalism as missing something, precisely because it cannot account for what he believes to be inexpressible ignorance. In light of this, one might say the following: Maudlin's position must incorporate liberal reference, Kaplanian indexicality (or alternatively descriptivist indexicality, if one is willing to secure reference to spacetime points not via indexicals *per se*, but rather via definite descriptions and the act of naming—something which we think *is* possible, as discussed earlier), and an assumption that such reference can fully furnish the characterization of a spacetime point. On the other hand, Dasgupta is free to choose his stance with regards to liberal reference and Kaplanian/descriptivist indexicality, but claims that indexicals cannot fully characterise a spacetime point in accordance with his intuition (and presumably—although he does not engage explicitly with such proposals—*mutatis mutandis* for our alternative proposal for securing reference to spacetime points via definite descriptions and the act of naming), leaving some form of inexpressible ignorance to be explored.

Focusing solely on the use of indexicals, descriptivist accounts do not *have* to be embraced by Dasgupta—but are perhaps more natural for him than Kaplanian accounts, where there has (at least restricting to the referential capacities of indexicals) to be an assumption that there is indeed something about location that one doesn't know. In a way, descriptivist accounts might be taken to afford a way of answering Perry's objection—recall: for Perry, Dasgupta must "provide some reason to believe there's inexpressible, in-principle ignorance *in this particular case*" (Perry, 2017, p. 233))—, since there is something clearly missing from the use of "here", but Dasgupta is still free to be a Kaplanian about indexicals if he so chooses. To be clear, however, and at the risk of labouring a point: in light of the alternative options for securing reference that we discussed earlier, we still do not think there being in-principle ignorance follows from the descriptivist account of indexicals—this must be considered an *additional* assumption instead.

1.4.2   *God's favourite point*

As mentioned earlier, Dasgupta presents a second motivation in "The bare necessities" against Maudlin's epistemological argument:

> Maudlin's view implies that whether something is detectable depends on factors that are, intuitively, entirely irrelevant to the matter. For on his view, whether our location in Newtonian space is detectable depends on whether or not we can ask open questions. So, for example, suppose that God had a favorite point in space. Then we would be able to ask open questions about our location in space, for example "Am I 3 feet or 6 feet from God's favorite point?" On Maudlin's view, it would then turn out that location is undetectable

after all. So on Maudlin's view, whether absolute position is detectable or not depends on whether God had a favorite point. And that seems clearly false: whether or not God has a favorite point is surely irrelevant to my epistemic situation *vis a vis* our position in space!

(Dasgupta, 2011, p. 146)

Again, this objection misses the mark. The reason, in this case, is that by introducing a privileged spacetime point, Dasgupta is *changing* the spacetime setting of NGT—in particular, he is augmenting Galilean/Newtonian spacetime with extra structure, effectively transforming it to what is sometimes referred to as *Aristotelian spacetime* (see Earman, 1989, pp. 34–35). If such a point were to exist, there would be questions which we could not answer, even in principle (much as in the case of the kinematic shift). This, of course (by a plausible and minimal Occamist norm—cf. Dasgupta, 2016), is precisely why we should *not* believe in the existence of such a point! The issue with Dasgupta's objection here, then, is that it merely moves the goalposts, and affords one no reason to think that there is any problem with the epistemological argument *per se*, when such a privileged spacetime point is *not* introduced.

In "Inexpressible ignorance," Dasgupta contemplates a very similar situation: a world of two-way eternal recurrence (divided into "epochs," e_i), but with one exception: there is a "special" epoch e_S, which "differs from the rest just in the fact that one electron is a little to the left of its counterparts in other epochs" (Dasgupta, 2015a, p. 448). On this case, Dasgupta writes that

> Maudlin's view ... implies that, were that electron a little bit over to the right, I would not be ignorant of which epoch I inhabit. And this is hard to take seriously: surely my ignorance of which epoch I inhabit cannot be cured by minute changes in far-off epochs!
>
> (Dasgupta, 2015a, p. 448)

In response to this, however, we side with Perry, who writes that

> It seems that Dasgupta is simply wrong here. There is no sense in which the change from [the world with e_S] to [an otherwise-identical world without e_S] merely removes expressibility. Everything we are expressibly ignorant of in [the former] is either false in [the latter], or is something we're also expressibly ignorant of in [the latter].
>
> (Perry, 2017, p. 232)

The overarching point is this: Dasgupta's various motivations for inexpressible ignorance by appeal to "perturbed" worlds (whether said perturbation is applied at the level of spacetime structure, or at the level of the material content of that world) do not succeed, for they illegitimately transfer expressible ignorance in the former case, to inexpressible ignorance in the latter case, while not recognizing the alternative: that there is simply no ignorance *at all* in the latter case.[24]

1.4.3   Parallels: linguistic responses to scepticism

There are interesting parallels between (i) the Dasgupta/Maudlin interaction on inexpressible ignorance, and (ii) a better-known debate in the contemporary philosophical literature—*viz.*, that regarding whether one can proffer a linguistic response to philosophical scepticism. Famously, Putnam argued that, since brains-in-vats can only ever refer to vat-facsimiles of real-world objects, they can never articulate the sceptical problem—and so, the issue is dissolved (see Putnam, 1981, for the original source, and McKinsey, 2018, for an elegant summary of notable responses). In reaction to this, Nagel maintained that the argument is too quick: perhaps brains-in-vats are simply *inexpressibly* ignorant of real-world facts—"Instead I must say, 'Perhaps I can't even think the truth about what I am, because I lack the necessary concepts and my circumstances make it impossible for me to acquire them!'" (Nagel, 1986, p. 73)—; or perhaps one can make use of non-referring terms to articulate the sceptical scenario—"I can use a term which fails to refer, provided I have a conception of the conditions under which it would refer" (Nagel, 1986, p. 72).[25]

Transparently, in these debates, Dasgupta parallels Nagel, insofar as the latter is to be read as endorsing the possibility of inexpressible ignorance—but who parallels Putnam? If one reads Putnam as stating that, because brains-in-vats are inexpressibly ignorant of real-world objects, they are in fact not ignorant of anything at all, then it would seem to be Dasgupta's final reading of Maudlin in Dasgupta (2015a) who is the appropriate analogue here.

Note, though, that if what we have said in the foregoing subsections is correct, then the response that Maudlin himself (rather than Dasgutpa's reading of Maudlin) would proffer to the sceptical problem is a little different. Suppose, to make the parallel with static shifts more explicit, that we consider various different sceptical scenarios, in which the data supplied to the brain-in-vat is the same, but the real-world facts are altered. Call these "sceptical shift scenarios." Then, on our first reading of Maudlin presented in §1.4.1.2, he would argue that these shifted worlds are genuinely distinct, and there are real-world facts of which we are inexpressibly ignorant, but that there is no epistemological problem, because we can identify indexically which of this class of worlds is our own. On our second reading of Maudlin presented in

§1.4.1.2, he would argue that there is in fact no inexpressible ignorance at all, for we can refer indexically to real-world objects.

Are either of these responses plausible in the sceptical shift scenario? In our view *no*, for one cannot identify indexically real-world objects as one can identify indexically one's own position—the situation is more analogous to the kinematic shift, or to the cases of gauge transformations in electromagnetism (for which see §1.5), than to the static shift. There is, however, one disanalogy between the sceptical shift scenarios and these latter two cases: in the sceptical case, if Putnam is correct, we are *inexpressibly* ignorant of the real-world facts; whereas in the latter two cases, we are *expressibly* ignorant. This difference stands in spite of the fact that we cannot identify indexically our world in either case.

Where, then, does this leave us? If Putnam is the analogue of Dasgupta's Maudlin, and (one aspect of) Nagel is the analogue of Dasgupta, then Maudlin himself finds neither a straightforward nor plausible parallel in the sceptical case—for it is *not* possible for a brain-in-vat to identify indexically real-world facts or objects in sceptical shift scenarios; this blocks both of our own readings of Maudlin in this context.[26]

1.5   Gauge in electromagnetism

In §1.3, we presented Maudlin's proposed solutions to both the hole argument and the static shift, and sought to explicate why Maudlin offers a different response in each case. In this section, we consider a structurally analogous problem of underdetermination (to both the static shift, and the hole argument in one of its guises) in electromagnetism, and the potential application of these solutions in that context—this comparison will afford some insight.

Solutions of electromagnetism are tuples $\langle M, \eta_{ab}, A^a, J^a \rangle$, where η_{ab} is the Minkowski metric field of special relativity, A^a is the electromagnetic vector potential, and J^a represents a current; dynamical equations for this theory are the Maxwell equations,

$$\nabla_a F^{ab} = J^b, \tag{1.4}$$

$$\nabla_{[a} F_{bc]} = 0. \tag{1.5}$$

(Here, the Faraday tensor F_{ab} is defined as $F_{ab} := \nabla_{[a} A_{b]}$, and ∇ is the derivative operator compatible with η_{ab}.) This theory is claimed to manifest gauge redundancy, in the sense that solutions of the theory in which the A^a field differs by a gradient term lead to the same Faraday tensor F_{ab}, and so to the same observable data (since the Faraday tensor encodes the electric and magnetic fields, which are taken to be the observable data of the theory).

In light of this gauge redundancy, the A^a field is often taken to be "unphysical"; rather, the physical content of electromagnetism is taken to be encoded in F_{ab}. Against this orthodoxy, in his response to Healey on the Aharonov-Bohm effect, Maudlin (1998) explores the possibility of taking the A^a field to be physical (for details on the philosophical import of the Aharonov-Bohm effect, and on associated issues of locality and separability, see Healey, 1997, 2007). He writes:

> The question is why gauge-invariance is a sine qua non for physical reality. Suppose, to be simplistic, one thought that the vector potential was real, and that there is ONE TRUE GAUGE which describes it at any time. [Footnote suppressed.] This would immediately render the explanation of the Aharonov-Bohm effect local and separable. What troubles would accrue for such a theory?
>
> One obvious trouble would be epistemological: since potentials which differ by a gauge transformation generate identical effects, no amount of observation could reveal the ONE TRUE GAUGE. This would be generally annoying and a real metaphysical/semantic problem for positivists. But one might be willing to pay this price, especially since the demand that every physically real quantity be accessible to experimental measurement seems defensible a priori only by a positivist of some stripe. One might also worry about problems of determinism: if different gauges really represent different states of affairs (if there is, as it were, an *active* reading of a local gauge transformation), and [*sic*] then *local* gauge transformations may lead to the sorts of problems with determinism that the hole-argument of John Earman and John Norton (1987) raised for General Relativity. These problems are certainly not trivial.
>
> (Maudlin, 1998, pp. 366–367)

Maudlin's point here is that, while taking seriously the representational capacities of the A^a field can ameliorate problems of non-locality and non-separability arising out of the Aharonov-Bohm effect, doing so comes with its own problems—*viz.*, the exact analogues of the underdetermination and indeterminacy problems faced in the hole argument. That this is so is clear: if the A^a field is physically real, it is nevertheless underdetermined *which* A^a field in a gauge equivalence class (related by gradient terms) is real—for, as discussed, the empirically observable data is encoded in the F_{ab} field, in which the gauge part of the A^a field is "washed out." The indeterminism problem also arises: consider a range of solutions to electromagnetism which agree on the data on some Cauchy surface Σ, but which differ thereafter by local gauge transformations on the A^a field: no observer with access only to data on Σ would be able to predict *which* solution correctly represents her world, *vis-à-vis* the value of the A^a field to the future of Σ—and indeed, the structure of electromagnetism does not specify which of these possibilities will occur.

One point is worth making at the outset here: Maudlin's proposal that there is "ONE TRUE GAUGE" is *merely* a proposal that we take the A^a field physically seriously; it is *not* a proposed solution to the problems of underdetermination and indeterminism which arise therefrom. One may, however, at this juncture seek to import Maudlin's proposed solutions to the hole argument and to the static shift—namely, metric essentialism and an epistemological argument—to this context. How would they fare?

Begin with the analogue of metric essentialism. In this case, this would be a "gauge essentialist" view, according to which only one element of the class of worlds represented by gauge-related solutions of electromagnetism is metaphysically possible—*a fortiori* physically possible.[27] As in the case of the hole argument, this would resolve immediately both the underdetermination and indeterminism concerns—although it is not at all obvious that Maudlin's motivations for metric essentialism presented in Maudlin (1988) carry over to this "gauge essentialism"—in which case, it is not at all obvious that this is a compelling view to hold.[28]

What of the epistemological argument? Recall that this argument succeeded (modulo the issues discussed in §1.4) against the underdetermination problem in both the static shift and the hole argument, and also against the indeterminism problem in the hole argument (at least in one sense—although the argument is impotent against the charge that the hole argument renders GR radically indeterministic). But how does the epistemological argument fare when it comes to the underdetermination and indeterminism problems in the context of gauge transformations in electromagnetism?

Our point in the case of gauge transformations in the standard A^a field formulation of electromagnetism as presented earlier is the following: the situation here is even worse for the epistemological argument, for in this case the response cannot successfully be levied against the underdetermination or the indeterminism problem. The reason for this is the following: while it is true that I *just know* that I am *here*, rather than *five metres to the left of here*, I do not likewise *just know* which gauge obtains—for internal gauge transformations upon the A^a field, unlike spatial transformations such as the static shift, are not identifiable indexically (cf. our earlier discussions of the sceptical shift scenarios—although as we noted, if Putnam is correct in that case, the ignorance there is not even expressible). Thus, the situation here is in fact more akin to the kinematic shift in NGT set in Newtonian spacetime than to the static shift: the lack of indexical identifiability means that the epistemological argument cannot get off the ground: just like absolute velocities, we are ignorant (albeit expressibly so) of the "ONE TRUE GAUGE."[29,30]

On these matters, Healey (2007, §4.4.3) seems to engage—albeit without explicit acknowledgement—with the Maudlin/Dasgupta debate over inexpressible ignorance, when he writes

> The epistemological problems with the localized gauge potential properties view have a semantic aspect. They are connected to the fact that it leads to unanswerable questions. Neither the theory itself, nor anything we can do when applying it, enables us to give determinate answers to questions about how the supposed localized gauge potential properties are distributed. This is not just a failing of language. We cannot even entertain a thought that they are one way rather than another, from among an infinity of gauge-related distributions. But an advocate of the localized gauge potential properties view may deny that this renders these properties problematic. It may be an epistemological defect in a theory to raise meaningful but unanswerable empirical questions. But the semantic features of the localized gauge potential properties view are such that there are questions it does not even permit one meaningfully to ask. And what appeared as an epistemological vice could be seen rather as a semantic virtue—the virtue of rendering metaphysical questions literally meaningless rather than empirically unanswerable!
>
> (Healey, 1997, pp. 122–123)

Healey begins here by suggesting—*à la* Dasgupta—that realists about the A^a field are committed to inexpressible ignorance regarding the values of that field; at the end of the passage, however, he suggests—*à la* Maudlin—that the other side of the coin here is to maintain that there are no unanswerable questions regarding the value of the gauge field A^a. In our view, both of these options are problematic: just as in the case of the kinematic shift, there does appear to be a plurality of possibilities for the value taken by the A^a field at any given spacetime point— and these different possibilities are all expressible! Thus, to repeat, we disagree with Healey, insofar as, in our view, the kinds considerations of inexpressible ignorance discussed at great length in the previous sections of this chapter simply do not come into play in (this formulation of) electromagnetism.

Why do we add the parenthetical qualification "this formulation of"? The reason is that there are, in fact, other formulations of electromagnetism, for which questions of inexpressible ignorance more plausibly apply. In recent years, a great deal of interest has arisen in what Dewar dubs the process of constructing "sophisticated" theories, in which one retains the same number of models, but "forgets" structure, such that the resulting models are isomorphic.[31] One example of this process—already discussed—is the move from NGT set in Newtonian spacetime to NGT set in Galilean spacetime.[32] Another example is moving from the A^a field formulation of electromagnetism to an alternative formulation in terms of fibre bundles (see Dewar, 2019, p. 501). Here is not the place to go into the (admittedly beautiful) mathematics of the fibre bundle formulation of electromagnetism; rather, we now simply present this version of the theory and discuss its upshots for inexpressible ignorance in the context of gauge theories.[33]

Models of the fibre bundle formulation of electromagnetism are given by tuples $\langle M, P, P_F, s, \omega \rangle$, where M is a differentiable manifold, P is a principal bundle and P_F is an associated bundle. s is a section of P_F representing material fields, and ω is a connection on P representing Yang-Mills fields. Letting the model $\mathcal{O} := \langle M, P, P_F, s, \omega \rangle$, implementing a gauge transformation on the material and Yang-Mills fields yields a new model $d_*\mathcal{O} := \langle M, P, P_F, d_*s, d_*\omega \rangle$, where d is a diffeomorphism. Technicalities aside, the important upshot here is easy to state: whereas in the A^a field formulation of electromagnetism, gauge transformations yielded distinct but non-isomorphic models, in the fibre bundle formulation of electromagnetism, gauge transformations yield distinct but *isomorphic* models, which differ by some diffeomorphism d.[34]

Since, in this formulation of electromagnetism, gauge-related models are isomorphic, the analogy with the static shift in NGT, and the hole argument in GR, is more exact. Given this, one might suggest that, in this formulation of electromagnetism, one *can* identify indexically the value of the gauge potential, just as one can (the claim from e.g. Maudlin goes) identify indexically one's absolute position in the spacetime case. Is this indeed the case? The answer is far from obvious to us, for recall that Yang-Mills fields are represented by *connections* on a principle bundle. As Jacobs writes, "While we can easily interpret a section as an assignment of field values to spacetime points, the same is not the case for the connection. The connection specifies relations between points of the principal bundle" (Jacobs, 2021, p. 193); as a result of this relational status, it is not *prima facie* clear whether it is the kind of thing which can indeed be identified indexically.

In Jacobs (2021, §7.4), Jacobs goes on to distinguish a "deflationary approach," according to which "neither the principal bundle nor the connection on it represent anything physical" (Jacobs, 2021, p. 193). from an "inflationary approach," which "reifies not the principal bundle but the bundle tangent to it" (Jacobs, 2021, p. 193). On the former approach, since the connection is not reified, one might claim that it is not (at least in any straightforward sense) amenable to indexical identification. On the latter approach, one reifies the "bundle of connections," which is the tangent bundle TP to the principal bundle P, quotiented by the latter's (Lie) group structure. In this case, one could—potentially!—apply (a version of) Maudlin's epistemological argument in order to maintain that one can identify indexically the value of the gauge potential at one's spacetime point. However, even having gone through this mathematical rigmarole, it is not obvious to us that the approach succeeds. The reason is that the approach still appears sensitive to one's preferred metaphysics of fibre bundles: if one is a "fibre bundle substantivalist" *à la* Arntzenius (2012), according to which the entire bundle is to be construed on the model of an extended spacetime, then perhaps there is room for the observer to maintain that they can identify indexically their position in the entire bundle-theoretic structure, including the bundle of connections, and thereby identify indexically the value of the gauge potential at their point in the fibre bundle. However, if one does not embrace this metaphysical picture, then one would have to give further arguments to the effect that one can identify indexically properties of objects (in this case, sections of the bundle of connections) at one's spacetime point. Absent further argumentation, it is not clear that this is the case: a sufficiently strong Hawthorne-Manley liberal about reference and singular thought might assent to it, but others might demur.[35]

The upshot of this section, then, is that there are important differences between external (spacetime) transformations and internal ("gauge") transformations. In order to resolve both the underdetermination and indeterminism problems in the case of electromagnetism (in its standard A^a field formulation), the epistemological argument will not suffice, and one must instead (absent some other solution!) appeal to (the perhaps somewhat implausible position of) gauge essentialism. While Healey maintains that there is inexpressible ignorance of the value of the A^a field at one's spacetime point in the standard formulation of electromagnetism, we remain unconvinced; to our minds, shifts to the gauge potential by a gradient term are better understood on the model of kinematic shifts in NGT set in Newtonian spacetime. That being said, there is greater room to claim that there is inexpressible ignorance—or, on the other hand, to run Maudlin's epistemological argument—when one moves to the fibre bundle formulation of electromagnetism, in which gauge-related models are isomorphic. However, the nature of these isomorphic models is very different here to in the spacetime cases of NGT or GR, and it is not obvious to us that arguments such as Maudlin's have any traction in such cases.

1.6   Close

In this chapter, we have undertaken a detailed study of Maudlin's preferred response to the static shift in NGT—namely, a certain kind of epistemological argument. We have distinguished "good" and "bad" versions of this argument (§1.2)—as in the case of (on one plausible reading) Maudlin on the spatial versus temporal static shifts, respectively—and have articulated why Maudlin does not offer this argument in response to the hole argument in GR (§1.3). In addition, we have addressed two objections to this argument due to Dasgupta (§1.4), with a focus on

Dasgupta's response based upon the notion of inexpressible ignorance; broadly speaking, we have sided with Maudlin and against Dasgupta in these disputes. Finally, we have addressed the possibility (or, as it turns out, impossibility) of the application of the argument to the parallel case of shifts to the vector potential of electromagnetism (§1.5).

Acknowledgements

We are especially indebted to Thomas Møller-Nielsen and Oliver Pooley for valuable discussions on this material, and are also grateful to Neil Dewar, Richard Healey, Mike Hicks, Caspar Jacobs, Joanna Luc, Jer Steeger, and the anonymous referee (who turned out to be Carl Hoefer) for very helpful feedback. We thank Trevor Teitel for pointing us to the work of Hawthorne and Manley. We are very grateful to Antonio Vassallo for inviting us to contribute to this volume. J.R. is supported by the John Templeton Foundation, grant number 61521.

Appendix: History of the epistemological argument

In this appendix, we seek to cast light on some of the history of the epistemological argument which has been the focus of this chapter. To our knowledge, the first mention of something like this argument is due to Horwich, who writes:[36]

> [I]t seems to me that Leibniz' argument is quite seductive. For if time does consist of a set of entities—the temporal instants—then it would (given some further plausible assumptions) be the case that there is a possible world, different from the actual one, in which every actual event occurs, but 10 seconds earlier than it does in the actual world. Yet one is reluctant to admit that such a possible world would really be different from the actual world.
>
> (Horwich, 1978, p. 407)

Horwich continues (we take the liberty of an extended quotation, since the passage is of enviable clarity):

> Why is this? I suspect it's because, not only would such a world be indiscernible, (in the weak sense) from the actual world, but also there could not be an epistemological issue concerning which of the worlds was actual. In general there may be cases of theories T_1 and T_2, which are observationally indistinguishable and present us with an epistemological problem. We don't know if we are in a world described by T_1 or a world described by T_2. But the supposed difference between W_1 and W_2 is such that no such epistemological problem can arise. Even if we had found

ourselves in W_2 we would have arrived at precisely the same theory of the world.

Now this epistemological equivalence does not immediately entail that every sentence true in W_1 is also true in W_2. For if there are instants, then we may introduce reference fixing definitions such as:

Let k be the instant at which event E occurs.

And then the sentence "E occurs at k" will be true in W_1 but false in W_2. However, these differences engender no epistemological problems because those sentences which are true in W_2, yet false in W_1, are known *a priori* to be false in W_1. More precisely: there is a certain set of sentences (such as "E is simultaneous with F") whose members are true in both W_1 and W_2. And any sentence which is true in W_1 but false in W_2 follows from the members of this set together with reference fixing definitions such as the one described earlier. Thus we can be quite sure that the sentences which describe W_2 do not all describe our world, even though they have the same observational consequences as those which do.

(Horwich, 1978, pp. 407–408)

For Horwich, the epistemological *discernibility*, but observational *indiscernibility*, of shifted worlds motivates the conclusion that they should not, after all, be regarded as being genuinely distinct. This, clearly, is a conclusion which Maudlin does *not* draw—and one might reasonably find Horwich's thought that the epistemological argument leads one to regard the shifted solutions as not being genuinely distinct as lacking positive motivation. In any case, Pooley is certainly correct when he writes on the difference between these two authors that "[h]ere Maudlin and Horwich are simply at loggerheads" (Pooley, 2015, p. 81).

Let us move on now to the second sustained appearance of the epistemological argument in the literature,[37] which (again, to our knowledge) is due to Field, who once again reaches a conclusion interestingly distinct from that of Horwich:

Now, one possible reply to [Leibniz's argument] (one that Horwich himself develops fairly persuasively in the article mentioned) is that the supposition [of statically shifted scenarios] isn't unreasonable in the way that it first appears, and that this is so because the possibility granted ... can't be used to generate epistemological problems about which sort of universe one is in.

(Field, 1984, pp. 76–77)

There are several points to be made here. As Pooley notes, Field misconstrues Horwich's point:

It is ironic that Field attributes essentially this response to the shift argument to Horwich for, in fact, Horwich's intuition propels him in the opposite direction.

(Pooley, 2015, p. 81)

That is, for Horwich, the epistemological discernibility (together with the observational indiscernibility) of shifted scenarios suggests that, in fact, these scenarios are *not* genuinely distinct. Field, however, reads Horwich in the opposite manner: he suggests that such epistemological discernibility is precisely what renders it permissible to regard shifted scenarios as genuinely distinct. Whatever one makes of this argument *per se*, it is clear that that it is inadequate as a reading of Horwich; it does, however (broadly speaking), align Field with Maudlin on this issue.

The final author deserving discussion in this section—and the third (to our knowledge) to write on the epistemological argument—is Teller (1987), who states the following:

Reference in a possible world is established by reference in the actual world. . . . In particular, if I talk about a counterfactual situation in which everything has been uniformly moved over from where it now actually is, I speak of a situation in which, for example, I have been moved over from where I actually am. . . . Thus, the counterfactual case in which everything has been uniformly moved over is, we can now see, distinguishable from the actual case; for in the counterfactual case, I (to pick an arbitrary example) am now at a different place from the place I can identify in the counterfactual situation as the place I now actually occupy. . . . This distinguishability is no "deep" metaphysical fact. It is just a reflection of facts about how language works in counterfactual contexts.

(Teller, 1987, pp. 443–444)

Teller, unlike Horwich but like Field and Maudlin, is comfortable with regarded statically shifted scenarios as being genuinely distinct. It is clear, then, that the epistemological argument should not be attributed to Maudlin alone. However, what is certainly novel about Maudlin's contribution on these matters is that he is the first to contrast the static shift with the kinematic shift—it is only with Maudlin that we find explicitly the point that the same kind of epistemological argument cannot be mustered in the latter context:

In sum, the *only* way that the static shift can be formulated is something like, "what if God had created the material universe oppositely oriented *to the way it is oriented now*?", and this is clearly a counterfactual situation. But we can ask "what if God created the material

universe at absolute rest?", not knowing whether we describe a counterfactual situation or not.

(Maudlin, 1993, p. 191)

Notes

1. Why do we write here "having fixed a representational context"? Being isomorphic, both of $\mathcal{M}$ and $d_*\mathcal{M}$ are equally apt to represent the same possibilities. However, if one of these models is elected to represent a possible
   world—thereby fixing a representational context—then the other model
   cannot be taken to represent that world, at least within the same representational context. This being said, even within a representational context, the
   worlds represented by such models are (almost) invariably taken to be empirically equivalent (the only author to deny this claim, to our knowledge, is
   Weatherall, 2018). For further discussion of all of these points, see Pooley
   and Read (2021).
2. This second point implies, according to Earman and Norton (1987), that GR
   is radically indeterministic. Note that this is a(n apparent) fact about GR,
   and can be divorced from epistemological considerations; this will be of relevance later.
3. Of course, a positive metaphysical picture of a general relativistic world in
   which the reality of the manifold is denied has yet to be offered: see
   Earman (1989, ch. 9) for one attempt to do this, which makes appeal to
   "Einstein algebras."
4. In fact, the situation is more subtle, in light of a critique of Maudlin's metric
   essentialism due to Norton (1988). Norton's challenge is to identify *which* of
   a class of isomorphic models can represent a possibility—why *this particular
   model*, and how can we distinguish this model from an "imposter"? We
   endorse Pooley's response to Norton:

   > Abstracting from the pragmatics of representation, all isomorphic models
   > *are* equally suited to represent the same spacetime. But, in practical situa-
   > tions, some model or other will be singled out, normally *quite arbitrarily*,
   > to represent a physical possibility. The advocate of [Maudlin's position]
   > claims only that, *relative* to such a choice of one model, the others must
   > be viewed either as representing impossible worlds (*per* the haecceitist essen-
   > tialist) or as representing nothing at all (*per* the anti-haecceitist).
   > (Pooley, 2002, p. 101)

5. Since $\mathcal{N}$ and $d_*\mathcal{N}$ are isomorphic, the same points as articulated in endnote 1
   apply in this case.
6. See Pooley (2015, p. 80) for a similar analysis (cf. also Pooley and Read,
   2021). We are in full agreement with Pooley when he writes the following
   of Maudlin's argument in the case of the temporal static shift:

   > Now this is not quite right. If we hear the counterfactual "things would look
   > awfully different if the big bang had occurred four minutes ago" as true, this
   > is because the time at which we place ourselves in the counterfactual sce-
   > nario is now. Things would indeed look awfully different if we could but
   > exist in such conditions to make any observations. But this counterfactual
   > scenario is not the one that the static shift asks us to consider. In another

billion years there is "someone who looks just like me writing a sentence that looks just like this" because, in the appropriately Leibniz-Shifted world, that's where I am temporally located. Had the world been as the shift scenario describes, things would have seemed to me to be just the same.

(Pooley, 2015, p. 80)

7. All notions of observational indistinguishability in the foregoing are "immanent": they are associated to "how things look" for an observer embedded in a world. There is also a "transcendental" notion of observational equivalence, which is associated to field values at spacetime points (different field values at the same point in different models being associated to transcendental observational distinguishability). By introducing this distinction, one can, perhaps, argue that Maudlin is making use of the transcendental notion of observational equivalence in the first of these passages on temporal static shifts—but would agree that the models remain immanently observationally indistinguishable. For more on these two notions of observational distinguishability, see Pooley and Read (2021).

8. Here, we are assuming for the sake of simplicity an anti-haecceitist ontology of spacetime points.

9. We thank Carl Hoefer for very clarifying remarks on this and the previous paragraph.

10. It is worth noting that metric essentialism will not work as a solution to the static shift problem in NGT, for in this theory the geometrical structure of Galilean spacetime is *fixed*, and only the matter content is altered on implementing a static shift. An obvious response here would be to appeal to a generalization of metric essentialism—e.g., (i) "matter density essentialism," according to which manifold points have essentially their matter density values given by ρ, or (ii) "gravitational potential essentialism," according to which manifold points have essentially their gravitational potential values given by φ. Whether such views have the same kinds of virtues that Maudlin adduces for metric essentialism in Maudlin (1988) is, however, unclear. (One way to motivate (ii) might be to argue that, given Knox's "spacetime functionalism"—see Knox (2014), and Read (2016) for relevant discussion—φ is best understood as part of the spatiotemporal content of the world.) Cf. also "gauge essentialism," discussed in §1.5.

11. Saunders' argument is anticipated in Stein (1977); for discussion of Stein on this matter, see Earman (1989, p. 55).

12. Our thanks to Richard Healey for discussions on the content of this paragraph.

13. In endnote 35, we suggest that a sufficiently strong liberal about reference and singular thought (on this liberalism, see later) may be able to refer to their absolute velocity, even in the case of NGT set in Newtonian spacetime. Note also that it may be possible to secure reference to one's absolute velocity by associating a name with a definite description (e.g.: "Let k stand for my worldline's tangent vector right now."), rather than by using exclusively indexicals: for more on this possibility, see our discussion on Maudlin's epistemological argument presented in §1.4.1.3. (Such securing of reference should be possible both in the case of NGT set in Newtonian spacetime and NGT set in Galilean spacetime—though, of course, there would remain something of which of is ignorant in the former case: namely, the magnitude of one's absolute velocity.) Our thanks to Carl Hoefer for discussion on this point.

14. Although Dasgupta, in the following, focuses upon NGT set in Galilean spacetime, identical issues would arise in the case of NGT set in Newtonian spacetime.

15 Dasgupta's vacillation between "open" and "unanswerable" at Dasgupta (2011, pp. 146–147) is unfortunate; nevertheless, his wholesale bypassing of "open" in Dasgupta (2015b) suggests that our reading is reasonable.

16. Dasgupta in fact offers a third, slightly distinct motivation that we term the "God's favourite point" argument. This argument is analyzed and refuted in §1.4.2.

17. Maudlin makes unspoken assumptions regarding indexicals in order to formulate this position; we discuss this in further detail in §1.4.1.3.

18 Regarding this reading, one might reasonably ask: "how can there be inexpressible *ignorance*, yet no epistemological challenge?" The point is that one might be (inexpressibly) ignorant of *something*, and yet still (by hook or by crook) be able to identify which of a class of world's is one's own (said ignorance notwithstanding). In virtue of this latter ability, there is (in the sense in which we intend the notion here) no epistemological challenge posed—though, of course, there are still epistemological challenges *tout court*.

19. To clarify, it is that nothing is inexpressible in this *particular* case; while this reading is predicated on rejecting the first premise, we remain sceptical that the second premise (that all ignorance is expressible) would be endorsed by Maudlin in full generality.

20. See Dasgupta (2011, p. 462) for Dasgupta's definition of "full essence," which builds upon a broadly Finean notion of essence—see Fine (1994).

21. There is, indeed, a precedent for such objections in the literature: see e.g., Teitel (2019), Wildman (2021).

22. We are very grateful to Carl Hoefer for suggesting to us that Maudlin's account may be reconcilable with a descriptivist conception of indexicals.

23. One might maintain that even if Kaplan's account of indexicals is endorsed, and one can thereby refer directly to the spacetime point which constitutes one's absolute position, there might yet remain inexpressible ignorance—*viz.*, of facts underwriting the haecceities of spacetime points. To this, we would reply that (a) this is changing the subject, for the question at hand is whether one can indexically identify one's own position, and (on these assumptions) this *is* the case, regardless of whether one can know (or even express) everything there is to know about the haecceities of the spacetime points constituting such positions. Moreover, (b) at this point, Perry's response to Dasgupta kicks in once more: the burden remains on the proponent of such a view to explain why we should believe in such facts to begin with. We discuss these matters in further detail later.

24. Note also that, even if one grants that there is inexpressible ignorance in such cases, this does not necessarily mean, given the (admittedly problematic) first reading of Maudlin presented in §1.4.1, that there exists here an epistemological challenge.

25. There is also a third option, raised by Hawthorne and Manley (albeit not in the context of philosophical scepticism): perhaps reference can be secured *accidentally*. On this, Hawthorne and Manley write that "one might fail to refer with a putatively logically proper name and not know it; and one might also refer without knowing that one has done so (there being a real danger of illusion)" (Hawthorne and Manley, 2012, p. 8). Although this is certainly a line which one might take towards scepticism in response to Putnam (who, one anticipates, would repudiate the argument), we set it aside in the following.

26. Perhaps one can identify indexically real-world facts or objects *as actual*—but, beyond that, one does not seem to be able to identify them indexically in sceptical shift scenarios. To be clear, note also that there are two senses of "real" at play in the foregoing: the first meaning "non-brain-in-vat," and the second meaning "an element of the ontology of the world." (A Lewisian modal realist would, of course, repudiate the claim that what is real is exhausted by the ontology of the actual world: we set this aside here.) In our discussion of the sceptical scenarios, we generally had the former in mind, as should have been evident from context.

27. Recall the issues discussed in endnote 4, which apply also in this case—though note that, since models of electromagnetism related by a gauge transformation are not isomorphic, one might (following Fletcher, 2020) argue that they do not have the same representational capacities.

28. Cf. endnote 10.

29 Models related by a static shift are isomorphic—unlike models related by a kinematic shift, or by a gauge transformation. Conjecture: epistemological arguments of the kind considered in this chapter. which proceed by appeal to the indexical identifiability of models, can be applied only in the former case.

30. One may be able to use definite descriptions and the act of naming in order to refer to the gauge field at one's spatiotemporal location—cf. our discussion of such a possibility in §1.4.1.3. Just as in the case of the kinematic shift in NGT set in Newtonian spacetime, however, one would remain ignorant of the *magnitude* of that quantity—see endnote 13.

31. See Dewar (2019) for the original article on sophistication, and Jacobs (2021), Martens and Read (2020) for further explorations and (sometimes critical) discussions of the procedure.

32. There are some subtleties here—see Martens and Read (2020, fn. 26).

33. Our presentation of the formalism in what follows tracks the elegant discussion of Jacobs (2021, ch. 7); see e.g., Weatherall (2016) for further discussions of this formalism.

34. For a rigorous presentation of this result, see Jacobs (2021, §7.4).

35. This being said, if one is willing to claim that one can identify indexically one's own velocity in the case of Galilean spacetime—recall our discussion of this point in §1.3—then perhaps one can maintain by analogy that one can identify indexically the value of the gauge potential at one's spacetime point. Note also that this sufficiently strong liberal might simply maintain that one can identify indexically the value of the A^a field at their spacetime point, even in the standard formulation of electromagnetism.

36. Note that Horwich's concern here is with temporal static shifts, rather than spatial static shifts.

37. Between Horwich and Field, there is van Fraassen, who writes in passing that "Note for example that 'I am here' is a sentence which is true no matter what the facts are and no matter what the world is like, and no matter what context of usage we consider. Its truth is ascertainable *a priori*" (Fraassen, 1980, p. 136).

Bibliography

Arntzenius, F. (2012). *Space, Time, & Stuff*. Oxford University Press.

Braun, D. (2017). Indexicals. In E. N. Zalta (Ed.), *The Stanford Encyclopedia of Philosophy* (Summer 2017 ed.). Metaphysics Research Lab, Stanford University. https://plato.stanford.edu/archives/sum2017/entries/indexicals/.

Dasgupta, S. (2011). The bare necessities. *Philosophical Perspectives* 25 (1), 115–160.

Dasgupta, S. (2015a). Inexpressible ignorance. *Philosophical Review* 124 (4), 441–480.

Dasgupta, S. (2015b). Substantivalism vs relationalism about space in classical physics. *Philosophy Compass* 10 (9), 601–624.

Dasgupta, S. (2016). Symmetry as an epistemic notion. *British Journal for the Philosophy of Science* 67 (3), 837–878.

Dewar, N. (2019). Sophistication about symmetries. *British Journal for the Philosophy of Science* 70 (2), 485–521.

Earman, J. (1989). *World Enough and Spacetime*. MIT Press.

Earman, J. and J. Norton (1987). What price spacetime substantivalism? The hole story. *British Journal for the Philosophy of Science* 38 (4), 515–525.

Field, H. (1984). Can we dispense with space-time? *Proceedings of the Biennial Meeting of the Philosophy of Science Association* 1984, 33–90.

Fine, K. (1994). Essence and modality. *Philosophical Perspectives* 8 (Logic and Language), 1–16.

Fletcher, S. C. (2020). On representational capacities, with an application to general relativity. *Foundations of Physics* 50 (4), 228–249.

Fraassen, B. C. V. (1980). *The Scientific Image*. Oxford University Press.

Hawthorne, J. and D. Manley (2012). *The Reference Book*. Oxford University Press.

Healey, R. (1997). Nonlocality and the Aharonov-Bohm effect. *Philosophy of Science* 64 (1), 18–41.

Healey, R. (2007). *Gauging What's Real: The Conceptual Foundations of Contemporary Gauge Theories*. Oxford University Press.

Horwich, P. (1978). On the existence of time, space and space-time. *Noûs* 12 (4), 397–419.

Jacobs, C. (2021). *Symmetries as a Guide to the Structure of Physical Quantities*. Ph. D. thesis, University of Oxford.

Kaplan, D. (1989). Demonstratives: An essay on the semantics, logic, metaphysics and epistemology of demonstratives and other indexicals. In J. Almog, J. Perry, and H. Wettstein (Eds.), *Themes From Kaplan*, pp. 481–563. Oxford University Press.

Knox, E. (2014). Newtonian spacetime structure in light of the equivalence principle. *British Journal for the Philosophy of Science* 65 (4), 863–880.

Kripke, S. (1980). *Naming and Necessity*. Blackwell.

Lehmkuhl, D. (2011). Mass-energy-momentum: Only there because of space-time. *British Journal for the Philosophy of Science* 62 (3), 453–488.

Malament, D. B. (2012). *Topics in the Foundations of General Relativity and Newtonian Gravitation Theory*. Chicago University Press.

Martens, N. C. M. and J. Read (2020). Sophistry about symmetries? *Synthese* 199 (1–2), 315–344.

Maudlin, T. (1988). The essence of space-time. *Proceedings of the Biennial Meeting of the Philosophy of Science Association* 1988, 82–91.

Maudlin, T. (1993). Buckets of water and waves of space: Why spacetime is probably a substance. *Philosophy of Science* 60 (2), 183–203.

Maudlin, T. (1998). Healey on the Aharonov-Bohm effect. *Philosophy of Science* 65 (2), 361–368.

Maudlin, T. (2012). *Philosophy of Physics: Space and Time*. Princeton University Press.

McKinsey, M. (2018). Skepticism and content externalism. In E. N. Zalta (Ed.), *The Stanford Encyclopedia of Philosophy* (Summer 2018 ed.). Metaphysics Research Lab, Stanford University. https://plato.stanford.edu/archives/sum2018/ entries/skepticism-content-externalism/.

Nagel, T. (1986). *The View From Nowhere*. Oxford University Press.

Norton, J. D. (1988). The hole argument. *Proceedings of the Biennial Meeting of the Philosophy of Science Association* 1988, 56–64.

Perry, J. (1979). The problem of the essential indexical. *Noûs* 13 (1), 3–21.

Perry, Z. R. (2017). How to be a substantivalist without getting shifty about it. *Philosophical Issues* 27 (1), 223–249.

Pooley, O. (2002). *The Reality of Spacetime*. Ph. D. thesis, University of Oxford.

Pooley, O. (2015). *The Reality of Spacetime*. Book Manuscript.

Pooley, O. (2017). Background independence, diffeomorphism invariance, and the meaning of coordinates. In D. Lehmkuhl, E. Scholz, and G. Schiemann (Eds.), *Towards a Theory of Spacetime Theories*, pp. 105–143. Birkhäuser.

Pooley, O. and J. Read (2021). On the mathematics and metaphysics of the hole argument. *The British Journal for the Philosophy of Science*. http://doi.org/10.1086/ 718274.

Putnam, H. (1981). Brains in a vat. In *Reason, Truth and History*, pp. 1–21. Cambridge University Press.

Read, J. (2016). *Background Independence in Classical and Quantum Gravity*. Master's thesis, University of Oxford.

Saunders, S. (2003). Indiscernibles, general covariance, and other symmetries: The case for non-reductive relationalism. In A. Ashtekar (Ed.), *Revisiting the Foundations of Relativistic Physics*, pp. 151–173. Springer.

Stachel, J. (1993). The meaning of general covariance. In J. Earman, A. Janis, and G. Massey (Eds.), *Philosophical Problems of the Internal and External Worlds: Essays on the Philosophy of Adolph Grünbaum*, pp. 129–160. University of Pittsburgh Press.

Stein, H. (1977). Some philosophical prehistory of general relativity. In J. Earman, C. Glymour, and J. Stachel (Eds.), *Foundations of Space-Time Theories: Minnesota Studies in the Philosophy of Science*, pp. 3–49. University of Minnesota Press.

Teitel, T. (2019). Contingent existence and the reduction of modality to essence. *Mind* 128 (509), 39–68.

Teller, P. (1987). Space-time as a physical quantity. In P. Achinstein and R. Kagon (Eds.), *Kelvin's Baltimore Lectures and Modern Theoretical Physics*, pp. 425–448. MIT Press.

Weatherall, J. O. (2016). Fiber bundles, Yang-Mills theory, and general relativity. *Synthese* 193 (8), 2389–2425.

Weatherall, J. O. (2018). Regarding the 'hole argument'. *British Journal for the Philosophy of Science* 69 (2), 329–350.

Wildman, N. (2021). Against the reduction of modality to essence. *Synthese* 198, 1455–1471.

2 Plugging holes in sophisticated substantivalism

Tomasz Bigaj

2.1 Introduction

The hole argument in general relativity does not cease to inspire and intrigue generations after generations of philosophers of physics. Since the argument rose from obscurity to fame thanks to the "three Johns" (John Stachel, John Earman and John Norton),[1] much ink has been spilled on the topic of its philosophical importance. One particular thread in the discussion that spans over thirty years and counting centres around the position known today as "sophisticated substantivalism." Sophisticated substantivalism promises to rewrite the general lesson from the hole argument, which is that spatiotemporal substantivalism is condemned to a radical form of indeterminism. The current chapter is directly inspired by a recent article due to Carolyn Brighouse, an erstwhile proponent of sophisticated substantivalism (Brighouse, 1994, 1997), who has lately come to doubt its intuitive appeal (Brighouse, 2020). Her worries stem from the fact that sophisticated substantivalism with its anti-haecceitistic component is unable to account for our intuitions regarding the Leibnizian shift, as well as the cases of symmetry-breaking indeterministic processes. Unwilling to fully embrace the opposite stance of haecceitism, Brighouse leaves us in limbo, waiting for some arguments that could break the impasse. While I am not foolhardy enough to think that I could rise to the challenge and solve the problem once and for all, I would like to point out that perhaps Brighouse's misgivings are a bit overstated. The situation for sophisticated substantivalism is not as dire as portrayed, as it has plenty of strategies at its disposal to deal with the troublesome cases.

Apart from a gentle polemic with Brighouse, I would like to address some other issues pertaining to the hole argument which have been relatively neglected. The hole argument is based on the mathematical concept of diffeomorphic transformations, but the role of these transformations is somewhat obscure. In particular, suggestions have been made to the effect that the hole argument could equally well be reconstructed using arbitrary permutations instead of diffeomorphisms (smooth transformations of manifolds). I believe that this view is basically incorrect,

DOI: 10.4324/9781003219019-4

and therefore I will try to delve deeper into the mathematical underpinnings of the hole argument to show why the assumption of the smoothness of the hole transformations is essential for the argument to go through. Another point worth considering is more philosophical in nature. It is the question of the role of some metaphysical presuppositions that are so often invoked in the debates on the hole argument and the fate of substantivalism. In particular, the philosophical stance of essentialism is typically used to boost the case for sophisticated substantivalism. And yet another recent article by Trevor Teitel casts doubts on this strategy (Teitel, 2019). At the end of this chapter I will try to explain why I think it is useful to appeal to some concepts from the metaphysics of modality, including the notion of essential properties, and in that way I will hopefully respond to Teitel's objections.

The chapter is structured as follows. Section 2.2 will begin with a slightly technical exposition of the concept of smooth transformations of differential manifolds. On the basis of this elementary analysis, I will explain why the smoothness of the hole transformations is required, and I will criticize an argument due to Joseph Melia (Melia, 1999) that the hole argument can be restated using arbitrary permutations. In section 2.3 I will delineate the stance of spatiotemporal substantivalism and categorize its variants using the concept of essential properties and generally the concept of representation *de re*. I will show how sophisticated substantivalism, based on the assumption that the metric properties of spatiotemporal points and regions are both necessary and sufficient to retain their identity in alternative scenarios, can pass a modified acid test loosely based on the Leibniz shift, as required by Earman and Norton. Section 2.4 is devoted to an analysis of cases where symmetry is broken in an apparently indeterministic fashion (e.g., the collapsing tower example). While on the face of it sophisticated substantivalism seems unable to deliver the right analysis of these examples, due to the fact that the alternative variants of the future considered there are globally qualitatively indistinguishable, I will show that there are actually as many as three strategies to overcome this obstacle. In Conclusion I will mention Teitel's argument to the effect that essentialism does not offer a satisfactory metaphysical explanation for anti-haecceitism. Rather than countering his argument directly, I will point out that the role of essentialism is to provide a clear characterization of the difference between substantivalism and relationism, as well as between different variants of substantivalism, as explained in Section 2.3.

2.2 Mathematics of diffeomorphism invariance

The mathematical underpinnings of the hole argument are typically presented in the philosophical literature in a much-simplified form.[2] This way of presentation gives a broadly correct picture of what is going on in the argument, but at the same time may give rise to some misunderstandings and misinterpretations. The usual presentation begins with identifying a structure of the form $\langle M, O_1, \ldots, O_n \rangle$ where M is a differential manifold representing spacetime, and $O_1, \ldots, O_n$ some geometric objects defined on points. Among the geometric objects we typically include the metric tensor g and the stress-energy tensor T. Then we introduce a particular transformation d of manifold M, $d:M \to M$, called a diffeomorphism. Roughly, a diffeomorphism of M is a smooth bijection defined on M (a more precise definition will be given later). Under the action of a diffeomorphism d the entire structure is transformed into a new one $\langle M, d^*O_1, \ldots, d^*O_n \rangle$, where d^* is the transformation on the set of geometric objects "induced" by d (sometimes d^*O_i are called "drag-along" or "carry-along" objects). A popular gloss over the notion of a drag-along object d^*O_i is that it is just object O_i which was initially attributed to point p, "moved" to the image point dp: $d^*O_i(dp) = O_i(p)$. However, later we will see that this is not an entirely accurate way of thinking about objects d^*O_i.

The gist of the hole argument is the comparison of the diffeomorphically-related models $\mathcal{M} = \langle M, O_1, \ldots, O_n \rangle$ and $d\mathcal{M} = \langle M, d^*O_1, \ldots, d^*O_n \rangle$. On the one hand, they are different in that they assign different geometric objects to the 'same' points in M. On the other hand, due to the general covariance (resp. diffeomorphism invariance)[3] of the equations of GR, both structures represent nomologically possible worlds. Given that it is possible to find a diffeomorphism which is the identity on all points in M outside a certain region H and a non-identity inside H, this leads to the conclusion that GR is a radically indeterministic theory (fixing all the facts outside H does not fix the facts inside H). This way of casting the hole argument, while correct in rough outline, leaves out certain important elements. In particular, it is unclear what precisely the role of diffeomorphisms and the diffeomorphism-invariance of GR is supposed to be in the argument.

At first sight it may seem that the models $\mathcal{M}$ and $d\mathcal{M}$ are simply isomorphic by design, regardless of whether d is smooth or not, and regardless of any property the GR equations may or may not possess (general covariance, diffeomorphism invariance).[4] The standard definition of isomorphism between structures in the model-theoretic sense is well-known. A model-theoretic structure $\mathfrak{R}$ has the form $\langle D, R_1, \ldots, R_n, f_1, \ldots, f_m \rangle$, where D is a non-empty set (the domain), R_i are relations on D of a given arity (interpreted as subsets of Cartesian products $D \times D \ldots \times D$ of the corresponding number of factors), and f_i are functions from $D \times D \ldots \times D$ to some sets of mathematical objects (e.g., numbers). Two models $\mathfrak{R} = \langle D, R_1, \ldots, R_n, f_1, \ldots, f_m \rangle$ and $\mathfrak{R}' = \langle D', R'_1, \ldots, R'_n, f'_1, \ldots, f'_m \rangle$ are isomorphic iff there if a bijection $\phi : D \to D$ such that $R'_i(\phi a_1, \ldots, \phi a_k)$ iff $R_i(a_1, \ldots, a_k)$ for all R_i and $a_1, \ldots, a_k \in D$, and $f'_i(\phi b_1, \ldots, \phi b_k) =$

$f(b_1, \ldots, b_k)$ for all f_i and $b_1, \ldots, b_k \in D$. Given our earlier interpretation of drag-along objects $d^* O_i$, it seems obvious that models $\mathcal{M}$ and $d\mathcal{M}$ satisfy the definition of isomorphic structures. But from a mathematical point of view isomorphic structures are just "notational variants" of the same underlying abstract structure, and therefore if one such structure is a representation of a possible world, so is the other. And apparently the fact of the isomorphy of $\mathcal{M}$ and $d\mathcal{M}$ does not depend on the assumption that d is a diffeomorphism. This gives rise to the suggestion, made by some, that the hole argument remains valid for any arbitrary permutation of manifold M, and that the argument has very little if anything to do with the particular features of General Relativity.[5] As a matter of fact, some claim that the essence of the hole argument is based on the assumption of general permutability, which is satisfied by any reasonable theory, not necessarily a space-time theory.[6] Roughly, this assumption states that a physical theory should be "blind" as to which object is which—any object can play the role of any other object as long as all their properties and relations are appropriately exchanged.

In what follows I will claim that this interpretation unnecessarily waters down the lesson from the original hole argument, making it an almost trivial observation (trivial at least from the perspective of a physicist—philosophers can always claim that an alternative in the form of some variant of haecceitism or primitive identity is conceivable). To see that, we will have to pay closer attention to some mathematical details of the diffeomorphic transformations of models of GR.[7] Let us start with a proper definition of a manifold (a smooth, or differentiable manifold). A manifold is not simply a set of bare, unstructured objects. It already possesses an important internal structure whose mathematical representation is given in the form of *charts*. More specifically, an n-manifold is a pair $(M, \mathfrak{C})$, where $\mathfrak{C}$ is a set of charts. A chart φ is a map from a subset $U \subseteq M$ to $\mathbb{R}^N$ such that φ is one-to-one and the image $\varphi[U]$ is an open set in $\mathbb{R}^N$ (φ can be thought of as assigning coordinates to spacetime points in the form of n-tuples of numbers). The set $\mathfrak{C}$ satisfies the following conditions:

i   all charts in $\mathfrak{C}$ are compatible with each other,
ii $\mathfrak{C}$ covers the entire manifold M, i.e., for every point $p \in M$ there is a chart whose domain contains p,
iii $\mathfrak{C}$ satisfies the condition of Hausdorff-separability.

Sometimes the fourth condition of maximality is added, stating roughly that $\mathfrak{C}$ is a maximal set satisfying (i)–(iii). We don't have to do that, since any set satisfying (i)–(iii) can be extended to a maximal one. As David Malament has shown (Malament, 2012, p. 3), for any manifold $(M, \mathfrak{C}_0)$

meeting the conditions (i)–(iii), if $\mathfrak{C}$ is the set of all n-charts compatible with all charts in $\mathfrak{C}_0$, then the manifold $(M, \mathfrak{C})$ is maximal. The details of conditions (i) and (iii) need not concern us. We can only mention that the compatibility of two charts ensures that if their domains intersect, they differ from each other on the common part of the domains in a "smooth" way. Hausdorff-separability, on the other hand, stipulates that for every two points there will be non-overlapping charts containing these points.

It is important to note that we can use the charts defined on M to define a topology on M in the form of open sets. A set $S \subseteq M$ can be characterized as open, if for every $p \in S$ there is a chart $\varphi : U \to \mathbb{R}^N$ such that $U \subseteq S$ (recall that by definition $\varphi[U]$ is open in $\mathbb{R}^N$) Speaking loosely, open sets in M are counterimages of open sets in $\mathbb{R}^N$ with respect to charts φ.[8] The existence of charts enables us also to introduce the concept of smoothness (i.e., differentiability) with respect to maps between manifolds. Suppose first that α is a map from M to $\mathbb{R}$. We will call α smooth, if for every chart $\varphi \in \mathfrak{C}$, $\alpha \circ \varphi^{-1}$ is smooth (i.e., C^∞-differentiable). This definition is appropriate, since $\alpha \circ \varphi^{-1}$ is a map from a subset of $\mathbb{R}^N$ to $\mathbb{R}$ and therefore it makes sense to apply to it the usual notion of differentiability. Next we can say what it means to be smooth for any map ψ from one manifold $(M, \mathfrak{C})$ to any other manifold $(M', \mathfrak{C}')$. A map $\psi : M \to M'$ is smooth, if for any smooth $\alpha : M' \to \mathbb{R}$, the composition $\alpha \circ \psi : M \to \mathbb{R}$ is also smooth. Finally, if a map $\psi : M \to M'$ has its inverse ψ^{-1}, and both ψ and ψ^{-1} are smooth, then ψ is called a diffeomorphism.

Next, I will briefly explain why the condition of smoothness is paramount in order to properly define the operation of "dragging along" with respect to geometric objects, such as vectors and tensors. Even though this is probably differential geometry 101, much too often philosophers tend to forget about this elementary fact, so it is worthwhile to spend some time to explain it. In order to determine how a smooth map $\psi : M \to M'$ can carry along geometric objects from M to M', we have to first remind ourselves how these geometric objects are introduced in the first place. The key concept here is that of a tangent space at point p (symbolized by M_p). There are several ways to introduce tangent vectors and tangent vector spaces at points for a given manifold. A slightly abstract but elegant way to do that is to define vectors as linear operators possessing the characteristic properties of derivative operators from the calculus. We begin by considering the set $S(p)$ of all smooth maps from an open set O containing p to $\mathbb{R}$. Now we can define a tangent vector ξ at p as a functional on $S(p)$ with the following properties:

i $\xi(f_1 + f_2) = \xi(f_1) + \xi(f_2)$,
ii $\xi(f_1 f_2) = f_1 \xi(f_2) + f_2 \xi(f_1)$
iii if $f = \text{const}$, then $\xi(f) = 0$.

Obviously, functionals ξ form a vector space (they are closed under addition $(\xi_1 + \xi_2)(f) = \xi_1(f) + \xi_2(f)$ and multiplication by numbers), which justifies calling them vectors.

The tangent vector space M_p serves as a basis for introducing higher-ranking geometric object, such as tensors. However, we will skip the well-known details how to do that, limiting ourselves to the case of contravariant vectors as an illustration. Suppose that we have two manifolds $(M, \mathfrak{C})$ and $(M', \mathfrak{C}')$, and that they are connected by a smooth map $\psi : M \to M'$ (which, at this point, need not be one-to-one or onto). The question is how ψ transforms contravariant vectors associated with points in M and M'. Let us select any vector ξ^a from M_p. It is natural to associate with ξ^a a vector from $M'_{\psi(p)}$ which acts upon maps from the set $S(\psi(p))$ the "same" way ξ^a acts on the "corresponding" maps from $S(p)$. That is, we define the "transformed" vector $\psi_*(\xi^a)$ as follows: for every $f \in S(\psi(p))$, $\psi_*(\xi^a)(f) = \xi^a(f \circ \psi)$ (see Figure 2.1). Note that if f is a smooth map on a subset of M', the composition $f \circ \psi$ is a smooth map on a subset of M thanks to the assumption that ψ is smooth. So the procedure is as follows: in order to define a tangent vector $\psi_*(\xi^a)$ at $\psi(p)$ in M' which is a ψ-counterpart of a given vector ξ^a at p in M, we take any smooth map f on an open subset of M' containing $\psi(p)$, we "pull" map f back to an open set in M containing p (using ψ as the link), and we stipulate that our new vector acts the same way on f as vector ξ^a acts on the pull-back map $f \circ \psi$.

There are two general observations that can be made on the basis of the careful mathematical analysis of the diffeomorphism-induced

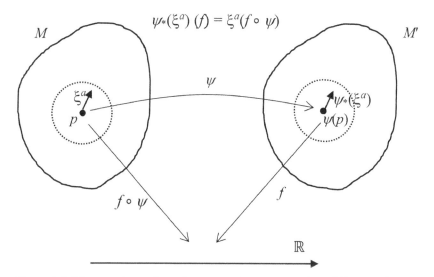

Figure 2.1 The transformation of contravariant vectors via a diffeomorphism.

transformations of tensor fields. Both observations point towards the suggestion, made at the beginning of this section, that the often-heard interpretation of diffeomorphically related models of GR as akin to models isomorphic under arbitrary permutations is misleading at the very least. First, observe that the assumption of the smoothness of the map ψ between two manifolds is necessary in order to properly introduce the transformations of tensors induced by ψ in the form of pullbacks ψ^* or pushforwards ψ_*. For instance, in order to identify the ψ-counterpart of a given contravariant vector ξ^a in M_p, we have to smoothly transform maps from $S(\psi(p))$ to $S(p)$. If ψ is not smooth, the composition $f \circ \psi$ containing an arbitrary smooth map f from $S(\psi(p))$ may not be smooth, and therefore may not belong to $S(p)$. As a result, the definition $\psi_*(\xi^a)(f) = \xi^a(f \circ \psi)$ ceases to make mathematical sense (the right-hand side of the equality is not well-defined when $f \circ \psi$ is not an element of $S(p)$). For instance, suppose that the map $\psi_{pq} : M \rightarrow M$ just permutes two points p and q, i.e., $\psi_{pq}(p) = q$, $\psi_{pq}(q) = p$ and $\psi_{pq}(x) = x$ for all other elements of M. In order to find a ψ_{pq}-counterpart of a selected vector $\xi^a \in M_p$ in M_p, we have to take an arbitrary smooth function f in $S(q)$, compose it with ψ_{pq} and apply ξ^a to this function. But function $f \circ \psi_{pq}$ is generally not continuous at p (except when $f(p) = f(q)$) since $f \circ \psi_{pq}(p) = f(q) \neq f(p)$, and $f \circ \psi_{pq}(x) = f(x)$ for all x in the immediate vicinity of p. Thus the action of derivative operators on f at p is not well defined.

Moreover, it can be further noted that fixing a smooth transformation $\psi : M \rightarrow M'$ at one point $p \in M$ does not fix the corresponding transformation ψ_* of contravariant vectors from M_p. Suppose that we have two distinct smooth transformations (diffeomorphisms) ψ and $\bar\psi$ connecting manifolds M and M' which happen to coincide at one point p: $\psi(p) = \bar\psi(p)$. In spite of that, the transformed contravariant vectors $\psi_*(\xi^a)$ and $\bar\psi_*(\xi^a)$ where $\xi^a \in M_p$, will be generally different. The appropriate definitions will be as follows: $\psi_*(\xi^a)(f) = \xi^a(f \circ \psi)$ and $\bar\psi_*(\xi^a)(f) = \xi^a(f \circ \bar\psi)$. Even though functions $f \circ \psi$ and $f \circ \bar\psi$ coincide at point p, they are different functions on the vicinity of p, and therefore the values of the differential operator ξ^a taken on these functions at p may differ. (Think about two different functions from $\mathbb{R}$ to $\mathbb{R}$, which intersect at a certain point but have different slopes at this point – the values of the derivatives of these functions at the selected point will be different).

Consequently, the hole argument cannot be extended to cover any arbitrary permutations of points. And yet Joseph Melia claims that this is precisely what happens: the argument remains valid even in the previously considered case in which the transformation of the original model merely swaps two points p and q, so we don't have to limit ourselves to diffeomorphic transformations (Melia, 1999, pp. 642–643). His argument, however, is based on an important modification of the original setup: rather than defining a map between a manifold $(M, \mathfrak{C})$ and itself, he effectively considers two distinct manifolds $(M, \mathfrak{C})$ and $(M, \mathfrak{C}')$ that share the "same" set of points M but differ in their charts. The difference is small but significant: every chart φ' in $\mathfrak{C}'$ is created by taking a chart φ from $\mathfrak{C}$ and switching the coordinates that φ assigns to points p and q: $\varphi'(p) = \varphi(q)$, $\varphi'(q) = \varphi(p)$, $\varphi'(x) = \varphi(x)$ for all the other points x.[9] Observe, moreover, that with this modification the permutation ψ_{pq} of points p and q *does* count as a diffeomorphism of manifolds $(M, \mathfrak{C})$ and $(M, \mathfrak{C}')$. The smoothness of the map ψ_{pq} can be quickly argued for as follows: take the composition $\varphi' \circ \psi_{pq} \circ \varphi^{-1} : \mathbb{R}^N \to \mathbb{R}^N$, for any chart φ. It is straightforward to verify that this composition is the identity. Speaking loosely, the map ψ_{pq} acting on 'bare' points, when cast in terms of the coordinates, turns out to be the identity, and thus very much smooth.

The second observation expands on the argument against Melia's claim that arbitrary permutations can be used to create alternative models of GR describing possible worlds in a way that threatens the thesis of determinism. Melia's position can be seen as flowing from an attractive and yet basically incorrect suggestion that models $\mathcal{M} = \langle M, O_1, \ldots, O_n \rangle$ and $d\mathcal{M} = \langle M, d^*O_1, \ldots, d^*O_n \rangle$ are isomorphic in the sense of isomorphism applicable to model-theoretic structures.[10] It is tempting to broadly characterize the "carried-along" geometric objects $d^*O_1, \ldots, d^*O_n$ attributed to point q as being identical with the geometric objects $O_1, \ldots, O_n$ associated with p, where $dp = q$ (and thus to adopt the identity $O_i(p) = d^*O_i(dp)$ for all points p). But if that were correct, then the resulting structure would be guaranteed to be isomorphic with the original one no matter what bijection d is (whether it is a diffeomorphism or any arbitrary permutation). The entire argument would become an exercise in general permutability rather than diffeomorphic invariance. But now we can see that this approach is misguided. The transformed objects $d^*O_i(q)$ (tensors) are not *identical* with the original ones $O_i(p)$. For starters, they "live" in different spaces—one built out of the tangent space M_p at p, the other built out of M_q at q. The transformation d^* should not be seen as connecting *the same* object attributed to different points, but rather as connecting an object defined at point p with its *d-counterpart* at q (and for that reason the commonly used terms "carry along," "drag along" or "shift" are misnomers—we are not "dragging" any mathematical object from p to q but rather identify among the objects already attached to q the one that can "play the role" of the original one at p). And the most important fact is that in order for this counterpart relation to make sense, the applied transformation must be a diffeomorphism. An arbitrary permutation won't do, for in that case the procedure of identifying an object corresponding to $O_i(p)$ among the objects attributed to q won't make mathematical sense.

2.3   From the hole argument to sophisticated substantivalism

Regardless of whether we agree or disagree that diffeomorphically-related models $\mathcal{M} = \langle M, O_1, \ldots, O_n \rangle$ and $d\mathcal{M} = \langle M, d^*O_1, \ldots, d^*O_n \rangle$ satisfy the formal, model-theoretic definition of isomorphic structures, we have to acknowledge that they represent *qualitatively indistinguishable* worlds. That is, all sentences formulated in a language without individual constants whose primitive vocabulary contains only empirically meaningful predicates should have the same logical value in both models. The only potential difference between the worlds represented by models $\mathcal{M}$ and $d\mathcal{M}$ can be expressed in statements about individual objects, i.e., particular spatiotemporal points and regions. That is, it may be for instance argued that there is a sentence of the form "Point (region) *a* has such-and-such metric properties" which is true in $\mathcal{M}$ but false in $d\mathcal{M}$. However, such a sentence does not constitute part of the qualitative characteristic of the possible worlds represented by $\mathcal{M}$ and $d\mathcal{M}$, due to its reference to the individual object *a*.

Still, many insist that statements about individual points and regions should be included in the language of the spacetime substantivalist. According to a popular view, the substantivalist is committed to acknowledging an objective difference between models of spacetime, one that is based purely on individual facts regarding which points and regions have which properties. As Earman and Norton put it, an "acid test" for substantivalism is provided by an assessment of the Leibniz shift in a Newtonian (or neo-Newtonian), flat spacetime (Earman and Norton, 1987, p. 521). The substantivalist should admit that two models of spacetime which differ only in that all the objects in the universe are moved ten feet in a given direction represent distinct physical possibilities in which same points and regions are occupied by distinct objects and fields, even though the resulting world is qualitatively indistinguishable from the original one. Consequently, the substantivalist must agree that the models of GR connected by a hole diffeomorphism represent distinct physical possibilities too, and thus determinism is violated.

A widely acknowledged reply to this argument is to insist that substantivalists are by no means obligated to treat diffeomorphism-related models as representing distinct physical possibilities. The key element here is the qualitative indistinguishability of these models. Even in a language whose variables range over spatiotemporal points and regions it is impossible to express a difference between *d*-related models $\mathcal{M}$ and $d\mathcal{M}$, as long as this language does not contain proper names for the spatiotemporal entities quantified over. Hence even those who acknowledge the existence of spatiotemporal objects may still insist that models $\mathcal{M}$ and $d\mathcal{M}$ are equivalent descriptions of the same underlying reality. The variant of substantivalism assuming that diffeomorphically related models (or, more broadly, any qualitatively indistinguishable models) represent the same reality, is known under the name "sophisticated substantivalism."[11]

However, the situation is a bit more complex than that. The division line between substantivalists and relationists in philosophy of spacetime does not separate the opponents simply according to whether or not they quantify over spatiotemporal points and regions. Rather, the dividing issue is whether they treat spatiotemporal locations as *independent* entities or as *derivative* ones. Relationism is commonly presented as the view that spatiotemporal locations are dependent on and derivable from material things and their relations, while substantivalism grants spatiotemporal points and regions the status of autonomous entities, capable of independent existence. There is no broad consensus as to how to interpret the crucial term "(in)dependence" in these contexts; however it is quite certain that it is supposed to be a modal term. (For instance, the independent existence of a spatiotemporal region occupied by a material object is often expressed in the statement that this very region *could* exist even if it *were not* occupied by the actually present object.) Thus the controversy between substantivalism and relationism cannot be properly formulated and discussed without some background in the metaphysics of modality (in particular modality *de re*).[12]

The key question here is how to identify actual objects in alternative, counterfactual scenarios. This question is usually formulated in the standard framework of possible worlds, which, even though controversial for philosophical reasons, turns out to be a useful tool in the currently considered context. Generally, a particular actual object can be identified in possible worlds either by some of its qualitative properties (intrinsic or relational), or without reference to any properties (haecceitistically). The qualitative properties used to identify objects in possible worlds are called *essential*. Thus a possible-world counterpart of a given object—such a counterpart is sometimes said to represent the original object *de re*—must possess all of its essential properties. Using this concept, it seems natural to characterize spacetime relationism as the position according to which the material contents of spatiotemporal points and regions constitute their essential properties. That is, no spatiotemporal location can exist without being occupied by the particular material things it actually houses. This characterization of relationism immediately invalidates the hole argument, since the argument presupposes that a given point can receive different material contents.[13]

On the other hand, substantivalism "disconnects" spatiotemporal regions from their material occupants. A particular region could host a different material field without losing its intrinsic identity. But this broad modal characteristic of substantivalism leaves us with (at least) two further options. First, we may insist that spatiotemporal locations do not possess any essential properties whatsoever. This option, which we may call *haecceitistic substantivalism*, is the one implicitly assumed by Earman and Norton, as they maintain that the substantivalist should admit that a given spatiotemporal point (region) could exist without any properties it actually possesses (whether strictly geometric, such as metric properties, or physical ones, encompassed in matter fields). Consequently, models connected by a hole-diffeomorphism represent situations in which some actual points and regions receive different metric and/or physical properties. However, this is by no means the only option available to the substantivalist. An alternative is to admit that spatiotemporal locations do possess essential properties other than the material things occupying them. Typically these essential properties are assumed to be metric properties, encompassed chiefly in the metric tensor. Thus a variant of substantivalism that may be dubbed *metric substantivalism* is possible.

Metric substantivalism can be further subdivided depending on how precisely we characterize the concept of essential properties. If we insist that possessing all the essential properties of a given object is a necessary but not sufficient condition for being a counterpart of this object (for representing this object *de re* in a possible world), then we are on a path to the variant of metric substantivalism proposed by Tim Maudlin (see Maudlin, 1988, 1990). In this approach, modeled on Kripke's variant of essentialism, we identify a given object a in a counterfactual scenario by stipulation, as in haecceitism, but in addition we have to make sure that this selected counterpart possesses the essence of a. For Maudlin the stipulation is given by the hole diffeomorphism $d : M \to M$; that is point p is identified with its image dp (the image dp represents *de re* point p in a possible world described by the transformed model dM). However, the assumption of essentialism demands that dp should possess the exact same metric properties (i.e., the metric tensor) as p. If this condition is not satisfied, the transformed model dM does not represent a metaphysical possibility. The only situation when dM represents a genuinely possible world is when d induces an isometry (when it preserves the metric properties of points), but this cannot lead to the conclusion of indeterminism as in the original hole argument.

Another option is to get rid altogether of all the vestiges of haecceitism in the form of the identifications $p \leftrightarrow dp$. We may simply assume that possessing the essential properties of object a is a necessary and sufficient condition for being a counterpart of a in a possible world. This of course implies, in contrast with Maudlin's version, that in some cases a particular object may have more than one counterpart, and that two actual objects can share their counterparts. Thus the counterpart relation ceases to be the identity. In the case of the hole transformation this approach leads to the conclusion that models M and dM represent the same possible world. This is so, because there can be no haecceitistic difference between M and dM, and qualitatively these models are indistinguishable, hence they must be identical. This conclusion confirms that we have thus reconstructed the position introduced earlier as sophisticated substantivalism.[14]

How about the acid test for substantivalism based on the Leibniz shift, as formulated by Earman and Norton? It is not difficult to observe that both haecceitistic substantivalism and Maudlin's version of metric substantivalism pass it with flying colors. A shift is represented by a transformation $d:M \rightarrow M$ of a flat-metric spacetime manifold M, where d is a translation by a constant vector field. Since $p \neq dp$ for all p, the resulting model dM is haecceitistically distinct from M, even though qualitatively indistinguishable from it. And because all points in M possess the same metric properties (metric tensor), d is guaranteed to induce an isometry and therefore model dM represents a genuine possibility according to Maudlin's essentialism. But sophisticated substantivalism in the form of strong metric essentialism fails the test. As no haecceitistic differences are admitted, the qualitatively indistinguishable models M and dM must be treated as representations of the same reality.

However, it is possible to devise a modified version of Earman and Norton's acid test which will correctly categorize sophisticated substantivalism as a genuine variant of substantivalism. Actually, it may be even claimed that this modified version better expresses our pre-theoretical intuitions regarding the Leibniz shift.[15] The new test is simple: instead of performing a static shift, we may imagine that the whole universe is moving uniformly in a particular direction. Since we don't know what the actual state of motion of the universe is, the test should consist in comparing two possibilities, none of which may be claimed to be actual. One possible scenario is that the centre of mass of the universe is stationary, whereas the other option is that it is moving uniformly in a given direction. The substantivalist should treat these two possibilities as distinct, since the same spatiotemporal locations will be generally occupied by different material objects. And the sophisticated substantivalist will have no problems with accepting that, since the two options will typically differ qualitatively as well (notwithstanding the fact that the difference will be empirically undetectable). For instance, in one possible world the qualitative sentence "There is a spatial point occupied by two bodies A and B in succession (where 'A' and 'B' are some qualitative characteristics uniquely defining two physical objects)" may be true, whereas in another world it will be false (for instance, if body A in this world is stationary).

Given that the possibility of a qualitative distinction between these options depends on some particulars of the mass distribution in both worlds, the new acid test should be formulated as flexibly as possible. We can achieve that by using the existential quantifier rather than the universal one: given any possible world whose centre of mass is stationary, *there is* a distinct possible world which moves uniformly in some direction. This is guaranteed to be satisfied, unless we consider a world with a uniform spatial distribution of mass (a world in which mass density is the same at any spatial point). Needless to say, the whole test is performed in the framework of a non-relativistic, Newtonian theory equipped with absolute space. But that is not problematic, since the test is a conceptual exercise of the "what if" kind. We don't need the assumption that the described scenarios are real, physical possibilities. In order to test whether a given metaphysical conception can be categorized as substantivalism, we ask the following question: is this conception capable of distinguishing between two imaginary scenarios of a Newtonian world in different states of motion? As I explained, sophisticated substantivalism has the wherewithal to do that, whereas relationism fails the test, since in each of the considered scenarios the spatial and spatiotemporal objects are derived from material bodies and their mutual relations, and these are the same regardless of the assumed state of motion of the entire universe.

2.4   Symmetry-breaking indeterminism

Consider the following scenario: a universe which consists of an infinite, featureless plane, and a solitary tower in the form of a perfect cylinder standing on the plane.[16] At a certain moment t the tower collapses under its own weight, but the direction in which it falls is not predetermined by the laws of the universe together with the initial conditions. Thus we have a clear case of indeterminism (or so it seems). And yet sophisticated substantivalism (or the conception of modality *de re* it presupposes) has troubles with accounting for this intuition. The basic explication of determinism is that in a deterministic world, for every complete specification of the past there is only one possible future that complies with the laws of the world. In the tower example it seems that there are an infinite number of potential futures for the same past: the tower has an infinite number of choices in which direction to crumple. Nevertheless all these futures are qualitatively indistinguishable from each other, hence according to sophisticated substantivalism all the scenarios of development describe one and the same possibility. Thus the tower case is incorrectly classified as deterministic.

The key feature of this example is the symmetry of the initial conditions and its breaking by a supposedly indeterministic process. The details regarding the tower, the plane etc., are not significant (even though they make for a convincing story). Analogous cases can be produced using entirely different setups, as long as they display the same symmetry-breaking feature. Thus we can consider a universe consisting of a number of indistinguishable black particles distributed symmetrically in space, and a law which prescribes that at a predetermined time one of the particles turns pink, but fails to identify which particle it will be. A more entertaining variant of this setup involves a group of identical bald philosophers sitting in a circle, and the process as a result of which one philosopher will grow a single hair on his head, but again no philosopher is preselected as the lucky one.[17] So each philosopher can hope that it will be him, but no one can be sure of that.

This challenge opens a Pandora's box with numerous proposals, solutions and commentaries. We will plunge into some of these shortly. But before we do that let us notice that there is something peculiar about the general setup, and in particular about the assumption of indeterminism presupposed in each of these scenarios. We are being told that the appropriate laws do not predict the exact outcome of the process (the direction in which the tower collapses, the identity of the lucky philosopher, and so on). But how could they, since there is no way to qualitatively identify any of the elements involved in the alternative scenarios (directions, individual philosophers)? There is no way any law could prescribe that *this* philosopher will grow a hair, since there is no possibility to explicate what "this" refers to without breaking the symmetry. It looks like the general setup already places determinism in an unwinnable position, since the assumed symmetry prevents us from even formulating a deterministic law, let alone accepting it as true.[18] The proponents of the symmetry-breaking example can shrug their shoulders and reply that in the end what counts is that determinism fails, regardless of whether it had a winning chance or not. But this victory seems too easy. It may be objected that the laws in these examples predict everything that can be predicted. What else should we expect?

Setting this problem aside for a while, let us move on to the typical responses sophisticated substantivalists can (and do) give to this challenge. One option is to question the methodology of counting possibilities used in the symmetry-breaking cases. While there is only one possible world containing the future evolution of the system, there may be distinct possibilities that do not correspond to different possible worlds. David Lewis famously introduced the distinction between possibilities and possible worlds in his conception known as "cheap haecceitism" (Lewis, 1986, pp. 230–235). He insisted that objects can possess counterparts in the actual world, and that these counterparts represent possibilities of "swapping" things without creating a new, qualitatively different reality. In the example with bald philosophers, each philosopher is a counterpart of every other philosopher. Thus for the philosopher who actually grew a hair, there is a number of his counterparts in the form of the remaining philosophers, and each counterpart represents the possibility that a different philosopher could grow the hair instead.

I have no room here to discuss various questionable features of cheap haecceitism, which I consider a highly controversial stance (but see Graf Fara, 2009 for some arguments). However, we should observe that this solution brings with it certain unwelcome consequences in the form of intuition-violating verdicts regarding other cases. Cheap haecceitism wrongly categorizes cases of "haecceitistic indeterminism" as genuinely indeterministic. Suppose that the universe contains two indistinguishable particles of the type we call "alfa," and that after a fixed amount of time each alfa particle turns into a beta-particle, with all beta-particles being indistinguishable as well. This universe seems to be perfectly deterministic, and yet according to cheap haecceitism there are two possibilities how the universe can evolve: one is the actual way, and the other involves the process with the beta-particles swapped. While the second option is not represented by a separate possible world (the existence of such a world would imply full-blown haecceitism), it nevertheless is represented by the existence of an actual-world counterpart for a given beta-particle.[19]

A different method of how to deal with the symmetry-breaking cases of indeterminism that has become the gold standard for sophisticated substantivalists is to make a distinction between global and local determinism. While in all of these examples the future of the entire universe is predetermined when cast in qualitative terms, we may distinguish parts of the universe for which there is more than one possible, qualitatively characterized evolution. Each section of the plane on which the tower can fall is an indeterministic system: the tower may or may not end up there. Similarly, for each of the four particles there is a chance that it will turn pink, but also a chance that it will remain black. And each bald philosopher by himself constitutes an indeterministic system in that it is not predetermined whether he will grow a hair. These facts supposedly explain why we have a strong intuition that symmetry-breaking cases involve some form of indeterminism. And sophisticated substantivalists have no problems with accounting for the violation of determinism at a local level without acknowledging haecceitistic differences or resorting to half-measures in the form of cheap haecceitism.[20]

The inclusion of local determinism can be done more formally in the way proposed by Belot and Melia (Belot, 1995; Melia, 1999). Following their lead, we will introduce the concept of a *duplication*, i.e., a function from one possible world (conceived as the set of objects) to another which connects objects with their perfect duplicates, i.e., objects that possess the same natural properties and stand in the same natural relations as the original ones. Let W be the set of all objects in the actual world, W'—the set of all objects in a possible world with the same laws as W, and W_t, W'_t—the appropriate subsets of objects, properties and relations occurring before time t. Let $f : W_t \to W'_t$ be a duplication of the history of W before t. Then standard determinism implies that there must be a 'global' duplication $g : W \to W'$ acting on the whole world. Thus if two worlds are duplicates up to time t, they must be duplicates everywhere. However, this definition is satisfied even in the case of symmetry-breaking indeterministic processes (i.e., it can be seen as a merely necessary but not sufficient condition for determinism). Whatever duplicate of the initial segment of the world with four black particles we consider, as long as the world containing this duplicate satisfies the actual laws, one particle will have to turn pink, and this secures the existence of a global duplication. In order to eliminate cases like these, Belot and Melia introduce the following modification of this condition of determinism and make the characterization also a sufficient condition: for any t, W_t and $W'_{t'}$ defined as earlier, and for any duplication $f : W_t \rightarrow W'_{t'}$, there exists a duplication $g: W \rightarrow W'$ such that the restriction of g to W_t is identical with f.

This new definition does the required job well—it classifies all the cases of symmetry-breaking processes as genuinely indeterministic. Let us use the example involving four black particles of which one turns pink to illustrate this. Let a, b, c and d be the names of the particles in the actual world. In addition, let us stipulate that it was particle a that turned pink at time t. Suppose now that we consider an exact copy (duplicate) of world W that contains particles a', b', c', d', and that in W' particle a' turns pink. The function $f : W_t \rightarrow W'_t$ such that $f(a) = b'$, $f(b) = c'$, $f(c) = d'$ and $f(d) = a'$ is obviously a duplication of W_t. However, when extended to cover the entire world W, function f is no longer a duplication, since a and b' are not duplicates after t: a turns pink while b' stays black. A duplication g of the entire world W must associate a with a', since these are the only objects that turn pink in their respective worlds. But such a function g obviously cannot be restricted to produce f as defined earlier. Thus the modified definition of determinism is not satisfied in this case. On the other hand, the definition is clearly satisfied in the case of alfa-particles decaying into beta-particles. Any permutation of alfa-particles can be extended to permutations of beta-particles (the reason being that beta-particles come into existence after t, and therefore their permutations are independent from the permutations of alfa-particles). Hence the case of merely haecceitistic indeterminism is categorized as (qualitatively) deterministic.

Brighouse admits that the strategy of modifying the definition of determinism in an appropriate way solves the problem of symmetry-breaking indeterminism. However, she complains that this solution lacks a deeper justification for the adopted criterion of determinism (Brighouse, 2020, pp. 8–9). She goes as far as to suggest that the tinkering with the original, well-justified expression of determinism has no motivation other than the desire to eliminate the cases of symmetry-breaking indeterminism. But if this is right, then the solution is spurious—we argue that sophisticated substantivalists may classify the troublesome cases as indeterministic simply because we have cooked up an arbitrary definition of determinism that achieves just that.

Still, I think that this complaint is a bit unfair. It seems that the modified characterization of determinism is motivated precisely by the distinction between global and local determinism, or—more specifically—by the wish to include the requirement of local determinism within the broad criterion of determinism. Melia correctly points out that we can't call a world deterministic if we can find in it systems whose past does not uniquely fix their future, even if the future of the entire world is fixed by its past. And the modified definition stated earlier makes sure that smaller systems will indeed behave deterministically. No matter what duplicate $f(a)$ of an object a before time t we consider, we are ensured that there will be a duplication g that extends f into the future—thus object $f(a)$ will have the same future as a. Hence Belot and Melia's corrected definition follows directly from the assumption that determinism must be preserved both globally and locally.

At this point we could pause, confident that one of the main challenges to sophisticated substantivalism has been dealt with. Yet there are still some curious facts regarding cases of symmetry-breaking indeterminism that deserve our attention. One intriguing observation that can be made here follows directly from Brighouse's apt remarks regarding the connection between one's ontology and the number of distinct possibilities one is willing to countenance (Brighouse, 2020, pp. 9–11). She compares two ontologies for the substantivalist: a two-category ontology including spatiotemporal points and objects occupying them, and a one-category ontology that accepts only points and their properties, without objects.[21] She then notes that the one-category ontology limits the number of distinct possibilities usually associated with shifting objects against spatiotemporal points. We may use this observation and apply it to the currently considered cases. As a result, we will see that switching from the dualist two-category ontology to the monist ontology of points and their properties has a surprising effect of undermining the appearance of indeterminism in the tower case and its cognates.

Let us again return to the case of four black particles, of which one is going to turn pink. How can we cast this situation in a language of points and their properties? Instead of talking about black particles, we may talk about points possessing the B-property (being occupied by a black particle—we may call these points "black-colored" for short) and the P-property (being occupied by a pink particle—"pink-colored"). Since all the particles are of the same type, there is no need to distinguish further subcategories of B-properties and P-properties. A duplication f of W_t in this case will map points onto points, while preserving their properties and relations. If the underlying spacetime is neo-Newtonian, then the only spatiotemporal relations between points that are frame-independent are temporal distance and spatial distance between points on a hyperplane of simultaneity. Thus a duplication f can shift points within any given hypersurface of simultaneity while mapping black-colored points onto black-colored points. Now, it is trivially true that any such duplication f can be extended to a global function g which is also a duplication. Function g can be simply the identity for all points whose temporal coordinate is greater than or equal t, while for times earlier than t, $g = f$. Therefore the situation is classified as deterministic in the light of the modified definition of determinism. Clearly, the reason for this is that our ontology no longer contains objects persisting in time (whether in the form of material things or spatial points), so fixing a duplicate for all times earlier than t hardly constrains our choice for duplicates after t.

But isn't this consequence in direct conflict with our intuition that the symmetry-breaking processes are indeterministic? Well, if our only goal were simply to recover this intuition in an appropriate formalization of determinism, then clearly the formalization based on the one-category ontology would not achieve that. However, I think that we may turn things around and take a second look at this supposedly watertight intuition. And it seems clear now that the intuition of indeterminism is strongly dependent on some background assumptions, in particular assumptions regarding the underlying ontology. Take the original tower example. It seems obvious that something is underdetermined here: given a particular section on the plane, it is not determined whether the tower will fall on it or miss it. But a section of a plane is a physical, material object which retains its identity over time (we may wait ten years, and then point at a particular section and state "This is the same section of the plane that we picked out ten years ago"). If we switch from the language of persisting objects to the description in terms of a one-category ontology, we can no longer connect past stages of particular sections of the plane with their future continuations. Consequently, the intuition regarding local indeterminism weakens its hold on us, as we no longer have persisting systems to which this intuition may be applied.

One may insist that even in the framework admitting only "momentary" objects ("occurents" in Peter Simons' terminology) it is still possible to defend the intuition of indeterminism in these cases. Even if a given section of the plane at the time of the tower's collapse is a momentary object that did not exist before, still it seems true that this particular region of spacetime could have a property different from the actually possessed while the entire past had been exactly the same. But consider the following scenario—a maximally austere universe consisting of empty and infinite spacetime, and a single localized flash occurring at a certain predetermined moment (let's say, five seconds after the initial moment of the universe's existence). Is this example a case of indeterminism? I can only speak for myself, but to me this scenario feels entirely deterministic, even though the flash might have occurred anywhere in the empty universe. Again, we can repeat the misgivings spelled out at the beginning of the section. If this situation is to be classified as indeterministic, what should be added to the laws of the universe and/or the description of the initial conditions in order to make the whole process deterministic? Any physical description which could identify particular regions inevitably breaks the presupposed symmetry. I don't want to press this point too strongly, but I would like to emphasize that the substantivalist has yet another option at her disposal in addition to the strategies described earlier, which is to question the intuition that symmetry-breaking "indeterministic" processes are indeed indeterministic. And the transition from the two-category ontology to the purely monist ontology of spacetime with no persisting objects seems to lend more credence to this strategy.

2.5   Conclusion

As we have seen, sophisticated substantivalism in the form of an appropriate variant of metric essentialism has the wherewithal to deal with the challenges of symmetry-breaking indeterminism, as well as with other, lesser challenges. However, before we wrap up the current discussion, I would like to mention a recent argument against spatiotemporal essentialism put forward by Trevor Teitel (Teitel, 2019). In his terminology sophisticated substantivalism is characterized as a variant of substantivalism that implies anti-haecceitism, which in turn is presented as the view that distinct metaphysical possibilities must differ qualitatively. Teitel notes that anti-haecceitism itself is capable of defusing the hole argument; however the assumption of essentialism is invoked in order to provide an explanation of why it is the case that only qualitative differences count towards the distinction between metaphysical possibilities. He argues in favor of the variant of essentialism he calls "sufficiency essentialism" as opposed to mere "necessity essentialism." Sufficiency metric essentialism in Teitel's approach states that for every actual spatiotemporal point and region, its conditional geometric role is essential to it, where the conditional geometric role of x in any metaphysical possibility w is expressed in the conditional statement that if the whole geometric structure of w is implemented at any w', then x plays the same qualitative role at w' as it does at w.

Teitel correctly points out that ordinary essentialism (for instance in the version proposed by Maudlin) is open to a modified hole argument, in which it is stipulated that the diffeomorphically transformed model $d\mathcal{M}$ contains some points and regions that are numerically distinct from the original ones. His sufficiency essentialism overcomes this obstacle, because it implies that if only the entire actual geometric structure is implemented in a given possible scenario, the points and regions occurring there must be identified with the original ones.[22] However, he stops short of announcing a complete victory of the essentialist solution to the hole argument. His main worry is that the metaphysical explanation of anti-haecceitism in the form of sufficiency essentialism is not entirely satisfactory, since the why-question can be repeated with respect to essentialism: why is it so that only objects with appropriate essences exist in the actual world? Why are metaphysical possibilities involving actual objects that lack their essential properties eliminated? Since no satisfactory answer to these questions can be supplied, Teitel concludes that we could equally well accept anti-haecceitism as a brute fact and forgo the whole business with essentialism.

I must admit that I am a bit mystified by the demand of "metaphysical explanations" with respect to some fundamental assumptions or theses. I am not sure what is expected from such a successful explanation—what formal and informal conditions it has to satisfy. I am tempted to reject Teitel's gambit by simply taking out the concept of metaphysical explanation altogether.[23] In my view the main question that we encounter when confronted with the hole argument is whether it is possible to offer a plausible variant of spatiotemporal substantivalism that would not commit us to the radical form of indeterminism. This challenge calls for further clarifications as to what counts as substantivalism. I offer an explication of the distinction between substantivalism and relationism in terms of the essential properties of spatiotemporal points and regions, as presented in section 2.3. This already commits me to the concepts of essences and essentialism, regardless of whether or not they can be used in dubious metaphysical explanations of anti-haecceitism. Is it possible to provide an alternative characterization of substantivalism and relationism without the notion of essence? Perhaps, but I don't know of any such characterization that I could accept. For sure it would be inappropriate to define substantivalism simply as a position which presupposes the existence of spatiotemporal points and regions, since even relationists can include them in their ontology.

Out of numerous non-equivalent variants of substantivalism explicated in essentialist terms, we have selected one that is based on the assumption that possessing all the essential properties of given spatiotemporal points and regions is both necessary and sufficient for representation *de re*. Again, this choice has nothing to do with any metaphysical explanations, but is a consequence of the fact that this variant of essentialist substantivalism seems to have, on balance, the best track record of dealing with various challenges, including the "mole" argument.[24] I have shown how to modify the "acid test" for substantivalism based on the Leibniz shift in order to make it possible for this radical variant of essentialist substantivalism to pass it. Finally, I have considered the most serious challenge to any position which implies the identification of qualitatively indistinguishable possibilities; that is the case of symmetry-breaking indeterministic processes. As it turns out, there are at least three strategies of how to deal with this challenge. We can adopt the method of counting possibilities

prescribed by Lewisian cheap haecceitism, or introduce a modified definition of determinism that pays attention not only to global but to local
determinism, or finally we can question the intuition that the considered
cases of symmetry breaking are indeed indeterministic. While the second
solution seems to be most appealing, notwithstanding Brighouse's latest
reservations, I am personally attracted to the third, rather controversial
approach. In support of this stance I offer an argument based on the
one-category ontology for spacetime theories. I suggest that when we
cast the symmetry-breaking cases in a language that does not refer to entities persisting in time, the intuition of indeterminism becomes shaky. Add
to this the fundamental impossibility of turning apparently indeterministic
laws into deterministic ones without breaking the initial symmetry, and we
have built a strong case against the presupposition that it is full-blown
indeterminism we are dealing with here.

In addition to these discussions, we have also addressed a slightly different issue related to the hole argument, namely the persistent opinion that
the hole-induced indeterminism is in fact a special case of permutation
invariance that affects virtually all scientific theories. I believe that this
claim is fundamentally wrong—the variant of indeterminism based on
general permutation invariance is a horse of a different, albeit related
color in comparison to the indeterminism implied by the diffeomorphism-
invariance of spatiotemporal theories. We have already seen that haecceitistic indeterminism, flowing from the unrestricted permutations of objects, is
much different from other variants of indeterminism, as it satisfies the modified definition of determinism presented in section 2.4 while the other variants don't. Moreover, I have argued in section 2.2 that the hole argument
cannot be based on arbitrary permutations of points, since mathematically
it does not make sense to introduce a transformation of geometric objects
defined on points (tangent spaces at points) without the assumption that
the underlying transformation of manifolds is smooth. I have refuted
Melia's argument to the contrary by showing that he applies the same operation twice: once in the form of the permutation of points, and for the
second time by swapping the coordinates of the permuted points. As a
matter of fact, diffeomorphism invariance is a much more specific feature
of spacetime theories than simple permutation invariance. Sophisticated
substantivalism insists that the inability of telling the difference between diffeomorphically related models simply means that the underlying spatiotemporal substance remains ontologically the same. My goal was simply to give
arguments that this position has a fighting chance in spite of mounting difficulties. I hope I have fulfilled this task satisfactorily.

Acknowledgements

I wish to express my gratitude to an anonymous referee for this volume
for making detailed comments and suggestions to an earlier version of

this chapter. The work on this chapter was financially supported by the
Polish National Science Center grant No. 2017/25/B/HS1/00620.

Notes

1. In Earman and Norton (1987) and Stachel (1989).
2. See also Norton (2019) for a non-technical exposition which includes a semi-complete bibliography to the problem.
3. The exact nature of the relation between general covariance and diffeomorphism-invariance is a matter of debate. See Norton (1993), Pooley (2010).
4. The isomorphy claim is repeated in many places, e.g., in Pooley (2013, p. 59).
5. See John Stachel's generalized hole argument in Stachel (2022) and its critical analysis in Pooley (2006). A similar suggestion is made in Melia (1999).
6. Compare Glick (2016) for an analysis of General Permutability in general relativity and quantum mechanics.
7. The subsequent presentation of the basic mathematical notions of differential geometry follows closely David Malament's excellent book (Malament, 2012).
8. Another, perhaps more common way of introducing differential manifolds is to assume from the outset that M is equipped with its topology, and then to stipulate that a chart must map open sets in M onto open sets in $\mathbb{R}^N$ (Malament, 2012, p. 4). Nevertheless, we will continue using Malament's definition of manifolds in terms of charts.
9. Note that charts φ and φ' defined that way are incompatible, unless $\varphi(p) = \varphi(q)$. Thus manifolds $(M, \mathfrak{C})$ and $(M, \mathfrak{C}')$ are genuinely distinct, in that they cannot be extended to the same maximal set of charts.
10. I am not suggesting that Melia himself derived his claim from this incorrect view, only that it is possible to do so.
11. The term "sophisticated substantivalism" has been introduced by Gordon Belot and John Earman in their critical analysis (Belot and Earman, 2000). The list of works whose authors subscribe to views close to sophisticated substantivalism includes Butterfield (1989), Brighouse (1994, 1997), Bartels (1996), Hoefer (1996), Pooley (2006, 2013), Bigaj (2016, 2020).
12. A similar view on the distinction between substantivalism and relationism has been recently developed and defended in North (2018). Jill North casts the two metaphysical options in terms of grounding and fundamentality, and she admits that these terms are modal in character, even though she doesn't provide a specific analysis of the terms within the framework of possible worlds and modalities *de re*.
13. Of course, we should keep in mind that by the hole argument we understand the "philosophical" argument that derives the negation of determinism from the existence of hole diffeomorphisms. The mathematical fact that such a diffeomorphism exists is not affected by any metaphysical considerations regarding essential properties.
14. We should perhaps clarify one issue here. Points p and dp, conceived as mathematical objects, are typically not identical, so diffeomorphisms are not assumed to be the identities on the manifold M. However, it may happen that p and dp represent the same physical points, if they are qualitatively indiscernible.

15. The modified version of the Leibniz shift available to sophisticated substantivalists directly responds to Brighouse's worry that "it *seems* to make sense to say on a substantivalist ontology that objects could have occupied different points, and seems no less sensible to maintain this even if the relative positions of objects remained unchanged" (Brighouse, 2020, p.4).

16. This example, originally formulated in Wilson (1993), was used in Belot (1995) as an argument against sophisticated substantivalism.

17. These examples are due to Melia (1999, p.650).

18. Would it help if we introduced individual names (labels) for the indistinguishable objects involved (directions, particles, philosophers)? I doubt it. It seems that the underlying laws should be invariant with respect to the permutations of non-qualitative labels, as is the case with the laws governing systems of "indistinguishable" quantum particles. Such permutation-invariant laws cannot predict for instance that it is particle number 3 that will turn pink.

19. Interestingly, Gordon Belot (who invented the example with alfa- and beta-particles) is of the firm opinion that haecceitistic indeterminism is a genuine type of indeterminism, and he even proposes a further modified variant of the definition of determinism that we give later in the text, in order to preserve this intuition. The variant considered by Belot includes the condition of uniqueness: for every duplication of the initial segment of the universe there is exactly one extension that is a global duplication (Belot, 1995, p. 193). I think that Belot's stance has a very unfortunate consequence in that it classifies all known deterministic theories (including Newtonian mechanics) as indeterministic.

20. Brighouse (2020) categorizes the proposal sketched in my Bigaj (2016, pp. 168–169) as a variant of cheap haecceitism, most probably due to the fact that according to this proposal the indeterministic behavior of the particle that turned pink is underwritten by the existence of qualitatively indiscernible (up to time *t*) particles that did not turn pink. However, in this approach the stress is put on the qualitative description of the situation. The cheap haecceitist says that *this* particle could have stayed black, because of the existence of its counterpart in the actual world that remained black after *t*. On the other hand, I insist only that an object possessing such-and-such qualitative features before *t* could remain black after *t*, and this modal statement is obviously made true by the existence of an actual black particle that never turned pink. To accept this, we don't need the assumption that one actual particle represents *de re* any other actual particle.

21. This second option is akin to the metaphysical conception known as supersubstantivalism. See Sider (2001); Schaffer (2009).

22. Teitel's "revised" hole argument was recognized in the literature before. David Glick in Glick (2016) presents it under the name of the "mole argument," crediting it to Jonathan Schaffer. Note also that Glick's suggested solution of the problem in the form of Minimal Structural Essentialism (MSE) bears uncanny resemblance to Teitel's sufficiency essentialism. For a critical evaluation of Glick's MSE, see Bigaj (2020), where an alternative view by the name of Radical Structural Essentialism is formulated and defended.

23. My reaction to Teitel's argument should not be construed as a wholesale attack on the "metaphysical" approach to the hole argument and the substantivalism–relationism debate. I merely wish to express my inability to capture the crucial concept of "metaphysical explanation" used by him in his criticism of the essentialist answer to the problem. My own take on this issue is that essentialism can be justified on methodological rather than "metaphysical" grounds. As I explain in the main text, by relying on

the modal concept of essential properties and relations we can easily make a distinction between relationism and substantivalism which otherwise would be difficult (if not impossible) to properly spell out.
24. For an extensive discussion, see again Bigaj (2020).

Bibliography

Bartels, A. (1996). Modern essentialism and the problem of individuation of spacetime points. *Erkenntnis* 45, 25–43.

Belot, G. (1995). New work for counterpart theorists: Determinism. *British Journal for the Philosophy of Science* 46, 185–195.

Belot, G. and J. Earman (2000). From metaphysics to physics. In J. Butterfield and C. Pagonis (Eds.), *From Physics to Philosophy*, pp. 166–186. Cambridge University Press.

Bigaj, T. (2016). Essentialism and modern physics. In T. Bigaj and C. Wüthrich (Eds.), *Metaphysics in Contemporary Physics*, pp. 145–178. Brill/Rodopi.

Bigaj, T. (2020). Radical structural essentialism for the spacetime substantivalist. In A. Marmodoro, D. Glick, and G. Darby (Eds.), *The Foundation of Reality: Fundamentality, Space, Time*, pp. 217–232. Oxford University Press.

Brighouse, C. (1994). Spacetime and holes. *Proceedings of the Biennial Meeting of the Philosophy of Science Association* 1994 (1), 117–125.

Brighouse, C. (1997). Determinism and modality. *British Journal for the Philosophy of Science* 48, 465–481.

Brighouse, C. (2020). Confessions of a (cheap) sophisticated substantivalist. *Foundations of Physics* 50, 348–359.

Butterfield, J. (1989). The hole truth. *British Journal for the Philosophy of Science* 40, 1–28.

Earman, J. and J. Norton (1987). What price spacetime substantivalism? The hole story. *British Journal for the Philosophy of Science* 38, 515–525.

Glick, D. (2016). Minimal structural essentialism. In A. Guay and T. Pradeau (Eds.), *Individuals Across the Sciences*, pp. 1–28. Oxford University Press.

Graf Fara, D. (2009). Dear haecceitism. *Erkenntnis* 70, 285–297.

Hoefer, C. (1996). The metaphysics of space-time substantivalism. *The Journal of Philosophy* 93, 5–27.

Lewis, D. (1986). *On the Plurality of Worlds*. Blackwell.

Malament, D. (2012). *Topics in the Foundations of General Relativity and Newtonian Gravitation Theory*. The University of Chicago Press.

Maudlin, T. (1988). The essence of space-time. *Proceedings of the Biennial Meeting of the Philosophy of Science Association* 1988 (2), 82–91.

Maudlin, T. (1990). Substances and space-time: What Aristotle would have said to Einstein. *Studies in History and Philosophy of Science* 21, 531–561.

Melia, J. (1999). Holes, haecceitism and two conceptions of determinism. *British Journal for the Philosophy of Science* 50, 639–664.

North, J. (2018). The structure of spacetime: A new approach to the spacetime ontology debate. In K. Bennett and D. Zimmerman (Eds.), *Oxford Studies in Metaphysics*, Volume 11, pp. 3–43. Oxford University Press.

Norton, J. (1993). General covariance and the foundations of general relativity: Eight decades of dispute. *Reports on Progress in Physics* 56, 797–858.

Norton, J. (2019). The hole argument. In E. Zalta (Ed.), *The Stanford Encyclopedia of Philosophy* (Summer 2019 ed.). Metaphysics Research Lab, Stanford University. https://plato.stanford.edu/archives/sum2019/entries/spacetimehole arg/.

Pooley, O. (2006). Points, particles, and structural realism. In S. Rickles, D. French, and J. Saatsi (Eds.), *The Structural Foundations of Quantum Gravity*, pp. 83–120. Oxford University Press.

Pooley, O. (2010). Substantive general covariance: Another decade of dispute. In M. Suárez, M. Dorato, and M. Rédei (Eds.), *EPSA Philosophical Issues in the Sciences: Launch of the European Philosophy of Science Association*, pp. 197–209. Springer.

Pooley, O. (2013). Relationist and substantivalist approaches to spacetime. In R. Batterman (Ed.), *The Oxford Handbook of Philosophy of Physics*, pp. 522–586. Oxford University Press.

Schaffer, J. (2009). Spacetime the one substance. *Philosophical Studies* 145, 131–148.

Sider, T. (2001). *Four Dimensionalism: An Ontology of Persistence and Time.* Oxford University Press.

Stachel, J. (1989). Einstein's search for general covariance. In D. Howard and J. Stachel (Eds.), *Einstein and the History of General Relativity*, pp. 63–100. Birkhäuser.

Stachel, J. (2022). 'The relations between things' versus 'the things between relations': The deeper meaning of the hole argument. In D. Malament (Ed.), *Reading Natural Philosophy: Essays in the History and Philosophy of Science and Mathematics*, pp. 231–266. Open Court.

Teitel, T. (2019). Holes in spacetime: Some neglected essentials. *The Journal of Philosophy* 116 (7), 353–389.

Wilson, M. (1993). There is a hole and a bucket, dear Leibniz. *Midwest Studies in Philosophy* XVIII, 202–241.

3 What represents spacetime? And what follows for substantivalism vs. relationalism and gravitational energy?

J. Brian Pitts

3.1 Four interrelated questions

This essay considers four interrelated questions that are not usually all explicitly considered simultaneously, but which have some subtle connections worth uncovering:

- What represents spacetime in GR?
- What is the status of gravitational energy(s) in GR?
- Substantivalism *vs.* relationism: is spacetime a substance?
- Is gravity exceptional, or just another field?

The first and third questions have been staples in philosophy for decades. Gravitational energy, controversial in physics since 1918, has become a fashionable topic in philosophy within the last decade. The fourth question, GR exceptionalism *vs.* particle physics egalitarianism, has rarely been discussed in philosophy or history (but see Kaiser, 1998; Pitts, 2017, 2020b), and is not often discussed *explicitly* in physics either, but disagreement on it lies near the core of rival research programs in quantum gravity (Rovelli, 2002; Brink, 2006; Smolin, 2006, chapter 6). Given that particle physics egalitarians rarely have been interested in philosophy until recently—the slogan "shut up and calculate" and Weinberg's chapter "Against philosophy" (Weinberg, 1992) come to mind—there might be some low-hanging fruit to pick in applying the egalitarian perspective in relation to these other issues. Feynman linked skepticism about gravitational energy to GR exceptionalism (Feynman et al., 1995, pp. 219–220). Curiel's recent work, discussed later in this chapter, also suggests this connection from the opposite direction (Curiel, 2019). Duerr also makes this connection (Duerr, 2021).

This essay proposes a particle physics-inspired answer to the question of what is spacetime in GR, relates that answer to (anti)realism about

DOI: 10.4324/9781003219019-5

gravitational energy, and provides another way to make gravitational energy(s) safe for relationism. This last conclusion coheres with some themes of the dynamical view of spacetime of Brown and (sometimes) Pooley (Brown and Pooley, 2006; Brown, 2005). Previously I have found support in particle physics for at least some of Brown's project *vis-a-vis* spacetime realist critiques (Freund et al., 1969; Pitts, 2011, 2016c, 2019). The result seems to be a coherent package of views that has not previously been seen clearly as a whole and in detail, and perhaps illustrates the fruitfulness of overcoming GR *vs.* particle physics divide—a divide that philosophers and historians have inherited, along with a GR exceptionalist bias, largely by default (but see Menon, 2021; Salimkhani, 2020).

3.2   What represents spacetime in GR?

The question of what represents spacetime in GR has been a standard topic in the philosophy since the 1980s. One obvious answer is $\langle M, g_{\mu\nu}\rangle$, the manifold with the metric. It is tempting, at least to philosophers, to think that one first has a manifold with identifiable points, and then lays a metric on top of it, a task that leaves freedom of choice. That path quickly leads to the hole argument, however, and is not required (Einstein, 1961, p. 155; Pitts, 2012; Menon, 2019). (If the points have some sort of essence that restricts the process of laying down the metric, that is another matter: cf. Maudlin, 1989.) Another influential answer to the question is the bare manifold M (Earman and Norton, 1987). A bare manifold presumably does not get its point individuation using promissory notes to be cashed out in terms of fields ($g_{\mu\nu}$ or otherwise) to be defined upon it. Point individuation appears to be primitive, and the hole argument lurks again. It is easy to feel dissatisfied, but harder to find an alternative. Iftime and Stachel's proposal to define spacetime by projection down from a bundle deserves consideration (Iftime and Stachel, 2006), as do old and new efforts to avoid individuals (Dasgupta, 2011).

Gravitational energy plays a key role in a foundational paper on the hole argument (Earman and Norton, 1987), motivating the classification of the metric as part of the contents of spacetime, as akin to matter (not, of course, in the technical GR sense of matter as non-gravitational):

> The metric tensor now incorporates the gravitational field and thus, like other physical fields, carries energy and momentum, whose density is represented by the gravitational field stress-energy pseudo-tensor ...[which] forces its classification as part of the contents of spacetime.
>
> (*ibid.*, p. 519)

This is surprising and perhaps too violent, one might think: should not one be able to think of spacetime in the absence of gravity, perhaps as something flat or trivial and in any case not energy-bearing, while energy-bearing gravitational waves are ascribed to gravity and not to spacetime—to contents and not to the container, as it were? Earman and Norton have considered that idea:

> We might consider dividing the metric into an unperturbed background and a perturbing wave in the hope that the latter alone can be classified as contained in spacetime. This move fails since there is no non-arbitrary way of effecting this division of the metric.
>
> (*ibid.*, p. 519)

Thus the claim that no principled split can be made plays a crucial role in forcing all of the metric, not just some energetic perturbation from triviality (flatness?), out of the spacetime category and into the material category for Earman and Norton. If a sufficiently principled split exists, then one might have an intermediate option between spacetime as the manifold and spacetime as the manifold with the metric.

The claimed non-existence of a non-arbitrary split into background and perturbation may suggest GR exceptionalist sympathies: distaste for background structures, for things that arguably act without being acted upon, along with a strong inductive claim over the trajectory of scientific progress. On the other hand, the background structures that one typically sees put forward as potentially formalisms to be interpreted (as opposed to mere mathematical conveniences), such as a background metric (Rosen, 1940) or an orthonormal basis (Møller, 1964) or even a background connection (Sorkin, 1988), have the disadvantage of introducing mathematical structure over and above what $g_{\mu\nu}$ requires (whether $g_{\mu\nu}$ is primitive or as derived), along with a new gauge "group" (perhaps a Brandt groupoid) denying *quantitative* significance to (most of) this extra structure and (in the case of a background metric or connection) bearing a striking resemblance to the coordinate freedom of GR. Such background structures were proposed to improve the status of gravitational energy in GR, but the Pickwickian character of the improvement, however, was not always adequately recognized: the advertised tensorial (coordinate-covariant) quality of the result was achieved at the expense of a less advertised new, comparably bad gauge dependence. What the right hand gives, the left takes away (Band, 1942; Pinto-Neto and Trajtenberg, 2000). With gravitational energy, the lump in the carpet has only been moved from coordinate dependence to gauge dependence.

This chapter, following a trajectory (Pitts, 2010, 2016a), avoids such traditional background structures and seeks a middle path. The background can be taken to be the constant scalar numerical matrix

$\eta =_{df} diag(-1, 1, 1, 1)$ (up to trivial notational variants such as $diag(1, -1, -1, -1)$ or $diag(-1, -1, -1, 1)$). This matrix encodes a sort of "vacuum" value for spacetime geometry, what remains (it is proposed) when not only all the fields standardly regarded as material are abstracted away, but also the gravitational field (in a sense to be made clearer). In special relativistic theories without gravity, or even with gravity as described by certain theories (such as massive scalar gravity: cf. Pitts, 2011, 2016c), $diag(-1, 1, 1, 1)$ encodes the spacetime geometry in the absence of gravity, as expressed in coordinates that are as good as any—I avoid the word "preferred" advisedly. In General Relativity this matrix often appears in weak-field expansions of GR and, more relevantly, in particle physics literature on quantum gravity in the tradition of covariant perturbation theory.[1] Neither the matrix's constancy nor its being a scalar obviously makes its being a physical field problematic, unless perhaps one objects to things that "act without being acted upon" (Pitts, 2006; Pooley, 2013). This same matrix also standardly appears in GR with spinors, where no one seems to find it objectionable or thinks that it radically changes the character of the theory relative to tensorial matter fields. (In James L. Anderson's analysis of absolute objects as amended and extended by Thorne, Lee, and Lightman, η is a confined object, not an absolute object—Thorne et al., 1973: it does not constitute the basis of a faithful representation of the manifold mapping group, or perhaps one should say, of coordinate transformations.) From these angles, η as a background representing spacetime in GR is neither very novel nor very objectionable.

From a particle physics egalitarian perspective, η plays a somewhat similar role to what $g_{\mu\nu}$ plays in a GR-exceptionalist view. In the latter, one can imagine abstracting away everything material (i.e., everything besides $g_{\mu\nu}$), and then seek to "build" a Lagrangian by coupling matter to $g_{\mu\nu}$ so as to give a dynamics to spacetime and to matter coupled to spacetime ($g_{\mu\nu}$); the Lagrangian should be a scalar density (up to a coordinate divergence) under coordinate transformations to give tensorial Euler-Lagrange equations when all fields are varied. For the particle physics egalitarian, the matter-like aspects of gravity (having energy-momentum, *etc.*, also emphasized by Rovelli, no particle physics egalitarian (Rovelli, 1997, p. 193) call for abstracting away one more field after the GR-exceptionalist stops: one abstracts away the gravitational potential out of $g_{\mu\nu}$, leaving only η. η, being numerically constant, does not individuate spacetime points; one has multiplicity without identity. (There might be an interesting resonance with the thin substantivalism of Dasgupta, 2011 and the Ramsification entertained briefly by Maudlin, 1989.) It might help to think of the world as akin to some device held together with screws, in which almost every piece (all but one or two, analogous to the matter fields as listed by the GR exceptionalist) is screwed into some base in multiple places (analogous to coupling to $g_{\mu\nu}$). The particle physics egalitarian claims that this base (analogous to $g_{\mu\nu}$) is itself composite: the gravitational potential $\gamma_{\mu\nu}$ is one more matter-like field that can be unscrewed from the true base, which (to continue the analogy) is the matrix $\eta = diag$ (−1, 1, 1, 1). Detailed features of the gravitational potential explain why it couples to everything in a universal way—why the other parts/ fields "screw into" both the gravitational potential and the base in the same fashion. If one imagines (re)assembling the world (or in any case its laws) by devising a suitable Lagrangian density, one has the spacetime background $\eta = diag(−1, 1, 1, 1)$, the gravitational potential $\gamma_{\mu\nu}$, and the (non-gravitational) matter fields as ingredients and needs to construct a Lagrangian density invariant (up to a divergence) under coordinate transformations. Why seek a Lagrangian density of *that* sort given these ingredients? The particle physics spin 2 derivations of Einstein's equations (with citations in Duff, 1975; Pitts, 2016a; Salimkhani, 2020) provide rather compelling guidance from very weak premises about avoiding violent instability, locality, (at least) Poincaré invariance, and the empirical fact of the bending of light.

The proposal about the gravitational potential overlaps a bit with a footnote by Pooley, who also raises several important challenges to taking the idea seriously. His view is that the best candidate (such as it is) for being the "gravitational field" is the

> *deviation of the metric from flatness:*h_{ab}, where $g_{ab} = \eta_{ab} + h_{ab}$. That this split is not precisely defined and does not correspond to any-thing fundamental in classical GR underscores the point that, in GR, talk of the "gravitational field" is at best unhelpful and at worst confused. The distinction between background geometry and the graviton modes of the quantum field propagating against that geometry is fundamental to perturbative string theory, but this is a feature that one might hope will not survive in a more fun-damental "background-independent" formulation.
>
> (Pooley, 2013, footnote 34, p. 539)[2]

In what sense is this split not precisely defined? Two senses come to mind. One sense, at least if η_{ab} is intended to be a flat metric *tensor* (not a numerical matrix), is the extra gauge freedom that arises in relat-ing the two metrics (Grishchuk et al., 1984), a gauge freedom that takes over much of the interesting role that coordinate transformations play in orthodox GR, thus separating the freedom to use spherical coordinates from notions of inertia and the like. Moving the lump in the carpet does not flatten the carpet, however; hence I propose not a flat metric tensor, but a numerical matrix $diag(−1, 1, 1, 1)$, leaving the coordinate freedom of GR to play its classic role. A second sense in which the devia-tion of the metric from flatness is not precisely defined involves field redefinitions. Why should we take the gravitational field (though I use the term "potential") to be the deviation $g_{ab} - \eta_{ab}$, and not $g^{ab} - \eta^{ab}$ (suppressing factors involving Newton's constant)? There is no compelling reason. Indeed the freedom to define the gravitational potential in different ways, many of them rather exotic (such as arbitrary powers of the metric, its inverse, or densitized relatives thereof—and beyond!) has been used as a resource to derive both Einstein's equations and infinitely many massive variants thereof (Ogievetskiĭ and Polubarinov, 1965; Freund et al., 1969; Pitts and Schieve, 2007; Pitts, 2016d) and to render the GR Lagrangian polynomial using either $(-g)^{\frac{5}{18}} g^{\mu\nu} - (-\eta)^{\frac{5}{18}} \eta^{\mu\nu}$ or $(-g)^{-\frac{5}{22}} g_{\mu\nu} - (-\eta)^{-\frac{5}{22}} \eta_{\mu\nu}$ (DeWitt, 1967). Ambiguity, however, is also present to some degree in geometrical GR. Why should we take the field variable to the metric $g_{\mu\nu}$,[3] rather than $g^{\mu\nu}$ (as one occasionally sees), or more plausibly $\mathfrak{g}^{\mu\nu}$ $(= \sqrt{-g}g^{\mu\nu})$, which has pride of place in writing wave equations (Papapetrou, 1948; Bruhat, 1962) and simplifies the Lagrangian (Goldberg, 1958)? Hence the problem of field redefinitions arises also in GR. Admittedly the various choices seem more natural and more closely related through natural matrix operations such as inverse and determinant, as opposed to power series expansions of (say) $(-g)^{\frac{1}{2}} g^{\mu\nu} - (-\eta)^{\frac{1}{2}} \eta^{\mu\nu}$ in terms of $g_{\mu\nu} - \eta_{\mu\nu}$ or *vice versa*. (I use bimetric notation, but the simplification to $\eta = diag(-1, 1, 1, 1)$ is obvious.) Viewing field redefinitions as choices of coordinates on a fiber might be an elegant approach for either the geometrical or the perturbative approach. The geometrical view also might render more natural the Lorentzian signature of $g_{\mu\nu}$ as a law, though perhaps nuts-and-bolts (quantum?) physics would prevent degeneracy or signature change from a particle physics standpoint. On the other hand, it remains unclear on what grounds one should or even can postulate "equal-time" or space-like commutation relations without a background notion of causality (Pitts and Schieve, 2004). Finally, while the gravitational potential is uniquely defined only up to field redefinitions, the background matrix η is unique (up to a conventional overall sign and a conventional choice to put "time," i.e., the sign that differs from the other three, first or last). Hence the ambiguity in defining the gravitational potential does not infect spacetime.

While one might see the appeal of a background of constant curvature (thus admitting a cosmological constant Λ), it might seem needlessly restrictive to specify a *flat* background. A background of constant curvature, however, apparently requires either explicit dependence on spacetime coordinates (unlike $diag(-1, 1, 1, 1,)$), which seems arbitrary, patchy (not attractive globally), and at best ugly, or the use of tensor calculus with a background metric *tensor* and consequent duplication of the gauge freedom. Is representing spacetime with the matrix $diag(-1, 1, 1, 1)$ hence too fragile? Perhaps it isn't, because Gia Dvali and collaborators argue that the need to accommodate theories requiring *S*-matrix formulations, including string theory, mandates a *flat* background, not one of constant (positive) curvature (Dvali and Gomez, 2016; Dvali, 2020). Hence there are modern independent plausible physical reasons for thinking that flat spacetime plays a fundamental role. If so, why shouldn't gravitational energy refer to it somehow?

3.3   Spacetime energy implies substantivalism?

Plausibly, if spacetime has energy-momentum, spacetime is substantival (Earman and Norton, 1987; Hoefer, 2000). This claim has been shared by Earman and Norton, who affirm gravitational energy's reality (but not localizability) and seem *prima facie* attracted to substantivalism (while encountering objections to it), and Hoefer, who denies gravitational energy and sees relationism-friendliness as an important benefit. Earman & Norton deny that $g_{\mu\nu}$ is part of spacetime to avoid spacetime energy:

> [GR] has made most compelling the identification of the bare manifold with spacetime. For in that theory geometric structures, such as the metric tensor, are clearly physical fields in spacetime. [footnote suppressed] The metric tensor now incorporates the gravitational field and thus, like other physical fields, carries energy and momentum ... in a way that forces its classification as part of the contents of spacetime.
>
> (Earman and Norton, 1987, p. 519)

In partial contrast, Hoefer denies gravitational energy and saves relationism:

> But if $T^{ab} = 0$ "empty space" can carry *genuine* energy-momentum of the gravitational field, then it (the empty space) should be counted as real also, and spacetime itself as represented by g^{ab} should be considered substantial and real.
>
> (Hoefer, 2000, p. 188)

> If empty spacetime need not be thought to possess genuine energy, at least one reason for considering it to be a substance is deflected.
>
> (*ibid.*, p. 196)

While granting the inference from spacetime energy to substantivalism, the view proposed here affirms gravitational energy, while denying that gravitational energy is spacetime energy. Thus one finds another way to avoid an argument for substantivalism from gravitational energy. One should consider, however, whether η savors of absolute space(time) before settling on a verdict about substantivalism (cf. Brown and Pooley, 2006).

It should be noted that the inference from spacetime energy to space-time substantivalism is intended to flow in only one direction. While authors of diverse views (noted earlier) seem to accept that spacetime energy is or would be sufficient for spacetime realism, there is no claim of a necessary condition. Duerr articulates a view in which spacetime is real without spacetime energy, because for Duerr (like Hoefer), gravitational energy does not exist (Duerr, 2020).

3.4 Particle physics egalitarianism *vs.* GR exceptionalism

Partly by default the philosophy of space & time since the late 1970s has leaned toward GR exceptionalism. Given the indifference (at best) of many particle physicists to philosophy, the only physicists who seemed relevant must have been general relativists. But the very community identification of general relativists—the sign on the door indicates that one is attached to a particular theory of a particular force, akin to identifying with Weberian electrodynamics—reflects and inculcates GR exceptionalism. GR exceptionalism *vs.* particle physics egalitarianism is rarely explicitly debated (but see Feynman et al., 1995; Duff, 1975; Kaiser, 1998; Rovelli, 2002; Smolin, 2006; Brink, 2006; Pitts, 2017, 2020b), but is clearly implicit in decades-long research programs in quantum gravity. Thus canonical quantum gravity was not merely intended by GR exceptionalists to use Hamiltonian methods to merge quantum mechanics and GR, but also to leave ample room for revolutionary and nonperturbative consequences not expected or even accommodated using particle physicists' perturbative techniques (Salisbury, 2020). This section aims both to illustrate the two attitudes and to indicate how the particle physics egalitarian attitude might affect discussions of gravitational energy.

Feynman taught a course at CalTech in 1962–3 in which he approached gravity as just another physical field, assumed like the others *a priori* except where empirical facts implied a distinction:

> [M]eson theorists ... have gotten used to the idea of fields, so that it is not hard for them to conceive that the universe is made up of twenty-nine or thirty-one other fields all in one grand equation; the phenomena of gravitation add another such field to the pot, it is a new field which was left out of previous considerations, and it is only one of the thirty or so; explaining gravitation therefore amounts to explaining three percent of the total number of known fields.
>
> (Feynman et al., 1995, p. 2)

Feynman, like various other authors (Gupta, 1954; Kraichnan, 1955; Weinberg, 1964; Deser, 1970; Pitts and Schieve, 2001; Pitts, 2016a), showed how to derive Einstein's equations with considerable rigor from very weak, plausible field theoretic postulates and a few empirical facts (such as the bending of light). The rigor greatly exceeds, say, Einstein's arguments about rotation and misconceptions about conservation laws in his process of discovery (Pitts, 2016b). One might expect that many physicists over decades would do better than one physicist in a few years having to invent many of the relevant tools. But the "spin 2" derivations of Einstein's equations still seem not to get the attention that they deserve in philosophy, at least until recently (Salimkhani, 2020). It was noticed recently that the spin 2 derivations make crucial use of a form of Noether's converse Hilbertian assertion, that the energy-momentum complex's being the sum of a term vanishing with the field equations and a term with automatically vanishing coordinate divergence implies general covariance (Pitts, 2016a).

Feynman explained later how his particle physics egalitarian views affected the debate on gravitational energy:

> What is the power radiated by such a [gravitational] wave? There are a great many people who worry needlessly at this question, because of a perennial prejudice that gravitation is somehow mysterious and different—they feel that it might be that gravity waves carry no energy at all. We can definitely show that they can indeed heat up a wall, so there is no question as to their energy content. The situation is exactly analogous to electrodynamics.
>
> (Feynman et al., 1995, pp. 219–220)

Feynman seems never to have given any very definite response to the usual worries about gravitational energy, but he seems committed to the existence of gravitational energy and to quasi-localization *avant la lettre*: heating up a wall implies that energy can be localized into small regions.

Curiel's recent work underlines the connection between GR exceptionalism and doubts about gravitational energy. As will be discussed, his GR exceptionalism motivates giving up on the idea of conservation altogether (Curiel, 2019). On the other hand, one might reconcile realism about gravitational energy with general relativist sympathies following Rovelli:

> A strong burst of gravitational waves could come from the sky and knock down the rock of Gibraltar, precisely as a strong burst of electromagnetic radiation could. Why is the first "matter" and the second "space"? Why should we regard the second burst as ontologically different from the second?
>
> (Rovelli, 1997, p. 193)

His book elaborates on matter-like aspects of gravity (Rovelli, 2004, p. 77).

3.5   Three questions about gravitational energy and a neglected option?

Recalling the views of Earman and Norton and of Hoefer on gravitational energy, one might consider three questions. First, is gravitational energy real? Second, should the (formal) gravitational (pseudo-?)energy be attributed just to $g_{\mu\nu}$, not to something else in addition or instead? Third, is $g_{\mu\nu}$ part of what represents spacetime? Four interesting sets of answers (perhaps among others) are:

- Gravitational energy is real and attributable just to $g_{\mu\nu}$, which is part of spacetime. (Yes, yes, and yes.)
- Gravitational energy is real and attributable just to $g_{\mu\nu}$, but $g_{\mu\nu}$ is not part of spacetime; only M is (Earman and Norton, 1987). (Yes, yes and no.)
- Gravitational energy is not real, but gravitational pseudo-energy is attributable just to $g_{\mu\nu}$, which is part of spacetime (Hoefer, 2000). (No, yes and yes.)
- Gravitational energy is real, but it is attributable to the gravitational field and not to $g_{\mu\nu}$. $g_{\mu\nu}$ is a composite entity made partly of non-energetic spacetime ingredient(s) and partly of the physical-material gravitational field. (Yes, no and no.)

Something like the fourth view was contemplated briefly by Earman & Norton; has the view been set aside deservedly? Let us recall:

> We might consider dividing the metric into an unperturbed background and a perturbing wave in the hope that the latter alone can be classified as contained in spacetime. This move fails since there is no non-arbitrary way of effecting this division of the metric.
>
> (Earman and Norton, 1987, p. 519)

While they do not provide an argument there, arguments had been given before and are provided elsewhere by Norton. A flat background metric introduces a whole new gauge "group" changing the flat background tensor while leaving the effective curved metric alone (Band, 1942; Grishchuk et al., 1984; Norton, 1994; Pinto-Neto and Trajtenberg, 2000; Petrov and Pitts, 2019). Even specifying a *flat* background metric isn't nearly enough to remove arbitrariness. But this chapter proposes not a background metric tensor, but a scalar matrix *diag*(−1, 1, 1, 1), the same at every point in every coordinate system—the matrix to which one could reduce the components of a flat background metric by adapting coordinates. The matrix $diag(-1, 1, 1, 1)$, however, is just 0s, 1s, and a -1. Absent a cosmological constant, this split crops up in weak field problems in GR and in perturbative treatments of gravity more generally, classical or quantum. This razor-thin background provides most benefits of a background metric or connection, with hardly any of the disadvantages. One has pure GR, or something extremely close. One can make sense of the background and the gravitational potential by recalling some lesser-known parts of the classical theory of geometric objects.

3.6   From tensors to geometric objects

Differential geometry generalized from tensors to "geometric objects" (Nijenhuis, 1952; Schouten, 1954; Yano, 1955; Aczel and Golab, 1960; Anderson, 1967; Friedman, 1973; Pitts, 2006, 2012; Read, 2022), a literature that made interesting progress into the 1960s. While more general notions existed, the basic idea was to generalize the tensor transformation law to any local algebraic transformation law built using derivatives of one coordinate system with respect to another. The transformation rule could involve higher derivatives (seen already with Christoffel symbols), be affine rather than linear (also seen with Christoffel symbols), or even be nonlinear. Few interesting examples of nonlinear or even affine geometric objects (except the connection or its projective and trace/volume pieces) were presented, however. That was historically contingent, because particle physicists were reinventing largely the same ideas at the same time (affine and nonlinear group realizations) and applying them to spinors in spacetime (DeWitt, 1949, 1950; Ogievetsky and Polubarinov, 1965; Ogievetskiĭ and Polubarinov, 1965; Pitts, 2012). For present purposes, the relevant post-tensorial geometric objects are affine with only first derivatives.

I suggest that the main culprit with the background structures traditionally used, is not the flatness or the background character, but the tensor character. How about a *constant scalar* background matrix $\eta = diag(-1, 1, 1, 1)$? This quantity has been common in covariant perturbation theory (Gupta, 1952; Feynman, 1963; Ogievetsky and Polubarinov, 1965; Ogievetskiĭ and Polubarinov, 1965; Veltman, 1981).[4] To consider a scalar background matrix in differential geometry, one can explore transformation rules beyond $O' = f\left(\frac{\partial x^{\mu'}}{\partial x^{\nu}}\right)O$ (Tashiro, 1950; Nijenhuis, 1952; Yano, 1955; Aczèl and Golab, 1960; Szybiak, 1963). Affine geometric objects have some nice properties: non-tensorial behavior gets excised in various contexts. The tensorial Lie derivative of a connection is a somewhat familiar but complicated example with second derivatives.

A less familiar example is a tensor (or tensor density) minus some constant(s), perhaps such as a "vacuum" value. Splitting $g_{\mu\nu}$ into a background matrix η and a gravitational potential $\gamma_{\mu\nu}$ (or building it from them, depending on what one takes to be primitive) provides a physically interesting example: the gravitational potential $\gamma_{\mu\nu}$. One could as easily put the indices up as down, and/or give any density weight, integral or not (Ogievetsky and Polubarinov, 1965; Pitts, 2016d).

Assuming a metric $g_{\mu\nu}$, we can define a gravitational potential (or "perturbation," though there is no assumption of smallness) by subtracting η, a sort of vacuum value, not necessarily in the QFT sense, but the default value when nothing interesting is happening and descriptive simplicity is employed. This procedure is routine in testing for stability, whether in the Higgs mechanism or in applied mathematics (Khazin and Shnol, 1991). Choosing the normalization to give the Lagrangian standard (non-geometric) dimensionality and depend on the gravitational potential in the standard way, one has $\gamma_{\mu\nu} = (g_{\mu\nu} - \eta)/\sqrt{32\pi G}$ (or the like under algebraic field redefinitions involving $\mathfrak{g}^{\mu\nu} - \eta$, etc., reasonable choices agreeing in their lowest order traceless parts in approximately Cartesian coordinates given suitable normalization, cf. Ogievetsky and Polubarinov, 1965). I (usually) avoid writing η as $\eta_{\mu\nu}$ to avoid the notational suggestion that η is a tensor; it is a matrix-valued scalar.

Before delving further into the technicalities, one can avert some tempting misconceptions. First, this definition is non-perturbative because $\gamma_{\mu\nu}$ is *not assumed small*. Neither the physics nor the coordinates are assumed "mild"; one could use spherical coordinates falling into a black hole or use approximately Cartesian coordinates in a "wrong" order such as $\langle x, t, y, z \rangle$ (albeit with a large effect on the gravitational potential value). No series expansion or restriction to achieve convergence is used. Second, unlike a flat background metric tensor, the matrix η imposes no topological restrictions. It is merely part of a change of variables: one takes the components of the metric tensor (or some similar quantity) and subtracts -1, 1 or 0. Third, the coordinates are completely arbitrary; no gauge fixing or coordinate condition, whether involving equations or inequalities (such as for time for a Hamiltonian) is employed. Fourth, the success or failure of quantum gravity programs that made use of this expansion perturbatively is irrelevant. The use is not intrinsically quantum, and one could quantize canonically or with a path integral or perhaps on some other way. This is pure GR, or within a hair's breadth thereof, with an innocent change of variables. Fifth, there is no extra gauge freedom, only the traditional coordinate freedom of GR. Nothing, or nearly nothing extra has been introduced, and no new gauge freedom to deny its physical meaning is present or needed. η appears in GR with spinors anyway and so cannot be very bad. There is little or no ontology for this razor-thin background. These remarks and those earlier are doubtless not the last word needed, but they might prime the pump.

With such misconceptions averted, we can explore the classical differential geometry of affine geometric objects. From $\gamma_{\mu\nu} = (g_{\mu\nu} - \eta)/\sqrt{32\pi G}$ and the tensor transformation rule

$$g'_{\mu\nu} = g_{\alpha\beta} \frac{\partial x^\alpha}{\partial x^{\mu'}} \frac{\partial x^\beta}{\partial x^{\nu'}}$$

one has:

$$\gamma'_{\mu\nu} = \gamma_{\alpha\beta} \frac{\partial x^\alpha}{\partial x^{\mu'}} \frac{\partial x^\beta}{\partial x^{\nu'}} + \frac{\eta_{\alpha\beta}}{\sqrt{32\pi G}} \left(\frac{\partial x^\alpha}{\partial x^{\mu'}} \frac{\partial x^\beta}{\partial x^{\nu'}} - \delta^\alpha_\mu \delta^\beta_\nu \right).$$

This is a local geometric object, with components at every point in every coordinate system covering that point and a transformation rule relating any two coordinate systems covering that point. It is an affine, first-differential-order geometric object, falling in between tensors and connections. Affine geometric objects have *linearly* transforming Lie derivatives (Tashiro, 1950; Nijenhuis, 1952; Yano, 1955): the non-tensorial part of the transformation rule is shaven off, making the Lie derivative a tensor:

$$\pounds_\xi \gamma_{\mu\nu} = \xi^\alpha \gamma_{\mu\nu,\alpha} + \gamma_{\alpha\nu} \xi^\alpha{}_{,\mu} + \gamma_{\mu\alpha} \xi^\alpha{}_{,\nu} + \frac{\eta_{\alpha\nu} \xi^\alpha{}_{,\mu} + \eta_{\alpha\mu} \xi^\alpha{}_{,\nu}}{\sqrt{32\pi G}} = \frac{1}{\sqrt{32\pi G}} \pounds_\xi g_{\mu\nu}.$$

Thus symmetries of the metric (Killing vector fields) are readily expressed as symmetries of the gravitational potential.

What of covariant differentiation? There is a little known formula for the covariant derivative of any first-order (even nonlinear) geometric object ω^N (Szybiak, 1963):

$$\nabla_a \omega^N = \partial_a \omega^N + \Gamma^\kappa_{a\lambda} \lim_{\frac{\partial x'}{\partial x} \to I} \frac{\partial \omega^{N'}}{\partial \left(\frac{\partial x^{\kappa'}}{\partial x^\lambda} \right)}.$$

Thus $\nabla_a \gamma_{\mu\nu} = \gamma_{\mu\nu,a} - \frac{1}{32\pi G} (\Gamma^\kappa_{a\nu} g_{\mu\kappa} + \Gamma^\kappa_{a\mu} g_{\kappa\nu})$, which agrees with the metric's covariant derivative up to a constant factor: $\nabla_a \gamma_{\mu\nu} = \frac{1}{\sqrt{32\pi G}} \nabla_a g_{\mu\nu}$. If the covariant derivative is taken using a $g_{\mu\nu}$-compatible connection $\Gamma^\kappa_{a\lambda}$, the result vanishes. Thus $\gamma_{\mu\nu}$'s Lie and covariant derivatives stand in for $g_{\mu\nu}$'s, as expected because $\Gamma^\kappa_{a\lambda}$ can be built from $\gamma_{\mu\nu}$ and η, and any derivative of η vanishes. One could do all of Riemannian geometry in terms of $\gamma_{\mu\nu}$ and η. Likewise one could derive the Euler-Lagrange equations and the gravitational energy-momentum pseudotensor(s) for GR in terms of $\gamma_{\mu\nu}$ and η.

The matrix η is no newcomer to gravitational energy, because both modern and classical works on gravitational energy have in some cases made use of a reference configuration. Such modern work includes Nester and collaborators (Chang et al., 2000; Nester, 2004) and overlapping teams including Grishchuk, Petrov, Katz, Bičák and Lynden-Bell (Grishchuk et al., 1984; Katz et al., 1997; Petrov and Katz, 2002). One then faces the question of the gauge freedom in relating the two metrics. On the other hand, the background matrix $diag(-1, 1, 1, 1)$ is unique up to conventional choices (actually $diag(-1, -1, -1, 1)$) in the Papapetrou-Belinfante pseudotensor (Papapetrou, 1948).[5] Papapetrou's interpretive concerns about his pseudotensor formalism, I suggest, are largely addressed *via* affine geometric objects (addressing his concerns about a "system of some 'auxiliary numbers'") and the multiplicity of gravitational energies (addressing gauge dependence) (Pitts, 2010). Papapetrou's eventual introduction of a flat metric *tensor* instead of a numerical matrix—the opposite of the move that I suggest—motivated him to gauge-fix the relation between the two metrics in order to avoid an infinity of distinct results, an infinity strikingly resembling coordinate dependence. But his gauge fixing differs only formally from Fock-style fixation of the coordinates with the harmonic condition in a single-metric formalism (Fock, 1959). If gauge-fixing is a satisfactory solution, then why not just fix harmonic coordinates in GR and declare gravitational energy to be localized in the true (e.g., harmonic) coordinates? It seems to me advantageous to accommodate $diag(-1, 1, 1, 1)$ and the metric perturbation within differential geometry and keep the coordinate freedom as it was.

3.7   What represents spacetime: A proposal

Now one can make more sense of the particle physics-inspired proposal for what represents spacetime. In between the standard suggestions of $\langle M, g_{\mu\nu}\rangle$ and just M, is the proposal: spacetime is $\langle M, \eta\rangle$. This suggestion might ring a bell by now: "We might consider dividing the metric into an unperturbed background and a perturbing wave in the hope that the latter alone can be classified as contained in spacetime." (Earman and Norton, 1987, p. 519) I suggest that a non-arbitrary division is $g_{\mu\nu} = \eta + \sqrt{32\pi G}\gamma_{\mu\nu}$ (*mutatis mutandis* with field redefinitions for the gravitational potential); hence η is the unperturbed background and $\gamma_{\mu\nu}$ (or some equivalent entity) is the perturbing "wave" (not necessarily wavy) contained in spacetime. Is this the golden mean, the Goldilocks zone? On this proposal, *much of* the metric is part of spacetime, as one might prefer. If nothing much is happening (weak fields), and coordinates are adapted to this situation (not far from Cartesian and with time in the temporal slot: cf. Bilyalov, 2002; Pitts, 2012), then the value of $g_{\mu\nu}$ is approximately η. On the other hand, η savors of Minkowski space-time, which by some lights is a relationism-friendly "glorious non-entity" (Brown and Pooley, 2006), though one might also consider a resemblance to traditional absolute space(time). Gravitational energy is not due to spacetime (though η might appear in it), but due to the gravitational potential $\gamma_{\mu\nu}$, likely quadratic in its first derivatives and possibly having some second derivatives (especially spatial or mixed). If gravitational energy is not spacetime energy, then one can believe in gravitational energy, even in infinitely many localized gravitational energies, with no pressure toward affirming substantivalism from such energy. With gravitational energy ascribed to $\gamma_{\mu\nu}$, not spacetime, one can do justice to the sensibility that the $g_{\mu\nu}$ is in many ways like a matter field (Rovelli, 2004, p. 77; Brown, 2005, chapter 9)—an idea linked to Rovelli's gravitational energy realism (Rovelli, 1997). As Hoefer said, "[i]f empty spacetime need not be thought to possess genuine energy, at least one reason for considering it to be a substance is deflected." (Hoefer, 2000, p. 196) Now this conclusion can be reached in another way, accepting gravitational energy but denying spacetime energy.

3.8 Curiel on gravitational energy

The idea of gravitational energy has received philosophical attention recently from the GR exceptionalist perspective as well. Curiel has underscored the non-existence of a local gravitational stress-energy tensor given certain assumptions and has suggested giving up the usual idea of conservation in favor of fungibility:

> I prove that, under certain natural conditions, there can be no tensor whose interpretation could be that it represents gravitational stress-energy in general relativity. It follows that gravitational stress-energy, such as it is in general relativity, is necessarily non-local.
>
> (Curiel, 2019, p. 91)

For Curiel the inference from non-tensoriality (or perhaps more generously, not being a geometric object) to non-localizability is rather direct. A local but non-tensorial object, such as the Noether operator (Schutz and Sorkin, 1977; Sorkin, 1977) or the closely related and more pedestrian pseudotensor(s), is not a candidate because it is not a physical quantity (Curiel, 2019, p. 97).

The fact that integrals over pseudotensors depend only on coordinates at the boundary (Chang et al., 1999) indicates that the temperature in Curiel's coffee is less coordinate-dependent than one might have feared. His concern about using a pseudotensor to ascertain how much a gravitational wave partly absorbed by a piezoelectric stick would warm his coffee, is unclear to me. If the concern is that it can be used but would give different answers in different coordinate systems, that seems unlikely because physical objects, such as coffee and thermometers, are not sensitive to a choice of coordinates. The piezoelectric stick does not need to "'know' which of Pitts's 'localized energies' it should draw on" because any of them will do. Recalling that a pseudotensor conservation law is logically equivalent to Einstein's equations (Anderson, 1967; Pitts, 2010), the conservation law is a way of expressing the content of GR that shows that a sum of material energy and gravitational (pseudo-?)energy satisfies the continuity equation.

Curiel has not specified the wavelength of the gravitational wave in question, but the ratio of this wavelength to the cup-detector affects the analysis (Schutz and Ricci, 2001). For wavelengths small compared to the detector, the problem is largely included in Schutz and Ricci's treatment. After discussing how one can treat gravitational waves on a gently varying background as a type of matter and use Isaacson's averaging over a few wavelengths to get a localization to that scale, they comment:

> In the textbooks you will find discussions of pseudotensors ... of Noether theorems and formulas for energy, and so on. None of these are worse than we have presented here, and in fact all of them are now known to be consistent with one another, if one does not ask them to do too much. In particular, if one wants only to localize the energy of a gravitational wave to a region of the size of a wavelength, and if the waves have short wavelength compared to the background curvature scale, then pseudotensors will give the same energy as the one we have defined here.
>
> (*ibid.*, p. 56)

Astrophysically plausible gravitational waves will tend to have wavelengths longer than a coffee cup, however, and a fundamental treatment ought to be able to handle all cases. Hence not all of Curiel's question is answered by Schutz and Ricci. My sense is that the question, if not already answered somewhere, is more like a puzzle than an anomaly (to borrow some Kuhnian concepts for an idea that hardly constitutes a paradigm sociologically): the question is worth answering, and there is no good reason to doubt that it can be answered.

Strikingly, symmetries of the action and Noether's theorem play no role for Curiel. Neither does the continuity equation appear, so Curiel does not entertain anything that could yield results such as $E = constant$. This absence is not an oversight, but a principled inference from the premises adopted:

> The formulation of the First Law [of thermodynamics] I rely on is somewhat unorthodox: that all forms of stress-energy are in

principle ultimately fungible—any form of energy can in principle be transformed into any other form [footnote suppressed]—not necessarily that there is some absolute measure of the total energy contained in a system or set of systems that is constant over time.

(Curiel, 2019, p. 96)

If coordinate-free GR doesn't have it, Curiel doesn't need it. Hence he does not need energy conservation, a view that some will consider bold, but a coherent view that, used carefully,[6] will never yield a false prediction (unless lower-brow approaches to GR would also).

One might wonder, however, whether fungibility is an adequate version of, or substitute for, the First Law of Thermodynamics. In the multi-trillion-dollar foreign exchange market, of which the most obvious manifestations at airports offer the clearest insights, one is given the opportunity to convert (say) USD to GBP at one rate, or the reverse at a very different rate. At airports this "spread" is enormous, so one could convert USD to GBP and back, leaving with far fewer USD than one started with, however. Hence mere fungibility, with no chance of recovering one's starting point, is not clearly a version of the First Law. Possibly it is something like a combination of the First and Second Laws at best: even if perhaps money is conserved, your money tends to decrease. But a clear analog of the First Law only is not to be found in mere fungibility without quantitative bookkeeping implying that the starting configuration is in some respects preserved.

One might also question the mathematical basis for requiring a *symmetric covariant* stress-energy tensor for gravity, in tension with what Lagrangian field theory offers. For a scalar field, Noether methods offer for the stress-energy a *mixed* (1, 1) weight 1 tensor *density* $\mathfrak{T}^\mu_\nu$ to give $\mathfrak{J}^\mu[\xi] = \mathfrak{T}^\mu_\nu \xi^\nu$. A displacement vector yields a conserved current that is a tangent vector *density* of weight 1, thus such that $\nabla_\mu \mathfrak{J}^\mu \equiv \partial_\mu \mathfrak{J}^\mu = 0$. Given that only a partial divergence $\partial_\mu \mathfrak{J}^\mu = 0$ has a chance of integration to $E = constant$ (Weyl, 1922, pp. 236, 269–271; Landau and Lifshitz, 1975, p. 280; Misner et al., 1973, p. 465; Lord, 1976, p. 139; Stephani, 1990, p. 141), this is the ideal case, so requiring in advance something different (a symmetric stress-energy tensor) shows a lack of interest in getting conserved quantities, the obtaining of which usually has been considered a core part of the issue. As noted earlier, conservation of angular momentum requires not symmetric stress-energy, but only rotational symmetry of the action, because angular momentum needn't be $x^{[\alpha} T^{\mu]\nu}$ (Bergmann and Thomson, 1953; Forger and Römer, 2004); there can be a spin contribution. For vector matter, similar Noether methods give a current $\mathfrak{J}^\mu[\xi]$ that is tensorial but not algebraic in ξ^ν. Hence one cannot peel off the vector field ξ^μ and get a stress-energy tensor, as one can with a scalar field. For GR, $\mathfrak{J}^\mu[\xi]$

either is non-tensorial or has higher derivatives, depending on which Lagrangian density one uses (Sorkin, 1977). If one asks GR what the Noether mathematics means instead of imposing requirements by hand, it (mostly?) makes sense: the Noether operator generalizing $\mathfrak{J}^{\mu}[\xi]$, depending differentially on ξ^{μ}, yields a pseudotensor if one sets $\xi^{\mu} = (1, 0, 0, 0)$. Abstaining from this sort of mathematics to remain pure of coordinates leads to giving up conservation laws that do exist. A criterion for physical quantities that excludes the Noether operator is not obviously appropriate.

It is also obscure what the rules are in Curiel's admittedly utopian quest for a gravitational stress-energy tensor $S_{\mu\nu}$. This hypothetical entity is supposed to be symmetric ($S_{[\mu\nu]} = 0$) and to have vanishing covariant divergence ($\nabla_{\mu}S^{\mu\nu} = 0$). If that were true, then one could add it to the material stress-energy tensor as follows:

$$\nabla_{\mu}(\sqrt{-g}S^{\mu\nu} + \sqrt{-g}T^{\mu\nu}) = \partial_{\mu}(\sqrt{-g}S^{\mu\nu} + \sqrt{-g}T^{\mu\nu}) + (\sqrt{-g}S^{\mu\alpha} + \sqrt{-g}T^{\mu\alpha})\Gamma^{\nu}_{\mu\alpha}$$
$$+(\sqrt{-g}S^{\alpha\nu} + \sqrt{-g}T^{\alpha\nu})\Gamma^{\mu}_{\mu\alpha} - (\sqrt{-g}S^{\mu\nu} + \sqrt{-g}T^{\mu\nu})\Gamma^{\alpha}_{\mu\alpha}$$

(using the covariant derivative of a density of weight 1 for the last term, cf. Anderson, 1967)

$$= \partial_{\mu}(\sqrt{-g}S^{\mu\nu} + \sqrt{-g}T^{\mu\nu}) + (\sqrt{-g}S^{\mu\alpha} + \sqrt{-g}T^{\mu\alpha})\Gamma^{\nu}_{\mu\alpha} = 0.$$

This result is *incompatible* with the continuity equation

$$\partial_{\mu}(\sqrt{-g}S^{\mu\nu} + \sqrt{-g}T^{\mu\nu}) \overset{?}{=} 0$$

because $(S^{\mu\alpha} + T^{\mu\alpha})\Gamma^{\nu}_{\mu\alpha} \neq 0$ typically. Hence the resulting equation would *prohibit* the existence of a conserved energy $E = constant$. What Curiel seems to regard as desirable given his heuristics but, alas, impossible would in fact be disastrous. Fortunately GR does permit $E = constant$ for asymptotically flat metrics because it admits the continuity equation, albeit for a quantity of which Curiel does not approve. Thus $S_{\mu\nu}$ wouldn't fit GR. Curiel agrees, but for different reasons involving his proof, which depend on premises not following from Einstein's equations.

> The existence of a gravitational stress-energy tensor, however, would necessarily entail that we modify our understanding and for-mulation of general relativity. That is why this argument is only *ex hypothesi*, and not meant to be one that would make sense in general relativity as we actually know it.
>
> (Curiel, 2019, p. 98)

If the discussion both assumes the supposed "spirit" of GR and contra-dicts GR's laws, it is unclear what the rules are or what proprietary question is being addressed instead of the usual gravitational energy localization problem in GR. I am not convinced that one should heed guidance from such an antinomian spirit.

3.9 Two standard worries about gravitational energy

I turn now to the two standard objections to pseudotensors and argue that both have gotten weaker in the last 20+ years. The two standard objections are that a pseudotensor depend essentially on coordinates, which nothing physically real would do, and that the pseudotensor is nonunique, which is also incompatible with physical reality. Pseudotensors relate weirdly to coordinates in at least two ways: false positives and false negatives. Schrödinger presented an early false-negative objection: Einstein's gravitational energy-momentum pseudotensor vanishes outside a round heavy body in some coordinates, but surely there should be gravitational energy outside a round heavy body, and whether there is gravitational energy should not depend on coordinates (Schrödinger, 1918; Cattani and De Maria, 1993). Bauer presented a mirror-image objection that same year, a false positive: Minkowski spacetime in (unimodular) spherical coordinates has a nonzero energy density and even infinite total energy (Bauer, 1918; Cattani and De Maria, 1993). The nonuniqueness objection is that there are many comparably good candidates, and they cannot all be real, so plausibly none of them is real. The coordinate dependence objection goes back to the 1910s, though there was interesting activity (such as by Møller among others) in the late 50s-early 60s, while the nonuniqueness objection seems to have become serious in the 1950s with the Landau-Lifshitz pseudotensor (Landau and Lifshitz, 1975), Goldberg's infinity of pseudotensors (Goldberg, 1958), and others. Thus most people have long since given up on localization. Some reject gravitational energy outright in light of its apparently inconsistent properties (Hoefer, 2000; Duerr, 2019). This is a principled view that makes at least as much sense as the standard view (Misner et al., 1973) that gravitational energy exists but is not localizable, a claim that Norton also finds obscure (Norton, 2014). I argue that both worries are inconclusive and getting weaker recently. I also note Read's recent functionalist defense of gravitational energy (Read, 2020).

I have explained previously how asking Noether's first theorem how many conserved energies to expect—one for each rigid symmetry of the action, hence infinitely many (Bergmann, 1958)—resolves Schrödinger's false-negative objection (Pitts, 2010). Lacking a coordinate transformation is not a bug (Read, 2022) In fact it is a feature: it permits the expression of infinitely many energies with only 10 or 16 components (Pitts, 2010).

To give an analogy, one might be puzzled by the inequivalence under translation (analogous to lack of a coordinate transformation rule) between "María es alta" (tall) and "Mary is short"—unless María ≠

Mary, in which case there is no reason to expect equivalent heights. If the comparative and context-dependent nature of "tall" and "short" are objectionable, then one can change the example to involve different unit systems. Perhaps María is n meters tall (in Spanish) and Mary is x feet y inches tall. But most of us cannot make such conversions sufficiently accurately without calculation, so the contradiction is not evident. Inequality implies $12x + y \neq \frac{100n}{2.54}$.

Bauer's false positive objection can be criticized on various technical grounds. First, it is unclear that unimodular spherical coordinates (or garden-variety spherical coordinates, for that matter) should be regarded as covering the whole manifold; by modern standards they don't. Second, such coordinates make Einstein's Γ–Γ action diverge. If the field equations and the canonical energy-momentum pseudotensor are derived from the action, and this coordinate system makes the action diverge, why admit them? Third, there is a little-recognized nonuniqueness in the Lagrangian density which renders the Einstein pseudotensor optional as the canonical energy-momentum pseudotensor even given metric variables (not to mention non-canonical pseudotensors). In Maxwell's electromagnetism, one could include the contraction $(\partial_\mu A^\mu)^2$ in the Lagrangian by adding a total divergence, and thus can write down a 1-parameter family of equivalent Lagrangian densities differing by a total divergence. GR admits a similar ambiguity at least linearly (Ohanian and Ruffini, 1994, p. 647). This ambiguity seems to disappear in an exact treatment in metric variables, but it surfaces using an orthonormal basis, as in Møller's work. If one wishes, one can gauge-fix the tetrad into the symmetric square root of the metric (DeWitt and DeWitt, 1952; Møller, 1964; Ogievetsky and Polubarinov, 1965; Pitts, 2010, 2012), which depends in a nonlinear way on the metric and on η. As long as one does not try to put a time coordinate in a spatial place or *vice versa* (roughly) (Pitts, 2012), one has an alternative GR Lagrangian built out of the metric and η with only first derivatives. Hence there is an apparently previously unrecognized 1-parameter ambiguity of GR Lagrangians *in metric variables*, leading to a similar 1-parameter ambiguity of canonical energy-momentum complexes, including the (gauge-fixed) Møller tetrad complex. But it gives 0 energy for Minkowski spacetime (Møller, 1964). Thus it is unclear that Bauer has used admissible coordinates, especially given his (Einstein Γ–Γ) action, and unclear that he has used the correct action, because an action exists that avoids Bauer's false positive objection even in his unimodular spherical coordinates. Recall that Møller's tetrad energy-momentum expression is tensorial (not a pseudotensor) under changes of coordinates, though dependent on the local Lorentz gauge. The local Lorentz gauge freedom can be fixed using the symmetric tetrad gauge condition to give a metric formalism with help from η (Pitts, 2010); some other ways of fixing the local Lorentz gauge give unreasonable results for the mass-energy (Mikhail et al., 1993). One might also take the view, already familiar for spherical symmetry, that one should make use of symmetries when they exist (Misner et al., 1973, p. 603). A plausible generalization is that one should adapt one's coordinates as far as possible to the largest set of commuting Killing vector fields when it is unique (Pitts, 2010). Then one should use Cartesian coordinates for Minkowski spacetime, which would resolve Bauer's objection even given the Einstein pseudotensor. Hence there are several resolutions of Bauer's objection.

Pseudotensoriality occurs in ordinary life, as in the colors of flags. In most places where English is the main language, "the flag is red, white and blue" is true. This claim admits translation into a true statement in France, because the French flag has the same colors. But it fails when translated into German or Spanish (at least in Spain) or most other languages. Failure of translation here is not mysterious: one is naturally referring to one's own national flag, so the statement in different languages has different referents (using the nation-state approximation to make the analogy vivid: one flag per language, which is more accurate in some places than others). In advanced countries, no normal person older than perhaps 5 years is unaware of the existence of other countries, so it is difficult not to know of the multiplicity of flags, so "the" flag will mean our flag. But if one somehow managed not to know that, while knowing multiple languages—perhaps one is part of an educational experiment in a totalitarian country—then one might only know of one flag and think that there is only one flag. Then one might read (perhaps due to a gap in censorship) apparently plausible but apparently incompatible statements such as "Die Flagge ist schwarz, rot und gold" and "the flag is red, white, and blue." The paradox would be resolved by learning that there are different countries with different flags. Flags and languages have a (more or less natural) relationship.[7] Perhaps energies and coordinate systems do as well, such as coordinates in which the corresponding displacement takes the form (1, 0, 0, 0) or the like.

Earlier it was noted that pseudotensors are economical: one is enabled to say infinitely many distinct things with a 10- or 16-component entity; the economy of pseudotensorial policies also has real-world examples, including the publishing industry and some multilingual academic writing. Whereas tensor calculus says the same old thing infinitely many times using the tensor transformation rule, no publisher feels obliged to publish translations of all of its books in every language. ("Coordinate-free" publication in terms of propositions or the language of thought, as opposed to publication in a language, is only science fiction at this stage.) In some academic fields, such as ancient or medieval philosophy, one finds untranslated Greek or Latin text in an article written in (usually) French, German or English, with along with untranslated quotations from the other two modern languages. While the audience able to appreciate such work fully is not large, it does exist. Perhaps gravitational energy also involves a sort of multilingualism to accommodate the natural connections between energies and coordinate systems.

3.10 Nonuniqueness objection and 3 or 4 possible answers

One would expect the pseudotensor to be unique if it represents real gravitational energy. But there are infinitely many candidates. So the pseudotensor does not represent real gravitational energy. This seems to be how the nonuniqueness objection runs. There are, however, four interesting replies to this objection.

An initial reply, which appeals to the widespread acceptance of material energy $T^{\mu\nu}$, is a *tu quoque* response (Pitts, 2010): gravitational energy is not qualitatively worse off than supposedly unproblematic material energy. Even a scalar field in flat spacetime suffers from nonuniqueness due to a multiplicity of comparably plausible candidates, due to the "improved" energy-momentum tensor, which has certain advantages (Callan et al., 1970).

A second reply comes from the work of Nester *et al.*, according to whom different pseudotensors describe different quasi-localizations with physical meaning tied to boundary conditions (Chang et al., 2000; Nester, 2004). Thus different pseudotensors are right in different contexts. Why should the same one be required in every context, given the close relationship between pseudotensors and boundaries?

A third reply is that there is a best One True pseudotensor. Perhaps it is the Papapetrou-Belinfante pseudotensor or a higher-tech relative thereof (Petrov and Katz, 2002). Clearly this third reply is incompatible with the second, but one can simply offer their disjunction, or even parts of each: maybe some pseudotensors are always wrong, but others are right in one context or another.

A fourth reply is rooted in old work of which the full import was perhaps not recognized (Bergmann, 1958): nonuniqueness is not a distinct objection, but only a repeat of coordinate dependence. Bergmann, perhaps working with a restricted class of pseudotensors, noted that "the totality of all conservation laws ... in one coordinate system is equivalent to one of them, stated in terms of all conceivable coordinate systems." (*ibid.*, p. 289). The "totality of all conservation laws" refers to different pseudotensors. He shows how to find the Einstein and the Landau-Lifshitz pseudotensors in his expression by choosing $\delta x^\sigma = k^\sigma$ (where k^σ is a set of constants) or $\delta x^\sigma = \mathfrak{g}^{\sigma\alpha} k_\alpha$. Especially if one has a reply to the coordinate dependence objection in terms of infinitely many energies, reducing the nonuniqueness objection to the coordinate dependence objection helps.

3.11   Conclusion

Given modern progress, there seems to be no reason to regard gravitational energy realism as doomed. If gravitational energy is real and localized, then material + gravitational energy-momentum is conserved: $\partial_\mu(\mathfrak{T}_\nu^\mu + \mathfrak{t}_\nu^\mu) = 0$, *really* (not just formally), which for isolated systems can give $E = constant$. If spacetime is $\langle M, \eta \rangle$, gravitational energy(s) isn't spacetime energy(s) and so do(es)n't imply substantivalism. Some classic and modern works on gravitational energy call for a reference configuration, while some work on quantum gravity argues that a (non-negative) constant curvature background geometry must be flat (Dvali, 2020). Given η, it is plausible to split $g_{\mu\nu}$ into $\gamma_{\mu\nu}$ and η. Hence there is coherence between the spacetime metric split and promising local representations of gravitational energy. This package of views seems plausible given particle physics egalitarianism, an option traditionally rarely entertained in philosophy or conceptual discussions of GR.

Notes

1. That covariant perturbation theory as a means of quantization did not work due to nonrenormalizability (infinities not removable without an infinite number of additional parameters) is not very relevant. Moreover, after its apparent demise in the 1970s, a possible revival in supergravity emerged (Bern et al., 2007, 2009). In any case one gets a quite satisfactory effective field theory (Donoghue, 2012).
2. The referee provided a timely reminder of this passage.
3. The abstract index notation, often associated with lower-case Latin letters for four-dimensional quantities, has been claimed to have "all the advantages of the component notation" while avoiding the disadvantage of obscuring the distinction between tensorial equations and equations for their components in some particular basis (perhaps adapted to the symmetries of some problem) (Wald, 1984, p.24). Wald. But what of expressive adequacy? Non-integral weight densities in the abstract index notation appear to be an unresolved problem—to say nothing of nonintegral powers of the metric (Ogievetsky and Polubarinov, 1965)—which in turn tends to restrict what can be thought.
4. Actually much of that work used for the background matrix $I = diag(1, 1, 1, 1)$ and $x^4 = ict$, which is even more striking, but a bridge too far, restricting the coordinates or admitting only a clumsy generalization.
5. Presumably no one will think that the matrix $diag(-1, 1, r^2, r^2 sin^2\theta)$, for example, is a plausible candidate to appear in laws of nature, although it is also, like $diag(-1,1,1,1)$, a matrix of components of a flat metric tensor in certain coordinates.
6. Elsewhere I have discussed how the supposed absence of conservation laws in GR encourages some people to object spuriously to the theory and others to think that energy non-conserving processes are thereby licensed (Pitts, 2010, 2020a).
7. Obviously politically fraught real-world issues are glossed over for the sake of the analogy. No views are intended about independence movements, recognition of minority languages/dialects, *etc.*

Bibliography

Aczél, J. and S. Golab (1960). *Funktionalgleichungen der Theorie der Geometrischen Objekte*. PWN.

Anderson, J. L. (1967). *Principles of Relativity Physics*. Academic.

Band, W. (1942). Comparison spaces in general relativity. *Physical Review* 61, 702–707.

Bauer, H. (1918). Über die Energiekomponenten des Gravitationsfeldes. *Physikalische Zeitschrift* 19, 163–165.

Bergmann, P. G. (1958). Conservation laws in general relativity as the generators of coordinate transformations. *Physical Review* 112, 287–289.

Bergmann, P. G. and R. Thomson (1953). Spin and angular momentum in general relativity. *Physical Review* 89, 400–407.

Bern, Z., J. J. Carrasco, L. J. Dixon, H. Johansson, D. A. Kosower, and R. Roiban (2007). Three-loop superfiniteness of N = 8 supergravity. *Physical Review Letters* 98, 161303. hep-th/0702112v2.

Bern, Z., J. J. Carrasco, L. J. Dixon, H. Johansson, and R. Roiban (2009). The ultraviolet behavior of N = 8 supergravity at four loops. *Physical Review Letters* 103, 081301. 0905.2326v1 [hep-th].

Bilyalov, R. F. (2002). Spinors on Riemannian manifolds. *Russian Mathematics (Iz. VUZ)* 46 (11), 6–23.

Brink, L. (2006). A non-geometric approach to 11-dimensional supergravity. In J. T. Liu, M. J. Duff, K. S. Stelle, and R. P. Woodard (Eds.), *DeserFest: A Celebration of the Life and Works of Stanley Deser*, pp. 40–54. World Scientific.

Brown, H. R. (2005). *Physical Relativity: Space-time Structure from a Dynamical Perspective*. Oxford University Press.

Brown, H. R. and O. Pooley (2006). Minkowski space-time: A glorious nonentity. In D. Dieks (Ed.), *The Ontology of Spacetime*, pp. 67–89. Elsevier. arXiv:physics/0403088 [physics.hist-ph].

Bruhat, Y. (1962). The Cauchy problem. In L. Witten (Ed.), *Gravitation: An Introduction to Current Research*, pp. 130–168. John Wiley and Sons. Translated by Stanley Deser.

Callan, Jr., C. G., S. Coleman, and R. Jackiw (1970). A new improved energymomentum tensor. *Annals of Physics* 59, 42–73.

Cattani, C. and M. De Maria (1993). Conservation laws and gravitational waves in General Relativity (1915–1918). In J. Earman, M. Janssen, and J. D. Norton (Eds.), *The Attraction of Gravitation: New Studies in the History of General Relativity*, Volume 5 of Einstein Studies, pp. 63–87. Birkhäuser.

Chang, C.-C., J. M. Nester, and C.-M. Chen (1999). Pseudotensors and quasilocal energy-momentum. *Physical Review Letters* 83, 1897–1901.

Chang, C.-C., J. M. Nester, and C.-M. Chen (2000). Energy-momentum (Quasi-) localization for gravitating systems. In L. Liu, J. Luo, X.-Z. Li, and J.-P. Hsu (Eds.), *The Proceedings of the Fourth International Workshop on Gravitation and Astrophysics: Beijing Normal University, China, October 10–15, 1999*, pp. 163–173. World Scientific. arXiv:gr-qc/9912058v1.

Curiel, E. (2019). On geometric objects, the non-existence of a gravitational stress-energy tensor, and the uniqueness of the Einstein field equation. *Studies in History and Philosophy of Modern Physics* 66, 90–102.

Dasgupta, S. (2011). The bare necessities. *Philosophical Perspectives* 25, 115–160.

Deser, S. (1970). Self-interaction and gauge invariance. *General Relativity and Gravitation* 1, 9–18. arXiv:gr-qc/0411023v2.

DeWitt, B. S. (1949). *I. The Theory of Gravitational Interactions. II. The Interaction of Gravitation with Light*. Ph. D. thesis, Harvard University. Supervisor Julian Schwinger.

DeWitt, B. S. (1950). *On the Application of Quantum Perturbation Theory to Gravitational Interactions*. https://repositories.lib.utexas.edu/handle/2152/9620.

DeWitt, B. S. (1967). Quantum theory of gravity. II. The manifestly covariant theory. *Physical Review* 162, 1195–1239.

DeWitt, B. S. and C. M. DeWitt (1952). The quantum theory of interacting gravitational and spinor fields. *Physical Review* 87, 116–122.

Donoghue, J. F. (2012). The effective field theory treatment of quantum gravity. *AIP Conference Proceedings* 1483, pp. 73–94. Sixth International School on Field Theory and Gravitation, Petropolis, Brazil, April 2012, http://doi.org/10.1063/1.4756964. arXiv:1209.3511 [gr-qc].

Duerr, P. M. (2019). Fantastic beasts and where (not) to find them: Local gravitational energy and energy conservation in general relativity. *Studies in History and Philosophy of Modern Physics* 65, 1–14.

Duerr, P. M. (2020). *Gravitational Energy and Energy Conservation in General Relativity and Other Theories of Gravity*. Ph. D. thesis, University of Oxford.

Duerr, P. M. (2021). Against 'functional gravitational energy': A critical note on functionalism, selective realism, and geometric objects and gravitational energy. *Synthese* 199, S299–S333.

Duff, M. J. (1975). Covariant quantization. In C. J. Isham, R. Penrose, and D. W. Sciama (Eds.), *Quantum Gravity: An Oxford Symposium*, pp. 78–135. Clarendon.

Dvali, G. (2020). S-matrix and anomaly of de Sitter. *Symmetry* 13 (1), 3. http://doi.org/10.3390/sym13010003; arXiv:2012.02133 [hep-th].

Dvali, G. and C. Gomez (2016). Quantum exclusion of positive cosmological constant? *Annalen der Physik* 528, 68–73. http://doi.org/10.1002/andp.201500216; arXiv:1412.8077 [hep-th].

Earman, J. and J. Norton (1987). What price spacetime substantivalism? The hole story. *The British Journal for the Philosophy of Science* 38, 515–525.

Einstein, A. (1961). *Relativity: The Special and the General Theory* (Fifteenth ed.). Crown Publishers.

Feynman, R. P. (1963). Quantum theory of gravitation. *Acta Physica Polonica* 24, 697–722.

Feynman, R. P., F. B. Morinigo, W. G. Wagner, B. Hatfield, J. Preskill, and K. S. Thorne (1995). *Feynman Lectures on Gravitation*. Addison-Wesley. Original by California Institute of Technology, 1963.

Fock, V. A. (1959). *The Theory of Space, Time and Gravitation*. Pergamon. Translated by N. Kemmer.

Forger, M. and H. Römer (2004). Currents and the energy-momentum tensor in classical field theory: A fresh look at an old problem. *Annals of Physics* 309, 306–389.

Freund, P. G. O., A. Maheshwari, and E. Schonberg (1969). Finite-range gravitation. *Astrophysical Journal* 157, 857–867.

Friedman, M. (1973). Relativity principles, absolute objects and symmetry groups. In P. Suppes (Ed.), *Space, Time, and Geometry*, pp. 296–320. D. Reidel.

Goldberg, J. N. (1958). Conservation laws in general relativity. *Physical Review* 111, 315–320.

Grishchuk, L. P., A. N. Petrov, and A. D. Popova (1984). Exact theory of the (Einstein) gravitational field in an arbitrary background space-time. *Communications in Mathematical Physics* 94, 379–396.

Gupta, S. N. (1952). Quantization of Einstein's gravitational field: General treatment. *Proceedings of the Physical Society A (London)* 65, 608–619.

Gupta, S. N. (1954). Gravitation and electromagnetism. *Physical Review* 96, 1683–1685.

Hoefer, C. (2000). Energy conservation in GTR. *Studies in History and Philosophy of Modern Physics* 31, 187–199.

Iftime, M. and J. Stachel (2006). The hole argument for covariant theories. *General Relativity and Gravitation* 38, 1241–1252. arXiv:gr-qc/0512021v2.

Kaiser, D. (1998). A ψ is just a ψ? Pedagogy, practice, and the reconstitution of General Relativity, 1942–1975. *Studies in History and Philosophy of Modern Physics* 29, 321–338.

Katz, J., J. Bičák, and D. Lynden-Bell (1997). Relativistic conservation laws and integral constraints for large cosmological perturbations. *Physical Review D* 55, 5957–5969. arXiv:gr-qc/0504041.

Khazin, L. G. and E. E. Shnol (1991). *Stability of Critical Equilibrium States*. Manchester University Press. Translation by Catherine Waterhouse with Arun V. Holden.

Kraichnan, R. H. (1955). Special-relativistic derivation of generally covariant gravitation theory. *Physical Review* 98, 1118–1122.

Landau, L. D. and E. M. Lifshitz (1975). *The Classical Theory of Fields* (Fourth revised English ed.). Pergamon, Oxford. Translated by Morton Hamermesh.

Lord, E. A. (1976). *Tensors, Relativity and Cosmology*. Tata McGraw-Hill Publishing Co.

Maudlin, T. (1989). The essence of space-time. *Proceedings of the Biennial Meeting of the Philosophy of Science Association* 1988 (2), 82–91.

Menon, T. (2019). Algebraic fields and the dynamical approach to physical geometry. *Philosophy of Science* 86 (5), 1273–1283.

Menon, T. (2021). Taking up superspace: The spacetime setting for supersymmetric field theory. In N. Huggett, B. Le Bihan, and C. Wüthrich (Eds.), *Philosophy beyond Spacetime: Implications from Quantum Gravity*, pp. 103–128. Oxford University Press.

Mikhail, F. I., M. I. Wanas, A. Hindawi, and E. I. Lashin (1993). Energy-momentum complex in Møller's tetrad theory of gravitation. *International Journal of Theoretical Physics* 32, 1627–1642. gr-qc/9406046.

Misner, C., K. Thorne, and J. A. Wheeler (1973). *Gravitation*. Freeman.

Møller, C. (1964). Momentum and energy in general relativity and gravitational radiation. *Matematisk-fysiske Meddelelser udgivet af Det Kongelige Danske Videnskabernes Selskab* 34 (3), 1–67.

Nester, J. M. (2004). General pseudotensors and quasilocal quantities. *Classical and Quantum Gravity* 21, S261–S280.

Nijenhuis, A. (1952). *Theory of the Geometric Object*. Ph. D. thesis, University of Amsterdam. Supervisor Jan A. Schouten.

Norton, J. D. (1994). Why geometry is not conventional: The verdict of covariance principles. In U. Majer and H.-J. Schmidt (Eds.), *Semantical Aspects of Spacetime Theories*, pp. 159–167. B. I. Wissenschaftsverlag.

Norton, J. D. (2014). What can we learn about the ontology of space and time from the theory of relativity? In L. Sklar (Ed.), *Physical Theory: Method and Interpretation*, pp. 187–228. Oxford University Press.

Ogievetskiĭ, V. I. and I. V. Polubarinov (1965). Spinors in gravitation theory. *Soviet Physics JETP* 21, 1093–1100.

Ogievetsky, V. I. and I. V. Polubarinov (1965). Interacting field of spin 2 and the Einstein equations. *Annals of Physics* 35, 167–208.

Ohanian, H. and R. Ruffini (1994). *Gravitation and Spacetime* (Second ed.). Norton.

Papapetrou, A. (1948). Einstein's theory of gravitation and flat space. *Proceedings of the Royal Irish Academy A* 52, 11–23.

Petrov, A. N. and J. Katz (2002). Conserved currents, superpotentials and cosmological perturbations. *Proceedings of the Royal Society (London) A* 458, 319–337.

Petrov, A. N. and J. B. Pitts (2019). The field-theoretic approach in GR and other metric theories. A review. *Space, Time and Fundamental Interactions* 4, 66–124. arXiv:2004.10525 [gr-qc].

Pinto-Neto, N. and P. I. Trajtenberg (2000). On the localization of the gravitational energy. *Brazilian Journal of Physics* 30, 181–188.

Pitts, J. B. (2006). Absolute objects and counterexamples: Jones-Geroch dust, Torretti constant curvature, tetrad-spinor, and scalar density. *Studies in History and Philosophy of Modern Physics* 37, 347–371. arXiv:gr-qc/0506102v4.

Pitts, J. B. (2010). Gauge-invariant localization of infinitely many gravitational energies from all possible auxiliary structures. *General Relativity and Gravitation* 42, 601–622.

Pitts, J. B. (2011). Massive Nordström scalar (density) gravities from universal coupling. *General Relativity and Gravitation* 43, 871–895. arXiv:1010.0227v1 [gr-qc].

Pitts, J. B. (2012). The nontriviality of trivial general covariance: How electrons restrict 'time' coordinates, spinors (almost) fit into tensor calculus, and 7/16 of a tetrad is surplus structure. *Studies in History and Philosophy of Modern Physics* 43, 1–24. arXiv:1111.4586.

Pitts, J. B. (2016a). Einstein's equations for spin 2 mass 0 from Noether's converse Hilbertian assertion. *Studies in History and Philosophy of Modern Physics* 56, 60–69. arXiv:1611.02673 [physics.hist-ph].

Pitts, J. B.(2016b). Einstein's physical strategy, energy conservation, symmetries, and stability: 'But Grossmann & I believed that the conservation laws were not satisfied'. *Studies in History and Philosophy of Modern Physics* 54, 52–72. arXiv:1604.03038 [physics.hist-ph].

Pitts, J. B. (2016c). Space-time philosophy reconstructed via massive Nordström scalar gravities? Laws vs. geometry, conventionality, and underdetermination. *Studies in History and Philosophy of Modern Physics* 53, 73–92. arXiv:1509.03303 [physics.hist-ph].

Pitts, J. B. (2016d). Universally coupled massive gravity, III: dRGTMaheshwari pure spin-2, Ogievetsky-Polubarinov and arbitrary mass terms. *Annals of Physics* 365, 73–90. arXiv:1505.03492 [gr-qc].

Pitts, J. B. (2017). Progress and gravity: Overcoming divisions between general relativity and particle physics and between science and HPS. In K. Chamcham, J. Silk, J. Barrow, and S. Saunders (Eds.), *The Philosophy of Cosmology*, pp. 263–282. Cambridge University Press. https://arxiv.org/abs/1907.11163.

Pitts, J. B. (2019). Space-time constructivism vs. modal provincialism: Or, How special relativistic theories needn't show Minkowski chronogeometry. *Studies in History and Philosophy of Modern Physics* 67, 191–198. Invited reviewed contribution to special issue on Harvey Brown's *Physical Relativity* 10 years later, edited by Simon Saunders; arXiv:1710.06404 [physics. histph].

Pitts, J. B. (2020a). Conservation laws and the philosophy of mind: Opening the black box, finding a mirror. *Philosophia* 48, 673–707.

Pitts, J. B. (2020b). Cosmological constant Λ vs. massive gravitons: A case study in General Relativity exceptionalism vs. particle physics egalitarianism. In A. Blum, R. Lalli, and J. Renn (Eds.), *The Renaissance of General Relativity in Context*, Volume 16 of Einstein Studies. Birkhäuser. arxiv:1906.02115.

Pitts, J. B. and W. C. Schieve (2001). Slightly bimetric gravitation. *General Relativity and Gravitation* 33, 1319–1350. arXiv:gr-qc/0101058v3.

Pitts, J. B. and W. C. Schieve (2004). Null cones and Einstein's equations in Minkowski spacetime. *Foundations of Physics* 34, 211–238. arXiv:gr-qc/0406102.

Pitts, J. B. and W. C. Schieve (2007). Universally coupled massive gravity. *Theoretical and Mathematical Physics* 151, 700–717. arXiv:gr-qc/0503051v3.

Pooley, O. (2013). Substantivalist and relationalist approaches to spacetime. In R. Batterman (Ed.), *The Oxford Handbook of Philosophy of Physics*, pp. 522–586. Oxford University Press.

Read, J. (2020). Functional gravitational energy. *The British Journal for the Philosophy of Science* 71, 205–232.

Read, J. (2022, forthcoming). Geometric objects and perspectivalism. In J. Read and N. Teh (Eds.), *The Philosophy and Physics of Noether's Theorems*. Cambridge University Press. http://philsci-archive.pitt.edu/18911/1/GONKv2.pdf.

Rosen, N. (1940). General relativity and flat space. I., II. *Physical Review* 57, 147–150, 150–153.

Rovelli, C. (1997). Halfway through the woods: Contemporary research on space and time. In J. Earman and J. D. Norton (Eds.), *The Cosmos of Science: Essays of Exploration*, pp. 180–223. University of Pittsburgh Press/ Universitätsverlag Konstanz.

Rovelli, C. (2002). Notes for a brief history of quantum gravity. In R. T. Jantzen, R. Ruffini, and V. G. Gurzadyan (Eds.), *Proceedings of the Ninth Marcel Grossmann Meeting (Held at the University of Rome "La Sapienza," 2–8 July 2000)*, pp. 742–768. World Scientific. arXiv:gr-qc/0006061.

Rovelli, C. (2004). *Quantum Gravity*. Cambridge University Press.

Salimkhani, K. (2020). The dynamical approach to spin-2 gravity. *Studies in History and Philosophy of Modern Physics* 72, 29–45.

Salisbury, D. (2020). Toward a quantum theory of gravity: Syracuse 1949–1962. In A. Blum, R. Lalli, and J. Renn (Eds.), *The Renaissance of General Relativity in Context, Volume 16 of Einstein Studies*, pp. 221–255. Birkhäuser. arXiv:1909.05412v1 [physics.hist-ph].

Schouten, J. A. (1954). *Ricci-Calculus: An Introduction to Tensor Analysis and Its Geometrical Applications* (Second ed.). Springer.

Schrödinger, E. (1918). Die Energiekomponenten des Gravitationsfeldes. *Physikalische Zeitschrift* 19, 4–7.

Schutz, B. F. and F. Ricci (2001). Gravitational waves, sources, and detectors. In I. Ciufolini, V. Gorini, U. Moschella, and P. Fré (Eds.), *Gravitational Waves*, pp. 11–88. Institute of Physics. arXiv:1005.4735 [gr-qc].

Schutz, B. F. and R. Sorkin (1977). Variational aspects of relativistic field theories, with applications to perfect fluids. *Annals of Physics* 107, 1–43.

Smolin, L. (2006). *The Trouble with Physics: The Rise of String Theory, the Fall of a Science, and What Comes Next*. Allen Lane.

Sorkin, R. D. (1977). On stress-energy tensors. *General Relativity and Gravitation* 8, 437–449.

Sorkin, R. D. (1988). Conserved quantities as action variations. In J. A. Isenberg (Ed.), *Mathematics and General Relativity: Proceedings of the AMS-IMS-SIAM Joint Summer Research Conference Held June 22–28, 1986 in Santa Cruz, California*, Volume 71 of Contemporary Mathematics, pp. 23–37. American Mathematical Society.

Stephani, H. (1990). *General Relativity* (Second ed.). Cambridge University Press.

Szybiak, A. (1963). Covariant derivative of geometric objects of the first class. *Bulletin de l'Académie Polonaise des Sciences, Série des Sciences Mathématiques, Astronomiques et Physiques* 11, 687–690.

Tashiro, Y. (1950). Sur la dérivée de Lie de l'être géométrique et son groupe d'invariance. *Tôhoku Mathematical Journal* 2, 166–181.

Thorne, K. S., D. L. Lee, and A. P. Lightman (1973). Foundations for a theory of gravitation theories. *Physical Review D* 7, 3563–3578.

Veltman, M. (1981). Quantum theory of gravitation. In R. Balian and J. Zinn-Justin (Eds.), *Les Houches Session XXVIII, 28 Juillet 6 Septembre 1975: Méthodes en Théorie des Champs/Methods in Field Theory* (Second ed.), pp. 265–327. North-Holland.

Wald, R. M. (1984). *General Relativity*. University of Chicago Press.

Weinberg, S. (1964). Photons and gravitons in S-matrix theory: Derivation of charge conservation and equality of gravitational and inertial mass. *Physical Review* 135, B1049–B1056.

Weinberg, S. (1992). *Dreams of a Final Theory*. Pantheon Books.

Weyl, H. (1922). *Space-Time-Matter. Methuen & Company*. Translated by Henry L. Brose from 4th edition of Raum-Zeit-Materie; reprinted by Dover, New York (1952).

Yano, K. (1955). *The Theory of Lie Derivatives and Its Applications*. NorthHolland.

4 The physics and metaphysics of pure shape dynamics

Antonio Vassallo, Pedro Naranjo, and Tim Koslowski

4.1 Introduction

The philosophical debate regarding the nature of space and time is one of the most long-lived disputes in the history of Western thought. The debate is usually presented in terms of the clash between *substantivalism* and *relationalism* (see, e.g., Hoefer et al., 2022a, b). In a nutshell, the doctrine of substantivalism maintains that spatial and temporal structures are ontologically self-subsistent—i.e., their existence is not parasitic upon the existence of matter. Relationalism, on the other hand, denies this claim.[1] Although the substantivalism/relationalism controversy has been pronounced dead several times, especially as far as physics is concerned (Rynasiewicz, 1996), it still seems to be alive and kicking. Furthermore, the debate has long transcended beyond the metaphysics of physics, branching out into the philosophy of mathematics and language (cf. the chapters by Cheng & Read and Bigaj in this volume). The present chapter aims to contribute to the debate by providing a quick yet up-to-date characterization of relationalism and its implementation in a modern physical theory of motion. The utility of such a characterization is evident: For example, Earman (1989) laments the fact that relationalism is an elusive doctrine, not least because:

> [T]here is no relationist counterpart to Newton's Scholium, the *locus classicus* of absolutism. Leibniz's correspondence with Clarke is often thought to fill this role, but it falls short of articulating a coherent relational doctrine and it even fails to provide a clear account of key points in Leibniz's own version of relationism.
>
> (*ibid.*, p. 12)

This remark highlights a quite uncontroversial historical fact. Substantivalism is a philosophical doctrine with a clear "birth date":

DOI: 10.4324/9781003219019-6

Once Newton laid down his Scholium, substantivalism became a well-established stance, with a solid physical backbone. The same thing cannot be said of relationalism. The construction of a philosophically as well as physically sound relational framework has taken centuries, and the work of many philosophers and physicists (see Barbour, 1982, for a historical overview of the conceptual development of relationalism). But even taken for granted that relationalism learned to walk when substantivalism was already running, this does not automatically imply that relationalism cannot be as conceptually clear as substantivalism. In fact, modern relationalism is not an elusive doctrine but, rather, a conceptually nuanced one.[2] The following sections will substantiate this claim.

We will start with a brief, non-technical overview of modern relational physics, with a special emphasis on the evolution of the approach in relation to the technical implementation of the main relationalist tenets (section 4.2). Having established the latest results in the quest for a totally relational theory, we will proceed with considering the possible metaphysical morals compatible with them (section 4.3). The discussion will make it apparent that the relational tenets go along with a surprisingly wide array of metaphysical positions, thus highlighting how a relationalist metaphysics is conceptually much richer than the simple ontological denial of space and time and the countenance to relative motion. The chapter will close with some speculations about the future of relational physics, especially in light of quantum physics.

4.2 From best-matching to pure shape dynamics

The goal of this section is to provide a concise portrait of the main recent developments that have occurred in the relational physics program towards the articulation of a "coherent relational doctrine"—paraphrasing Earman's quotation.

4.2.1 *The modern relationalist tenets*

The birth of modern relational dynamics can be traced back to three seminal papers by Julian Barbour and Bruno Bertotti (Barbour, 1974; Barbour and Bertotti, 1977, 1982), which led to the implementation of what is today known as the *Mach-Poincaré principle*:

> **Tenet I (Mach-Poincaré principle—classic version)** *Physical, i.e., relational initial configurations and their (intrinsic) first derivatives alone should uniquely determine the dynamical evolution of a closed system.*

The motivation for tenet I stems from the relationalist desire to eliminate redundant structure from physical theories, which typically arises because of the presence of various symmetries in the physical system under scrutiny. This aversion to redundant structure may be dressed with heavy empiricist overtones by claiming that *all structures whose variation amounts to no empirically observable difference should be banned from physics*. Not surprisingly, Newton's space and time are the redundant structures *par excellence* for the relationalist due to their dubious causal efficacy and unobservable nature. Indeed, the Mach-Poincaré principle follows from two well-known anti-substantivalist theses:

> **Tenet II (Spatial relationalism—classic version)** Lengths, be they distances or sizes, must be defined relative to physical systems, not spatial points.
>
> **Tenet III (Temporal relationalism)** Temporal structures, such as chronological ordering, duration, and temporal flow, must be defined only in terms of changes in the relational configurations of physical systems.

Moreover, the modern relationalist places a particular emphasis on the fact that every measurement simply amounts to the comparison of two physical systems. Formally, this means that only *ratios* of physical quantities carry objective information.[3] Going back to the case of space and time, this comparativist attitude implies that the relationalist should let go of any notion of size. What remains is simply the *shape* of the system.[4] To illustrate the idea, let us consider 3 particles in Newtonian space, which may be thought of as placed in the vertices of a triangle. Let x_a be the particle positions and $r_{ab} \equiv |x_a - x_b|$ the inter-particle separations. The empirical fact mentioned implies that we should take one of the r_{ab} as a unit of length and compare the remaining two against it, yielding the two ratios which are objective. The upshot of this procedure is the fact that only the shape of the triangle matters, for only two independent angles are needed. We can hence modify tenet II to read:

> **Tenet IV (Spatial relationalism—modern version)** The only physically objective spatial information of a physical system is encoded in its shape, intended as its dimensionless and scale-invariant relational configuration.

Mathematically, the procedure of systematically getting rid of redundant structure associated with some symmetry is commonly known as *quotienting out*. Schematically, if $\mathcal{Q}$ is the relevant configuration space of a given system and $\mathcal{G}$ is the symmetry group, the quotienting out yielding the space carrying truly physical information is $\mathcal{Q}_{ss} := \mathcal{Q}/\mathcal{G}$, whose dimension is simply $\dim(\mathcal{Q}_{ss}) = \dim(\mathcal{Q}) - \dim(\mathcal{G})$. In the case of N classical particles, $\mathcal{Q}$ is standard configuration space, $\mathcal{G} = \mathsf{Sim}(3)$, the joint group of Euclidean translations T, rotations R and dilatations S (called *similarity group*), and $\mathcal{Q}_{ss} = \mathcal{Q}/\mathsf{Sim}(3)$ is referred to as the *shape space* of the system. It is easy to see why such a reduced configuration space is called like this: In this simple scenario, the objective spatial information of the N-particle system is encoded in the shape of the N-gon defined by the N particles. Hence, in this case, each point on shape space represents an N-gon's shape (i.e., an equivalence class of N-gons under T, R and S transformations).[5]

In Barbour and Bertotti (1982), the effective implementation of tenet I in the context of classical particle dynamics is carried out by means of the so-called *best-matching*, namely, a sort of intrinsic derivative that allows one to clearly establish a measure of *equilocality*, i.e., to be able to know when two points are located in the same position at different times provided relational data alone are given. Importantly, Barbour-Bertotti consider the *relative configuration space*, arrived at after quotienting by translations T and rotations R. The desire to build a scale-invariant theory in compliance with tenet IV led to Barbour (2003), where also dilatations S are taken into account: It was at this point that shape space entered the modern relational program.[6] However, the introduction of scale transformations comes at a price: In order to account for dynamics and match empirical data, a non-shape degree of freedom has to be introduced—the ratio of dilatational momenta.[7] The upshot is a relational formulation of classical dynamics as far as the system in its entirety is considered, but which becomes the usual Galilean-invariant dynamics when only subsystems are taken into account (more on this at the end of subsection 4.2.2). This is a clear vindication of Mach's idea that the fully relational character of the dynamics can be attained only when the whole universe is considered.

It is illuminating to reflect a bit on this issue, for it is central to any relational description of physics. It was a well-known fact to pioneers of relational approaches that observable initial configurations and their first derivatives alone do not suffice to uniquely determine the evolution of a system. The culprit is the total angular momentum $\mathbf{J}$. It had been known since Lagrange's work that the value of $\mathbf{J}$ of a system cannot be obtained from said initial data.[8] One of the remarkable consequences of best-matching is the vanishing of $\mathbf{J}$ for a *closed* system—like the whole Universe. This, of course, does not prevent subsystems from having a non-zero angular momentum. Thus, by seriously considering the holistic character of relational physics, *closed* systems alone are elevated to genuinely distinct systems.

The search for a theory of gravity that fully complies with the relational tenets of Barbour (2003) eventually led to *Shape Dynamics*

(SD).[9] SD is a theory of conformal 3-geometries, as opposed to standard Einstein's gravity, which is a theory of Riemannian 4-geometries. In other words, the dynamical variables of the theory are only the parts of a Riemannian 3-metric that determine angles. Note how, also in this case, a conformal 3-geometry is what is left of a Riemannian 3-geometry after translations, rotations, and scale degrees of freedom are quotiented out. The core approach of SD represents a conceptual continuity with Barbour and Bertotti's original relational framework in the following sense: Given two configurations (i.e., two conformal 3-geometries), find the curve that minimises, by means of best-matching, their intrinsic distinctness. Crucially, though, the same caveat mentioned earlier in the particle case carries over to dynamical geometry: The physical arena of SD is not conformal superspace, as one would have thought for an allegedly scale-invariant theory: An additional degree of freedom must be introduced, which is associated with the total volume of space.[10]

An immediate worry surfaces at this point. Given that SD is a theory of conformal 3-geometries, isn't it the case that it conflicts with tenet II/IV about spatial relationalism? After all, there is nothing in a conformal 3-geometry that may ground the existence of such a shape as a relational configuration of *something* material. If such a shape is indeed a web of relations, the *relata* seem to be something akin to geometric points. If that is the case, isn't this a sophisticated form of substantivalism in disguise?[11] There are two possible replies to this objection. The first is to go for an "egalitarian" interpretation of the gravitational field as yet another physical field—more precisely, a massless spin-2 field. Under this reading, a conformal 3-geometry is just a mathematical description of something as material as, say, the electromagnetic field. And, indeed, the modern relationalist has the resources to incorporate entities with an infinite number of degrees of freedom in her framework (SD being the case in point). In the gravitational case, the material *relata* would then be field magnitudes rather than spatial points.[12] The problem with this response is that the "egalitarian" interpretation of the gravitational field in general relativistic physics is problematic and hugely controversial (see, e.g., Pitts' chapter in this book to catch a glimpse of the debate). A second response may be to show that matter can be incorporated in SD, thus constituting the "backbone" of a conformal 3-geometry. The problem with this response is that, at the moment, there is no easy way to couple matter to the purely gravitational sector of SD (see Gomes and Koslowski, 2012, for a preliminary attempt at doing so). As things stand, this is still an open issue.

4.2.2 The pure shape dynamics program

After this brief tour on the development of relational approaches to dynamics, we finally get to the latest refinement of the theory, dubbed

Pure Shape Dynamics (PSD; see Koslowski et al., 2021, for a general technical introduction to the framework). In a nutshell, the qualifier "Pure" means that PSD describes *any* dynamical theory exclusively in terms of the intrinsic geometric properties of the *unparametrized* curve γ_0 traced out by the physical system in shape space $\mathcal{Q}_{ss}$, hence ensuring that there are no external reference structures nor clock processes necessary to describe γ_0 in $\mathcal{Q}_{ss}$. In other words, the PSD project is highly innovative: It aims at rewriting the whole spectrum of theories of dynamics that characterized and will characterize the development of physics in fully relational, scale-invariant terms. For any "standard" physical system, be it a Newtonian N-particle system, a general relativistic cosmological model, a de Broglie-Bohm particle system, etc., PSD proposes the same recipe: Quotient out the relevant symmetries from the system and then fully geometrize its dynamical development. To further stress this insistence on the intrinsic geometric properties of the curve associated with a physical system, one speaks of its equation of *state*, in contrast to its equation of motion. We may slightly modify tenet I to match PSD "intrinsically geometric" nature as follows:

Tenet V (Mach-Poincaré principle—modern version) *Physical, i.e., relational initial configurations and their (intrinsic) derivatives alone should uniquely determine the dynamical evolution of a closed system.*

The key innovation brought about by tenet V is to consider in general *higher-order* derivatives of the curve, thereby allowing us to describe the dynamics of a physical system in terms of the curve alone, without the need of any additional non-shape parameters as in standard SD. Clearly, there is a sense in which tenet V is *weaker* than tenet I, for the former requires more initial data than the latter (higher-order versus first derivatives) to describe a dynamical system. This weakening is anything but a drawback from a modern relationalist standpoint since the upshot is the elimination of any non-shape degree of freedom from the dynamical description of a system—thus delivering a truly relational dynamics cast in terms of the degrees of freedom intrinsic to the system only.

Given an already "quotiented out" physical system, we shall express the equation of state of the unparametrized curve γ_0 in its associated shape space $\mathcal{Q}_{ss}$ as follows:

$$
\begin{aligned}
dq^a &= u^a(q^a, \alpha_I^a), \\
d\alpha_I^a &= A_I^a(q^a, \alpha_I^a),
\end{aligned}
\tag{4.1}
$$

and demand that the right-hand side be described in terms of dimensionless and scale-invariant quantities, whose intrinsic change is obtained employing Hamilton's equations of motion. In (4.1), q^a are points in shape space, namely they represent the universal configurations of the system, u^a is the unit tangent vector defined by the shape momenta p_a:

$$u^a \equiv g^{ab}(q) \frac{p_b}{\sqrt{g^{cd}p_c p_d}} , \qquad (4.2)$$

which allows us to define the direction ϕ^A at q^a. It is through the unit tangent vector and the associated direction that the shape momenta enter Hamilton's equations, which are in turn used in the intermediary steps leading to the equation of state (4.1). Finally, α_I^a is the set of any further degrees of freedom needed to fully describe the system. It is this set α_I^a that includes higher-order derivatives of the curve and spares us of the need of additional non-shape degrees of freedom. For consistency, the elements in α_I^a must exhaust the set of all possible dimensionless and scale-invariant quantities that can be formed out of the different parameters entering a given theory. Note the unifying nature of (4.1): In principle, the whole of relational dynamics, classical, relativistic, and quantum, boils down to the dynamical structure encoded in this equation of state.

The essential difference with standard SD is that PSD exclusively relies on the intrinsic geometric properties of the dynamical curve in shape space, as given in (4.1), whereas standard SD does not share this prominent role of said curve. In particular, this insistence on intrinsic properties is best exhibited by the unparametrized character of the curve in PSD, which, recall, guarantees that no external reference structures nor clock processes are needed to describe γ_0, the α_I^a alone being responsible for this job.

As already stressed, there is an important reason for considering PSD an improvement of the original SD framework. In short, the dynamical evolution as described by PSD does not fundamentally rely on any notion of parametrization whatsoever and, hence, it is a genuinely intrinsic description of a physical system. Remarkably enough, PSD is capable of reproducing known physics despite its decidedly intrinsic nature (see Koslowski et al., 2021, for the $E = 0$ N-body problem; the case of dynamical geometry will be the subject of another paper, currently in preparation). On the other hand, the original formulation of SD needs some monotonically increasing parameter—be it the ratio of dilatational momenta or York time/spatial volume, as discussed earlier—to be defined on a dynamical curve in order to make sense of the physical evolution. Such a parameter, although representing a much weaker structure than Newtonian time, still represents something "external" to the system.

This brings us to one of the key aspects of PSD, namely its account of dynamics. Given that we deal with unparametrized curves, the challenge is to find an intrinsic feature of the system that serves the purpose of a physically meaningful labeling of change. To this extent, we shall exploit the idea put forward in Barbour et al. (2014a) that a shape contains structure encoded in stable records.[13] Thus, the evolution is towards configurations (i.e., shapes) that maximize the complexity of the system. The result provides the desired ground for introducing a direction of change, which boils down to the direction of accumulation of the already mentioned stable records.[14] Clearly, the first thing we must provide is a natural definition of complexity, along with a suitable measure of it, which, to comply with the central tenet of PSD, should be given in terms of the intrinsic geometry of the unparametrized curve in shape space.

Given that the only fully worked-out physical system exhibiting generic formation of stable records is the N-body system, we shall use it as an example to motivate our approach to dynamics. It is worth pointing out that promising results come from the vacuum Bianchi IX cosmological model, where a natural candidate exists for a measure of shape complexity in geometrodynamics (see Barbour et al., 2013, section 3.5). However, the extension of our arguments to (the shape dynamical version of) full general relativity and quantum mechanics is still a work in progress. For current purposes, complexity is essentially the amount of clustering of a system, with a cluster being a set of particles that stay close relative to the extension of the total system. Next, we demand that the complexity function grow when (i) the number of clusters do, and (ii) the clusters become ever more pronounced, namely when the ratio between the extension of the clusters to the total extension of the system grow.

Given these premises, perhaps the simplest measure of complexity for the N-body system is

$$\mathsf{Com}(q) = -\frac{1}{m_{\mathrm{tot}}^{5/2}} \sqrt{I_{\mathrm{cm}}} V_N = \frac{\ell_{\mathrm{rms}}}{\ell_{\mathrm{mhl}}}, \tag{4.3}$$

where I_{cm} is the centre-of-mass moment of inertia, V_N is Newton's potential and ℓ_{rms} and ℓ_{mhl} account for the greatest and least inter-particle separations, respectively. Thus, their ratio, (4.3), measures the extent to which particles are clustered.

The important thing about the complexity function (4.3) is that it has a minimum (dubbed *Janus Point*) and grows in either direction away from it (Barbour et al., 2014b, a). It can be argued that this direction of increasing complexity be identified with the so-called *arrow of time* for internal observers (ones within one of the two branches at either side of the Janus Point). Thus, we arrive at a description of the experienced arrow of time in terms of purely intrinsic properties of the unparametrized curve on shape space.

Not only is the experienced arrow of time, or duration, accounted for by means of the curve in shape space, so too is length. This is achieved by the so-called *ephemeris equations*.[15] For the purposes of this discussion, we need only point out the general expression:

$$F_i = G_i(q^a, \alpha_l^a), \tag{4.4}$$

where F_i stands for a function of either length or duration (i = 1, 2) and G_i is a function that solely depends on intrinsic properties of the curve, $\{q^a, \alpha_l^a\}$. Hence, the PSD framework promises to deliver a general formal mechanism for the appearance of "everyday" spatial and temporal structures from the fundamentally unparametrized dynamics of the theory.

Finally, let us consider how PSD recovers the standard dynamics for subsystems of the universe. In the classical N-body case, the dynamics features generic solutions which break up the original system into subsystems, consisting of individual particles and clusters, that become increasingly isolated in the asymptotic regime (Marchal and Saari, 1976). Such almost isolated subsystems will develop approximately conserved charges, namely the energy E, linear momentum $\mathbf{P}$ and angular momentum $\mathbf{J}$. Within the dynamically formed subsystems, there are pairs of particles that may function as physical rods and clocks. These are referred to as *Kepler pairs* because their asymptotic dynamics tends to elliptical Keplerian motion.

In conclusion, PSD represents a natural evolution of Barbour and Bertotti's original ideas, which provides a robust formal framework for a full "Machianization" of physics. Of course, much work still has to be done but, as the previous discussion shows, the technical foundations of the theory are already laid down.

4.3   A fascinating metaphysical diversity

In the previous section, we have seen the fundamental theoretical elements that enter the dynamical description of a system according to PSD. Simply speaking, what is strictly needed to start constructing this dynamical description amounts to two elements: The shape space of the system and the dynamical law(s) (4.1). The former represents the topologically structured collection of all constructible intrinsic configurations q^a of a given system, while the latter generates the dynamical curve of the system, which is nothing but an ordered series of relational configurations.

For a scientific realist, this description represents only half of the story. What is missing at this point is the characterization of a metaphysical link between the formalism of PSD and the physical world. Otherwise said, if we believe that PSD tells us something about the nature of the world, we should point out what it is that the formalism refers to. In order to address this point, two distinct but deeply interrelated questions should be asked: *What* there is in the world, and *how* the world is according to PSD. The first question concerns the *ontology* of the theory, while the second concerns the best *metaphysics* for the physical world compatible with the framework. We take this distinction as a mere working hypothesis, glossing over the actual relation between ontology and meta-physics—we will just assume that metaphysics does not (entirely) depend on ontology. This is, in fact, a controversial matter, largely orthogonal to our discussion (see, e.g., Varzi, 2011, who instead defends the conceptual priority of ontology over metaphysics).

The question regarding the ontology of PSD is rather complex and tricky. Indeed, the ontological characterization of PSD depends on the actual implementation of the quotienting out procedure presented in the previous section and, hence, on the features of the starting theory subjected to such a procedure. For example, in the case of an *N*-particle Newtonian model, the quotienting out procedure "transforms" an ontology of material particles inhabiting a Newtonian background space into one of a "web" of particles related by Euclidean spatial relations (i.e., a Newtonian shape). The same applies to the general relativistic case, where an ontology of conformal 3-geometries is arrived at once the quotienting out procedure is performed on the space of Riemannian 3-geometries (see Gryb and Thébault, 2016, for a comparison of the ontologies of SD and general relativity). From this point of view, the quotienting out procedure can be seen as a *reconceptualization* that renders a starting fundamental ontology fully Leibnizian/Machian.

This discussion begs the question as to whether PSD's ontology is model-dependent, i.e., whether we should ask ontological questions only in the context of the particular model considered. A positive answer to this question would point at some sort of perspectival realist attitude, in the sense that it would imply that PSD cannot deliver a unique and objectively true "fundamental inventory" of the world: Each model has "its own" ontology, so to speak (see, e.g., Crețu, 2020, for a characterization of perspectival realism). Such an attitude, however, may be seen as dramatically weakening one of the strongest motivations for pursuing the PSD program, i.e., to provide a genuinely unifying frame-work for physics—and not just an unrelated collection of "reconceptual-ized" physical models.

In order to deny that the ontology of PSD is model-dependent, two strategies are available. The first is to take all the models at once as pro-viding the "ontological bedrock" of the world. This choice does not seem ideal since it would inflate the fundamental ontology while simulta-neously depriving it of explanatory power (e.g., no explanation of the fact that Newtonian gravitation is an approximation of general relativ-ity). The second option is instead to argue that there is in fact a class of models that faithfully represent the actual world (in the sense of being fully empirically adequate), while the remaining models should be seen either as physical *possibilia* or as providing ontologies that are derivative on (in the sense of supervening on, or even being reducible to) the fundamental one. Such a choice is ontologically more parsimoni-ous, and has the advantage of explaining how different physical domains are related by way of approximations or limiting procedures. Of course, this option will become available only if, and when, a full characterization of PSD will be available, which will include classical, quantum, and quantum-gravitational physics. Moreover, a consistent characterization of the dependence relations between the "fundamental" and the "derivative" models should be provided.

It is now clear why answering the "What is there?" question is not straightforward in the context of PSD, given the work-in-progress status of the framework. Hence, we leave it at that for the time being. In the following, we will just take for granted that a shape is a configuration of *something*, without inquiring further into what this something is. We will instead inquire into the type of metaphysics that is best suited to characterize a world made of shapes behaving according to (4.1), i.e., we will consider some possible answers to the question "How is the world, according to PSD?"

4.3.1   Shape space realism

Let's start from what is probably the most straightforward metaphysical reading of the theoretical framework of PSD, that is, *shape space realism*. According to this reading, (4.1) is just an algorithm that singles out a curve in shape space when it is fed certain initial conditions $\{q_0^a, \alpha_{0I}^a\}$. However, there is nothing ontologically significant that privileges some initial conditions over others; ditto for the ensuing curves. In other words, all points (and sequences thereof) in shape space have to be taken ontologically on a par. Let's unpack this last statement. First, we have to spell out what a point in shape space represents, then we have to specify in what sense these "things" must be considered equal from an ontic perspective.

The first task is quite easy to accomplish, recalling the quotienting out procedure discussed in the previous section. The outcome of such a procedure is to strip a universal spatial configuration of elements of reality—that is, a universal snapshot of material particles, or field magnitudes, or anything else the original theory is about—of their unobservable degrees of freedom that are usually associated with the symmetries of the system. This is the way the starting non-relational configuration space of the system is translated into the fully intrinsic configuration space that we call shape space. Hence, each point in shape space represents an "instantaneous" universal relational configuration of elements of reality. For example, starting from a Newtonian N-particle system, we get a space where each point represents a configuration of N particles related through the Euclidean spatial relations that survive the quotient by the similarity group. In this case, shape space is the structured set of all the possible shapes that can be constructed with these ingredients. This idea can be easily generalized, e.g., to the general relativistic case. In this case, shape space is the collection of all Riemannian 3-geometries invariant under spatial diffeomorphisms and conformal transformations.[16]

Now that we have clarified what shape space is "made of," we should take up the second task, i.e., specifying the sense in which all shapes featuring in this space are ontologically on a par. One may be tempted to say that shape space is the collection of all (physically) *possible* shapes (according to the relational ingredients inherited from the quotienting out procedure). This modal characterization,[17] however, would introduce a possible source of ontological asymmetry in that it may be argued that a subset of these shapes must represent the *actual* world, that is, the one curve in shape space which is generated when the initial conditions obtaining at our world are fed into (4.1). The way out of this challenge is to claim that in shape space there is no "actual" configuration as opposed to a "possible" one. The dynamics of the theory makes it the case that we get to experience some configurations rather than others, but this has nothing to do with the reality of each of them. In other words, all shapes are actual. In this way, shape space becomes the maximal collection of all the universal relational configurations *simpliciter*. Otherwise said, for shape space realists, shape space *is* the actual world.[18] This, of course, does not mean that shape space is some sort of "necessary being." Indeed, a shape space realist does not deny that the structure of shape space might have been different, which means that such a different shape space represents a possible world.

We have finally reached the metaphysical picture of the physical world according to shape space realism: All the universal "snapshots" simply exist all at once in a radically timeless and changeless sense. The dynamics encoded in (4.1), and the ensuing constructions (4.3) and (4.4), serve the only purpose of explaining the illusion of there being "time" and "change" in terms of the formation of stable records—including mental records—in an unparametrized sequence of configurations generated by certain (arbitrary) initial conditions. In this way, shape space realism is able to provide a straightforward metaphysical link between the theory's formalism and this timeless and changeless physical world. Simply speaking, all the mathematical structures entering PSD dynamics directly refer to geometric features of shape space *qua* physical space. In this sense, shape space realism can be seen as a close cousin of the configuration space realism proposed in the context of quantum mechanics (see, e.g., Albert, 2013).

As straightforward as it may be, shape space realism as a metaphysical stance is far from being a naive metaphysical reading of the PSD formalism. In fact, it exhibits some peculiar features that are not usually ascribed to relationalism. First of all, it is at the same time a radically anti-realist (even idealist) and a radically realist position. It obviously denies the reality of much of what we usually call "manifest image" of the world, such as the passing of time, change—including motion in ordinary space—, but also the notion of identity of an object—which is usually intended as identity *over time*. At the same time, however, it affirms the existence of each and every relational configuration compatible with the quotienting out procedure. The way it supports this "ontological democracy", so to speak, is by claiming that these configurations are parts of a structured substance. Such a substance bears topological as well as metrical properties and, hence, it is some sort of "space" akin to the 3-dimensional physical space we are acquainted with—even though it is much more structured than a simple Euclidean space.

This is possibly the most surprising feature of this stance. Although the whole starting point is the usual relationalist one, the end result is a position that is undoubtedly realist with respect to *some sort* of space. One may retort that such a space is not a collection of "points" in the usual geometric sense but, rather, it is "made of" relational configurations in the plain metaphysical sense of "being composed by."[19] However, this would be a partial truth. In fact, composing a space out of configurations requires not only that said configurations "be there," but also that they be ordered in a precise topological sense. The shape space realist cannot explain such a structure using facts inhering into the configurations themselves. The shapes are topologically ordered as a primitive fact of the matter, and this primitive topological structure represents the nature of shape space itself.

An immediate objection comes to mind at this point. Isn't it the case that shape space displays the hallmark of an absolute structure? After all, it is an immutable structure onto which we, for some reason, happen to project the illusory image of an ever-changing 3-dimensional world. This means that, as for the case of Newtonian absolute space, shape space is just there, indifferent to the physics encoded in (4.1). It is easy to realize, however, that this analogy cuts no ice. While in fact Newtonian space is an arena where material facts happen, in shape space nothing really happens: This space is some sort of "block universe" where all physically allowed universal snapshots are given all at once, timelessly. So, even if it is true that no physical happening can influence shape space, this is just because in the shape space realist picture there is no physical happening at all.

This discussion prompts a second and more compelling objection. The starting point for this objection is that, usually, the point of contact between theoretical predictions and empirical observations can be traced back to material objects inhabiting physical 3-space at a certain time. This is, after all, the kind of "stuff" labs are made of. Hence, in order to be empirically testable, a theory should provide a coherent story that links some theoretical terms to observable objects localized in ordinary space and time. The objection can hence be articulated as follows: How does shape space realism account for empirical observations, given that it regards ordinary space and time as some sort of illusion? This is, perhaps not surprisingly, a problem analogous to that encountered by configuration space realists (see, e.g., Maudlin, 2019, chapter 4, for an illuminating discussion of the issue at stake). This is not to say that PSD *qua* theoretical framework does not have the resources to provide a technical account of such a story—recall again the mechanism behind equations (4.3) and (4.4). However, adherence to shape space realism strips these technical reconstructions of ordinary space and time of metaphysical significance, raising the question of how we get to perceive these illusory constructs in the first place. However, at the present stage, this remains an open problem for those willing to defend shape space realism.

4.3.2 Shape monism

One may object that shape space realism is not, in fact, a straightforward metaphysical reading of the PSD framework. This objection is based on the fact that the characteristic physical traits of PSD are not to be traced back to the quotienting out procedure that leads to shape space, but to the dynamical laws (4.1). Hence, it makes more sense to be realist towards the curves generated by (4.1), rather than to shape space as a whole. But what does this possibly mean? How should we conceive of a curve in shape space as (part of) the physical world?

The answer to this question involves a two-step argumentative strategy similar to that discussed at the beginning of subsection 4.3.1. First, a characterization of a shape is called for. Here it is possible to repeat what was already argued for, i.e., that shapes are "instantaneous" universal relational configurations of elements of reality represented by points in shape space. Curves in shape space, then, are a mathematical depiction of a dynamical sequence of shapes.

Secondly, and differently from the shape space realism case, it has to be argued that not all points (or curves) in shape space are to be taken ontologically on a par. Such a move implies that the physical world is a much more constrained structure than a vast block of kinematically allowed shapes. The world-building constraint is, in fact, (4.1). The effect of this tenet is to deconstruct shape space into a myriad of worlds represented by the integral curves of (4.1), each of which generated by a particular set of initial conditions. Among this huge set of possible worlds, the actual one is trivially represented by the curve generated by the initial conditions actually obtaining (whatever they are).

The worlds generated by (4.1), as already said, are much smaller and more ordered than shape space as a whole. They are basically a linear sequence of universal configurations, so that now the "snapshot" metaphor makes more sense: A solution of (4.1) is analogous to a film strip showing the entire history of a universe all at once. Note that, also under this view, there is nothing inherently temporal in the fundamental picture of reality. Given that the curves generated by (4.1) are unparametrized, there is no fundamental sense in which the ordering is directed: There is no forward or backward direction in which the film stock unwinds. What is fundamental is the sequence of shapes given all at once in a timeless sense. From this point, the challenge is to come up with an explanation of how we get to observe a universe where (directed) time and change play an important role. We will defer this task to future work, being content to note that, given the resources mentioned in subsection 4.2.2 (i.e., complexity function, ephemeris equations, asymptotic Keplerian motions), it is much easier for the "film stock" view of relational dynamics to recover the manifest image of the world, rather than the shape space realist.

Of course, just saying that a world is a film stock whose frames are shapes is not enough to metaphysically characterize such a world. Indeed, even before starting to reflect on how to recover time from the film stock, a story is needed about the extent to which the film stock depicts a *unique* evolving subject and it is not just a collection of pictures of different and unrelated subjects. This is a compelling problem. Each shape, in fact, is a universal relational configuration of "stuff", but it is not possible to claim just that each shape represents *the same* stuff at different times. Without a solution to this problem, it is difficult to conceive of a curve in shape space as an organic whole representing the evolution of a single world (possible or actual). So, at this point, shape space realism seems to have the upper hand in that it works without needing any notion of trans-shape identity whatsoever.

But, if the problem with trans-shape identity comes from the difficulty to establish the identity of the elements of a starting configuration throughout the stages of dynamical evolution, a first step towards solving this conundrum may be to challenge the very assumption that the identity of a shape comes from the stuff (*relata* as well as relations) it is made of. If facts about the identity of a shape *as a whole* come before—in some appropriate sense of "before"—facts about the identity of its parts, then it seems *prima facie* simpler to argue that a curve generated by (4.1) depicts a unique relationally evolving subject.

The most straightforward way to implement this strategy is to claim that a shape is not *made of* stuff. In other words, the relations and *relata* do not compose a shape, in the sense that they are not metaphysically prior to the whole represented by a shape. The notion of metaphysical priority can be best understood in terms of grounding: For example, Socrates is metaphysically prior to the singleton {Socrates}, because the existence of {Socrates} is grounded in the existence of Socrates. Otherwise said, since grounding can be intended as a partial ordering with respect to fundamentality, "being ontologically prior to" basically means "being more fundamental than." Hence, the proposed way out is to claim that the shape as a whole is a fundamental object, with its proper parts—represented by relations and *relata*—being metaphysically dependent upon, or posterior to, it. This is, in a nutshell, the main claim of *shape monism*.[20]

The main justification for shape monism is evident: The very notion of shape is that of an organic, integrated, universal whole that does not need anything external to itself in order to be characterized. This justification is supplemented with the subversion of the standard mereological tenet that the whole depends on its parts. This is an important point to be reiterated: Shape monism does not claim that a shape has no parts, but that the whole is prior to its parts. Hence, the talk of subsystems of the universe makes perfect sense also under shape monism. Subsystems—even the most simple ones—are perfectly real entities which, however, are not fundamental (i.e., they metaphysically depend on the whole).

At first sight, shape monism may look bizarre from a relationalist perspective. After all, this type of monism is far from asserting the fundamentality of spatial relations. To the contrary, these relations are parasitic upon the whole shape. But is this really so embarrassing for the relationalist? Maybe this initial discomfort results unwarranted upon closer inspection. First, shape monism salvages the core of the relationalist doctrine, i.e., the denial of space and time as ontologically independent substances. Indeed, a shape is not something that is placed in an external space; rather, it "weaves up" a suitable embedding space through the appropriate local constructions (4.4). Second, it may be argued that shape monism perfectly captures the deep interdependence of all the elements of a shape, in the sense that each and every single element should be characterized relationally, not intrinsically. Barbour (1982) beautifully summarizes this point as follows:

> [L]eibniz and Mach suggest that if we want to get a true idea of what a point of space-time is like we should look outward at the universe, not inward into some supposed amorphous treacle called the space-time manifold. The complete notion of a point of space-time in fact consists of the appearance of the entire universe as seen from that point. Copernicus did not convince people that the earth was moving by getting them to examine the earth but rather the heavens. Similarly, the reality of different points of space-time rests ultimately on the existence of different (coherently related) viewpoints of the universe as a whole.

> (*ibid.*, p. 265)

In short, all problems between monism and relationalism disappear once we realize that the former is not a thesis about the metaphysical priority of relations but, rather, about the relatedness of all the parts of the cosmos. To sum up, shape monism is a viable option for a metaphysics of PSD.

This is, however, part of the story. The problem remains as to how to make sense of a curve as a succession of dynamical stages of a *single* shape. The most simple solution is to point out that the identification of all shapes as different configurations of a unique "cosmic object" is just assumed as a primitive fact. Recall that the quotienting out procedure discussed in subsection 4.2.1 was introduced to strip a single, well-defined global system (e.g., a Newtonian N-particle system) of unobservable degrees of freedom. So, although the procedure deletes any "label" that *relata* and relations had (since a fully relational theory should include permutations as one of the symmetries quotiented away, see endnote 5), still it does not impact on the primitive identity of the whole object being rendered fully relational.

The argumentative strategy sketched here, obviously, needs further refinement. First of all, claiming that, as a brute fact of the matter, there is a "universal object" whose dynamical evolution is captured by an integral curve of (4.1) looks more like a trivialization of the problem of trans-shape identity rather than a real solution. Secondly, it is doubtful whether it is a solution of the problem *at all*. Shape monism just asserts the metaphysical priority of the whole over its parts, but it does not say much about how the identity of these parts should be carried over a dynamical curve. This renders it difficult to trace the evolution of a subsystem of the universe: Couldn't it be the case that different dynamical stages of a shape represent different "carvings" of this shape into smaller parts? If that was the case, then the identity of any subsystem would be destroyed in the transition from one dynamical stage to the other. This is not by itself a fatal flaw of the position, but just an indication that still a lot of work lies ahead for the shape monist.

4.3.3 Shapes as ontic structures

Shape monism is not the only way to take equation (4.1) metaphysically seriously. An immediate option suggests itself once we recall that (4.1) is more of an equation *schema* that encompasses the "dynamical gist" of relational physics according to PSD, rather than a full-fledged dynamical law. In other words, it is necessary to specify the system under consideration in order to fill in the details of (4.1). Obviously, (4.1) is not a general dynamical scheme by accident. To the contrary, it represents a compact description of what is common to the relational dynamics of any possible physical system that can undergo the quotienting out procedure. But what can these "common traits" be? To answer this question, let's consider the general issue stemming from theory change in physics.

In a nutshell, this issue arises whenever we want to give a realist understanding of a physical theory, and amounts to pointing out that such a physical theory is subject to scientific revision and may in the future be dramatically altered or even discarded. How can we take such a theory as telling us something compelling about the physical world if it may soon be replaced by a more empirically accurate theory? For example, an early 19th-century physicist may have entertained the idea that heat was in fact a substance called *caloric*, but a modern physicist regards such an ontic commitment as obsolete and inaccurate under the light of modern statistical mechanics.[21] Among the possible responses to this challenge, one of the most famous—and interesting in the context of this chapter—was given by Henri Poincaré (Poincaré, 1900, p. 15).

According to Poincaré, physical theories are unable to successfully grasp the fundamental ontology of the natural world—what he calls the "real objects"—, because this aspect of reality is unknowable to us ("eternally hidden from us by nature"). This, however, does not mean that we should regard physical theories as recipes to get empirical predictions, in a somewhat instrumentalist fashion. There is, according to Poincaré, something real about the physical world that empirically adequate theories are successful in capturing. These real features of the world are nothing but the relations between physical objects. So, for example, what 19th-century physicists called "caloric" can be identified with what we today call "heat" insofar as both designations are intended to refer to some intimate interactive processes underlying material substances. The gist of Poincaré's response is the core tenet of modern *structural realism* in philosophy of science.[22] Hence, for structuralists, it is not anymore a problem if there is a change in the theoretical entities postulated by physical theories, since they maintain that it is the way such entities are related that is preserved through theory change. The empirical successes of different theories are then explained by the fact that they capture some structural aspects of the world.

Having the structuralist response in mind, it is now easy to answer our starting question. (4.1) is a compact way to convey the relational aspects of the dynamics that are common to all possible systems to which the quotienting out procedure can be applied. Whatever the fundamental ontology of the physical world might be (material particles, field magnitudes, strings, or something even more exotic), the key relational aspects of the dynamics are already captured by (4.1). These aspects are exactly the features of reality that physical theories like Newtonian mechanics and general relativity get right. Then, the fact that, say, general relativity is more empirically accurate than Newtonian mechanics can be traced back to the actual implementation of (4.1) in the two cases—the general relativistic implementation of (4.1) being more accurate than the Newtonian counterpart.

This certainly represents a step forward in giving a metaphysical reading of (4.1), but it is not sufficient to do the job. Indeed, structural realism as briefly characterized earlier looks more like an epistemic thesis than a metaphysical one, i.e., it is about what we can know rather than how the world is. This impression is reinforced by Poincaré's skeptical attitude towards the possibility of fully metaphysically informing our knowledge of the world. Luckily enough, structural realism can be turned into an ontic stance by simply arguing that the world is in fact *made of* structures (see, e.g., Ladyman, 1998). This response dissolves Poincaré's worry in that it denies that the fundamental ontology of the world is constituted by individual objects whose identity and features are *intrinsic* to them. Instead, what there is is a "web" of relations and *relata*, with the identity and features of the latter being determined by their "position" in the structure. The ontic structural realist may go even further by arguing that the *relata* are entirely ontologically parasitic on the relations: Fundamentally, there are only relations. Here we will blur the distinction between the moderate and the radical versions of ontic structural realism, since this difference does not hinge on the conclusions we are going to draw. Hence, from now on, we will remain agnostic towards the degree of ontic dependence that *relata* bear to relations in a structure.

This ontic variety of structural realism seems to be tailor-made for PSD.[23] It is in fact easy to argue that a shape is a structure in this specific sense. Hence, under an ontic structural realist reading, (4.1) generates an ordered (i.e., structured) sequence of spatial structures. The challenge for the structuralist, from this point on, is to provide a sound metaphysical picture (relying on the constructions (4.3) and (4.4) of how space, time, and subsystems bearing their identity over time appear out of the "structure of structures" described by (4.1). Let's try to sketch a possible strategy that the structuralist can follow to this extent.

First of all, the structuralist may repeat the reasoning that led the monist to claim that a solution of (4.1) given certain initial conditions represents the entire "history" of a physically possible world (the actual world being described by the solution generated by the initial conditions being realized). From this point of view, the structuralist can exploit the monist's metaphor of a dynamical curve being a film stock that depicts the different stages of evolution of a cosmic structure. At this point, the challenge arises again, as to how to connect together the different "frames" of the film, i.e., the different shapes making the curve. Contrary to the monist, the structuralist cannot invoke some sort of primitive identity of the cosmos as a unique *individual* object. In the present case, the metaphysical accent is put on structures, and it is evident from the formalism of PSD that there is a plurality of them— whereas the monist may argue that they are just different ways to "ontologically carve" the cosmos, which remains a singular entity.

A simple way out of this issue is to bite the bullet and accept that the shapes making up a curve in shape space are distinct entities, i.e., there are no primitive facts about trans-shape identity of relations and *relata* holding at a world. This obviously does not mean that these facts cannot be shown to *supervene* on the succession of shapes laid down by (4.1). PSD has in fact the formal resources to back up this supervenience picture. For example, having in mind that (4.1) orders configurations in terms of their similarity, and considering the construction underlying (4.3) and (4.4), it is possible to argue—perhaps in a Humean fashion, as done for example by Huggett (2006) in the case of Newtonian dynamics—that there is a privileged embedding of the dynamical succession of shapes into a 4-dimensional Riemannian geometry. The different configurations would then become different spacelike slices of this embedding "spacetime." From this point on, the usual picture of individual objects stretching over spacetime in continuous trajectories could be recovered, with the identity over time of such objects being provided exactly by the fact that they are temporal stages of these spatiotemporal trajectories. Note that nothing in this construction would be assumed *a priori* or, worse, would enter the fundamental structural ontology. Indeed, this construction would be a mere descriptive tool to cast the totally relational dynamics encoded in (4.1) in more familiar terms.

The strategy sketched earlier seems to be a viable solution to the issue of finding a trans-shape identity criterion, as well as finding a story for the emergence of everyday space and time from PSD's relational picture. As a matter of fact, this construction has already been proposed in the Newtonian as well as in the general relativistic cases (see Vassallo and Esfeld, 2016; Vassallo et al., 2017, as well as Lazarovici, 2018, section 4), although a concrete implementation in the PSD case is still an open question.

4.4 Conclusion

The previous sections highlight the tremendous progress that relationalism has undergone in the last few decades. It all started from some bold conjectures on how to technically implement Leibniz and Mach's ideas, and ended up in an equally bold program that seeks to subsume physics in its entirety under the relational flag. However, as the discussion in section 4.2 makes it clear, this program has still a long way to go. One of the most pressing issues, from this point of view, is finding a concrete implementation of quantum physics, including quantum gravity.[24] This task is formidable, but it is nonetheless within the PSD's program abilities.

Given that PSD's main tenet is to describe dynamics solely in terms of the intrinsic properties of the unparametrized curve in shape space, it should be expected that the prominent feature of quantum dynamics, i.e., the linearly evolving wave function, should be reduced to such intrinsic properties as well. Just to give a glimpse of the current line of research in this direction, we just point out that an effective reduction of the wave function to the geometric properties of the curve potentially requires that the family of *all* dynamical curves, for all possible initial conditions, be considered *at once*, unlike the single curve of classical dynamics. This is analogous to the way the path integral formulation of quantum mechanics generalizes the action principle of classical mechanics, i.e., by replacing the unique classical trajectory for a system with a multiplicity of quantum-mechanically possible trajectories. This "path multiplicity" clearly cannot be readily made reconcilable with the deterministic nature of the Mach-Poincaré principle. One may envisage, at this point, the rise of some relational counterparts of the interpretations of quantum mechanics, for example with respect to the ontic status of these trajectories (all existing in parallel, or just one of them being "actualized"). And, perhaps, some benefits for the standard debate in quantum foundations could come from moving to a fully relational arena.

In conclusion, the road to quantum relational physics is full of hurdles, but it nonetheless points to an exciting and promising direction.

Acknowledgements

We are very grateful to an anonymous referee and the guest editor Andrea Oldofredi for their enlightening comments on a previous version of this chapter. P.N. and A.V. acknowledge financial support from the Polish National Science Centre, grant nr. 2019/33/B/HS1/01772.

Notes

1. There is obviously more to this oversimplified story. Earman (1989, chapter 1, section 3), for example, distinguishes between an ontological and a semantic version of the anti-substantivalist claim, and points out that the logical relation between the anti-substantivalist claim and the commitment to the only existence of relative motions is not that straightforward. Here we lump together these fine-grained theses in a unique coarse-grained one, since this level of accuracy is sufficient to drive our point home.
2. One may object that we are just being evasive on relationalism's elusiveness by preferring from the outset a particular strand of the doctrine (what Pooley, 2013, calls *Machian* relationalism) over the others. A defence of our choice would require a paper on its own. For the time being, we just submit that Machian relationalism is the only stance so far codified in a well worked out theoretical framework, whereas the other variants are better seen as "interpretations" of standard, non-relational theories such as Newtonian mechanics.
3. Since we are here providing a brief overview of modern relationalism, we refrain from commenting further on this point. See Dasgupta (2013, 2020); Baker (2020), for a discussion on the possible consequences of comparativism towards physical magnitudes.

4. Actually, parameters like masses and charges play an important role in the dynamics, even if they only enter as dimensionless ratios. Strictly speaking, then, what remains after symmetries are removed is the shape of the system weighted by said parameters. For simplicity's sake, we shall leave them out of the discussion, since the essential tenets go through regardless.

5. In order to strip each particle of any intrinsic identity—which would constitute a non-relational feature, one may add the permutation group to the symmetries to be quotiented out.

6. This is not to say that shape space methods in physics were not known before Barbour (2003). To the contrary, the notion of shape space had already had very useful applications in, e.g., molecular physics (see Kendall et al., 1999, for a textbook on the subject.)

7. The dilatational momentum is defined as $D \equiv \sum_a^N \mathbf{r}_a^{cm} \cdot \mathbf{p}_{cm}^a$. Given its monotonicity, the ratio D/D_0, with D_0 some arbitrary choice, has arguably been used as a physical time variable (Barbour et al., 2013, 2014a, b).

8. Poincaré was of course well aware of—and bothered by—this fact, and this had a crucial importance in his formulation of what we have called the Mach-Poincaré principle (the *locus classicus* for this discussion is Poincaré, 1905, chapter VII).

9. It is beyond the scope of this short survey to go into the technical details of SD. The interested reader is advised to look at the nice introduction in Barbour (2012) as well as the comprehensive, yet pedagogical book by Mercati (2018).

10 Mathematically, let us denote Riem the set of Riemannian 3-geometries and Diff(3) the group of spatial diffeomorphisms. Then, Superspace = Riem/Diff(3). Further, let S be conformal superspace S = Superspace/conformal transformations. Finally, the physical space of SD is $S_V \equiv$ Superspace/VPCT = S $\times \mathbb{R}^+$, where VPCT is the group of *volume-preserving* conformal transformations and $\mathbb{R}^+$ represents the spatial volume or its conjugate variable, the so-called *York time*.

11. This objection is put forward, e.g., in Pooley (2001).

12. We note *en passant* that this is one of the reasons that lead Rynasiewicz to claim that the substantivalism/relationalism debate is trivialized in modern physics: The fact that it is possible to switch from a "spacetime" to a "gravitational field" talk in general relativistic physics without changing the physical gist of the theory is a sign that the dispute is merely verbal (cf. Rynasiewicz, 1996, section V).

13. Although Barbour and collaborators worked in the context of standard SD, their main results can be easily carried over to PSD, as the ensuing discussion will clarify.

14. See the exchange between Zeh (2016) and Barbour et al. (2016) for a critical discussion of this point.

15. Ephemeris equations are heavily model-dependent. Explicit expressions exist for the N-body system (Koslowski et al., 2021) and Bianchi IX cosmological model (Koslowski et al., 2018).

16. In principle, the same reasoning can be applied, *mutatis mutandis*, to quantum theories. Recall from the discussion of tenet IV, that the quotienting out procedure works whenever the starting theory exhibits a symmetry group $\mathcal{G}$ whose elements represent redundant degrees of freedom. There is no reason to doubt that this works in the quantum domain as well. For example, in the case of the de Broglie-Bohm theory, we expect the corresponding PSD models to be very similar to the Newtonian ones given that both theories are about material particles inhabiting a background Newtonian space (the quantum motion on shape space will substantially differ

from the classical one, obviously). The fate of the wave function in PSD is an entirely different issue (see section 4.4).

17. To simplify the discussion, we are subsuming modal talk under "possible worlds" parlance, being aware that the notion of "possibility" and that of "possible world" may not coincide, depending on the particular metaphysics of modality adopted (see, e.g., Lewis, 1986, pp. 230–232, on this score).

18. Note that this kind of proposal goes back to the early stages of the Machian program, even before SD was born. Indeed, Barbour (1994, 1999) proposed to call the substantival relational configuration space "Platonia," and the configurations with higher probability to be experienced "time capsules."

19. Another option may be to argue the other way around, i.e., things being shapes in virtue of them being points of shape space. Such an option, if defensible, would constitute a supersubstantivalist take on relational physics (see Lehmkuhl, 2018, for a recent review of supersubstantivalism). We will set aside this possibility since, ironically, there is not enough space to discuss it here.

20. The general characterization of monism provided here heavily relies on Schaffer (2010). The reader is invited to check this source for an exhaustive discussion of this position.

21. Here we are simplifying the discussion a lot. The interested reader can start to dig deeper into this issue in the philosophy of science by taking a look at Psillos (2018).

22. This is not to say that Poincaré was the only "responsible" for the rise of scientific structuralism. See Worrall (1989), for the first articulation of this stance in a modern context, and Ladyman (2020), for an introduction to the debate.

23. Note, however, that structuralism does not mitigate the issues with establishing a proper ontology for PSD discussed at the beginning of the section. While, in fact, it washes away the need to settle for fundamental *objects*, the problem is shifted to that of finding the fundamental *relations*.

24. Rovelli's covariant loop quantum gravity (Rovelli and Vidotto, 2015) is an already well-established quantum gravity program that is based on relational considerations. However, Rovelli's relationalism is markedly non-Machian in that it denies any "global" structure to the world. For Rovelli, the world is a constellation of unrelated events consisting of the local interactions between physical systems.

Bibliography

Albert, D. (2013). Wave function realism. In D. Albert and A. Ney (Eds.), *The Wave Function*, Chapter 1, pp. 52–57. Oxford University Press.

Baker, D. (2020). Some consequences of physics for the comparative metaphysics of quantity. In K. Bennett and D. Zimmerman (Eds.), *Oxford Studies in Metaphysics*, Volume 12, pp. 75–112. Oxford University Press.

Barbour, J. (1974). Relative-distance Machian theories. *Nature* 249, 328–329.

Barbour, J. (1982, September). Relational concepts of space and time. *British Journal for the Philosophy of Science* 33 (3), 251–274.

Barbour, J. (1994). The timelessness of quantum gravity: II. the appearance of dynamics in static configurations. *Classical and Quantum Gravity* 11, 2875–2897.

Barbour, J. (1999). *The End of Time. The Next Revolution in Physics*. Oxford University Press.

Barbour, J. (2003). Scale-invariant gravity: Particle dynamics. *Classical and Quantum Gravity* 20, 1543–1570. http://arxiv.org/abs/gr-qc/0211021v2.

Barbour, J. (2012). Shape dynamics. An introduction. In F. Finster, O. Müller, M. Nardmann, J. Tolksdorf, and E. Zeidler (Eds.), *Quantum Field Theory and Gravity, Proceedings*, pp. 257–297. Birkhäuser Basel. https://arxiv.org/abs/ 1105.0183[gr-qc].

Barbour, J. and B. Bertotti (1977). Gravity and inertia in a Machian framework. *Il nuovo cimento B* 38 (1), 1–27.

Barbour, J. and B. Bertotti (1982). Mach's principle and the structure of dynamical theories. *Proceedings of the Royal Society A* 382, 295–306.

Barbour, J., T. Koslowski, and F. Mercati (2013). A gravitational origin of the arrows of time. *arXiv:1310.5167v1*. https://arxiv.org/abs/1310.5167v1.

Barbour, J., T. Koslowski, and F. Mercati (2014a). Identification of a gravitational arrow of time. *Physical Review Letters* 113, 181101.

Barbour, J., T. Koslowski, and F. Mercati (2014b). The solution to the problem of time in shape dynamics. *Classical and Quantum Gravity* 31 (15), 155001.

Barbour, J., T. Koslowski, and F. Mercati (2016). Janus Points and arrows of time. *arXiv:1604.03956v1*. https://arxiv.org/abs/1604.03956.

Creţu, A.-M. (2020). Perspectival realism. In M. Peters (Ed.), *Encyclopedia of Educational Philosophy and Theory*. Springer.

Dasgupta, S. (2013). Absolutism vs comparativism about quantity. In K. Bennett and D. Zimmerman (Eds.), *Oxford Studies in Metaphysics*, Volume 8, pp. 105–148. Oxford University Press.

Dasgupta, S. (2020). How to be a relationalist. In K. Bennett and D. Zimmerman (Eds.), *Oxford Studies in Metaphysics*, Volume 12, pp. 113–163. Oxford UniversityPress.

Earman, J. (1989). *World Enough and Space-time. Absolute Versus Relational Theories of Spacetime*. The MIT Press.

Gomes, H. and T. Koslowski (2012). Coupling shape dynamics to matter gives spacetime. *General Relativity and Gravitation* 44 (6), 1539–1553. http:// arxiv.org/abs/1110.3837v2.

Gryb, S. and K. Thébault (2016). Time remains. *British Journal for the Philosophy of Science* 67 (3), 663–705. http://arxiv.org/abs/1408.2691.

Hoefer, C., N. Huggett, and J. Read (2022a). Absolute and relational space and motion: Classical theories. In E. Zalta (Ed.), *The Stanford Encyclopedia of Philosophy* (Spring 2022 ed.). https://plato.stanford.edu/entries/spacetimetheories-classical/.

Hoefer, C., N. Huggett, and J. Read (2022b). Absolute and relational space and motion: Post-Newtonian theories. In E. Zalta (Ed.), *The Stanford Encyclopedia of Philosophy* (Spring 2022 ed.). https://plato.stanford.edu/entries/space time-theories/.

Huggett, N. (2006). The regularity account of relational spacetime. *Mind* 115, 41–73.

Kendall, D., D. Barden, T. K. Carne, and H. Le (1999). *Shape and Shape Theory*. John Wiley & Sons LTD.

Koslowski, T., F. Mercati, and D. Sloan (2018). Through the big bang: Continuing Einstein's equations beyond a cosmological singularity. *Physics Letters B* 778, 339–343.

Koslowski, T., P. Naranjo, and A. Vassallo (2021). Pure shape dynamics: General framework. *arXiv:2112.14101* [gr-qc]. https://arxiv.org/abs/2112.14101.

Ladyman, J. (1998). What is structural realism? *Studies in History and Philosophy of Science* 29, 409–424.

Ladyman, J. (2020). Structural realism. In E. Zalta (Ed.), *The Stanford Encyclopedia of Philosophy* (Winter ed.). Metaphysics Research Lab, Stanford University. https://plato.stanford.edu/archives/win2020/entries/structuralrealism/.

Lazarovici, D. (2018). Super-Humeanism: A starving ontology. *Studies in History and Philosophy of Modern Physics* 64, 79–86.

Lehmkuhl, D. (2018). The metaphysics of super-substantivalism. *Noûs* 52 (1), 24–46.

Lewis, D. (1986). *On the Plurality of Worlds*. Blackwell.

Marchal, C. and D. Saari (1976). On the final evolution of the n-body problem. *Journal of Differential Equations* 20, 150–186.

Maudlin, T. (2019). *Philosophy of Physics: Quantum Theory*. Princeton University Press.

Mercati, F. (2018). *Shape Dynamics*. Oxford University Press.

Poincaré, H. (1900). Relations entre la physique expérimentale et de la physique mathématique. In C.-E. Guillaume and L. Poincaré (Eds.), *Rapports présentés au congrès international de physique*, Tome 1, pp. 1–29. Gauthier-Villars.

Poincaré, H. (1905). *Science and Hypothesis*. Walter Scott.

Pooley, O. (2001). *Relationalism Rehabilitated? II: Relativity*. Manuscript. http://philsci-archive.pitt.edu/221/. Accessed September 2021.

Pooley, O. (2013). Substantivalist and relationalist approaches to spacetime. In R. Batterman (Ed.), *The Oxford Handbook of Philosophy of Physics*, pp. 522–586. Oxford University Press.

Psillos, S. (2018). Realism and theory change in science. In E. Zalta (Ed.), *The Stanford Encyclopedia of Philosophy* (Summer 2018 ed.). Metaphysics Research Lab, Stanford University. https://plato.stanford.edu/archives/sum2018/entries/realism-theory-change/.

Rovelli, C. and F. Vidotto (2015). *Covariant Loop Quantum Gravity*. Cambridge University Press.

Rynasiewicz, R. (1996, June). Absolute versus relational space-time: An outmoded debate? *The Journal of Philosophy* 93 (6), 279–306.

Schaffer, J. (2010). Monism: The priority of the whole. *Philosophical Review* 119 (1), 31–76.

Varzi, A. (2011). On doing ontology without metaphysics. *Philosophical Perspectives* 25, 407–423.

Vassallo, A., D.-A. Deckert, and M. Esfeld (2017). Relationalism about mechanics based on a minimalist ontology of matter. *European Journal for Philosophy of Science* 7 (2), 299–318. http://arxiv.org/abs/1609.00277.

Vassallo, A. and M. Esfeld (2016). Leibnizian relationalism for general relativistic physics. *Studies in History and Philosophy of Modern Physics* 55, 101–107. https://arxiv.org/abs/1608.07257.

Worrall, J. (1989). Structural realism: The best of both worlds? *Dialectica* 43, 99–124.

Zeh, H. (2016). Comment on the "Janus Point" explanation of the arrow of time. *arXiv:1601.02790v1*. https://arxiv.org/abs/1601.02790.

5 Rotating black holes as time machines: An interim report

Juliusz Doboszewski

5.1 Introduction

Kerr spacetimes are a family of exact solutions to the vacuum Einstein's field equations which can be used to represent the exterior of a spinning body, in particular a rotating black hole. They are of crucial importance for observational and theoretical astrophysics (where it is highly plausible that, when considered as isolated systems, all black holes in our universe are Kerr black holes), and for a number of central open problems in mathematical relativity. Philosophical interest in the Kerr spacetime comes from the observation that due to the rich causal structure of the maximally extended[1] Kerr spacetime, some topics of traditional (within last thirty or so years) philosophical interest show up jointly in a physically relevant class of solutions. These include failure of determinism, possibility of relativistic hypercomputation, and possibility of time travel, or time machines. Since all of these properties rely on the presence of a Cauchy horizon in the deep interior of a black hole, two central philosophical questions arise. First, should our commitments to the empirical adequacy of Kerr exterior for some compact astrophysical sources imply additional commitment to the analytic extension through Cauchy horizon as the appropriate black hole interior solution? Second, do we have good reasons for believing that Kerr's Cauchy horizon is a plausible feature of approximately Kerr-like subregions of our universe? This, in turn, amounts to deciding (a) whether and which features of the Kerr spacetime will carry over to spacetimes "in the vicinity" of Kerr, and, in particular, (b) whether we can expect that Cauchy horizons occur in generic situations, or if they are merely a feature of a highly symmetric solution.

I will not argue that rotating black holes are definitely time machines, nor that they are definitely not, but advocate for a more cautious view. Although in the context of rotating black holes existence of time machines (understood as a philosophical problem) to some extent reduces to well-defined mathematical problems on which a lot of progress has recently been made, and although numerous recent results do begin to point in a consistent direction, many concerns remain about whether these results

DOI: 10.4324/9781003219019-7

will generalize. And so, at this stage, abstaining from deciding either way remains the most prudent option.

Nevertheless, this topic is ripe for a philosophical analysis, and my goal here is to provide the reader with a sense of the foundational and philosophical issues arising, as well as a map of the most important topics and results. Older questions—such as interpretation of global properties of exact solutions, or roles of the causal conditions in the initial value formulation of general relativity—remain important, but novel themes emerge, such as physical significance of low regularity (that is, less than twice continuously differentiable) spacetimes. All of those are of crucial importance when considering whether the available evidence is sufficiently strong to rule out Cauchy horizons (and, through that, time machines) in this class of spacetimes.

The chapter is organized as follows. Sections 5.2.1 and 5.2.2 briefly recall the main features and astrophysical applications of the Kerr spacetime. Sections 5.2.3 and 5.3 cover philosophically interesting aspects of the Kerr spacetime and clarify the sense in which it can be thought of as a time machine spacetime. Classical forms of cosmic censorship (section 5.4), quantum effects (section 5.5), and a recent thermodynamical argument (section 5.6) are then discussed. To summarize the overall picture, many recent results begin to constrain the viability of Kerr black holes as time machines. The first line of argument, of section 5.4, is the strongest, but some doubts concerning approximation schemes used in these calculations may be raised; the second line of evidence, in section 5.5, relies on the semi-classical approximation, but (somewhat speculatively) singularity resolution in a more fundamental theory of quantum gravity might play out either way; the third line of ruling out time machines is, as I am arguing in section 5.6, not very convincing.

5.2 Basic information about the Kerr spacetime

Let's warm up by recalling some of the basic features of Kerr spacetime.

5.2.1 Form of the metric and its causal structure

This section serves merely as a reminder and a pointer to proper sources; it is impossible to fit a complete presentation of the Kerr spacetime on a few pages. The most careful discussion I am aware of can be found in O'Neill (2014); see also Wiltshire et al. (2009) (especially the list of the most important coordinate reprentations of the solution in the first chapter).

Kerr spacetimes are a family of stationary, asymptotically flat, vacuum solutions to the Einstein's field equations, with a ring-shaped curvature singularity around the axis of rotation. The family can be thought of as spanned by two parameters, mass m and spin a. For the subextremal part of the family $(0 < a^2 < m^2)$, its maximal analytic extension is an infinite series of three spacetime regions: asymptotically flat and globally hyperbolic region I (representing the exterior of a rotating black hole) and two interior regions, II and III.

Taken jointly, regions I+ II are globally hyperbolic,[2] and separated by the black hole event horizon: no signal from region II can reach an observer at infinity in region I. Region III is connected to region II by a Cauchy horizon.[3] This region is not chronological, i.e., there are distinct points p and q that are related by the chronological precedence relation $p \ll q$ (which holds iff there is a future directed timelike curve from p to q) for which the relation "$\ll$" is symmetric; a timelike curve connecting p with itself is called a closed timelike curve (CTC). See also Figure 1 in the first chapter of Wiltshire et al. (2009) for another useful illustration of these features.

The mathematical importance of the Kerr family stems from its place in a number of central conjectures. Stated rather colloquially, these conjectures include:

- Weak Cosmic Censorship hypothesis: in generic situations, gravitational collapse leads to formation of event horizons;
- Strong Cosmic Censorship hypothesis (SCC): generic initial data cannot be extended beyond their domains of dependence, which, for Kerr, means that the Cauchy horizon will not be present in near-Kerr spacetimes (in other words, occurrence of Cauchy horizon is an unstable feature of the Kerr spacetime);
- Kerr stability: small perturbations of the exterior lead to spacetimes similar to Kerr;
- Kerr rigidity: a generalization of the black hole no hair property (stating that stationary black holes are uniquely characterized by the numbers specifying their location and three additional quantities: mass, spin, and charge);
- The Final State Conjecture: that the endstate of gravitational collapse converges to Kerr.

A plausible expectation is that all of these conjectures hold. But these problems are open, hard, and subtle, as even a brief and partial overview of what is known about the SCC in section 5.4 demonstrates. For a recent pedagogical overview of these issues in the context of the Penrose 1965 singularity theorem (and pointers to the appropriate literature, which is too vast to cover here) see Landsman (2021).

5.2.2 *Observational evidence*

In an astrophysical context, the Kerr hypothesis states that compact objects (both supermassive and stellar size) thought to be black holes, such as the central object in Saggitarius A (*SgrA**) in the centre of our galaxy, indeed are Kerr black holes known from general relativity. Three main independent lines of observational evidence are available, and, so far, in agreement with the Kerr hypothesis.

The first line comes from the detection of gravitational waves from binary systems by the LIGO-Virgo collaboration (Abbott et al., 2016, and subsequent detections). In this case the Kerr hypothesis implies that the waveform after the merger phase converges to a stationary state given by the Kerr metric. This could be phrased as a text of the no-hair property of black holes; see Isi et al. (2019) for a discussion using the event $GW150914$, and Cardoso and Pani (2019) for an up to date overview of alternative models, including so-called exotic compact objects (which differ from black holes in not having an event horizons).

The second comes from optical observations of the central object *SgrA** within our galaxy, in particular, bright sources such as the start *S2* orbiting around it (Abuter et al., 2018a; Abuter et al., 2018b). The Kerr hypothesis is the main explanation of the mass and simple structure of *SgrA**. See also Eckart et al. (2017) for a discussion of philosophical issues raised by such observations.

The third comes from high resolution imaging of the black hole photon ring based on Earth-size virtual Very Long Baseline Interferometry array operating at 1.3 mm radio frequency (The Event Horizon Telescope Collaboration et al., 2019a); published results concern sources $M87^*$ and *SgrA**. And, again, the conclusion (The Event Horizon Telescope Collaboration et al., 2019b) is that the shape of their photon ring is consistent with the Kerr hypothesis.

None of these lines of evidence is decisive, in part because there are good prospects for significant improvements in these measurements: in the stellar-mass cases, by increasing sample size in future observational runs; for the supermassive candidate *SgrA**, by more precise measurement of motions in the optical spectrum as well as Event Horizon Telescope observations; and for both $M87^*$ and *SgrA**, by higher resolution measurements with extensions of the Event Horizon Telescope array.

Obviously, all of this evidence might only inform us about the exterior region of a black hole. So although observational evidence summarized is in agreement with the Kerr hypothesis, the choice of the interior solution remains unconstrained by it. Naturally, one could try to interpret empirical adequacy of the Kerr spacetime in the exterior as at least suggestive of its empirical adequacy as an interior solution as well. On that basis, one could further argue that chronology violating regions of Kerr may indeed be plausibly realized within our universe. Such a possibility should be seriously considered, and, if possible, constrained. In the absence of empirical access to interiors of those sources, conceptual analysis and mathematical methods are the only means of progress available.

5.2.3 *Philosophically interesting features of the maximal extension*

We are now in a position to make some observations concerning features of the Kerr spacetime which are of particular philosophical interest. These include failure of determinism, the Malament-Hogarth property, and a possibility of time machines.

The first of those arises because the domain of dependence (which consists of regions I+II) of Kerr initial data is extendible, and any such extension contains a Cauchy horizon. This is a failure of determinism, because the behavior of physical fields in I+II fails to uniquely fix their behavior in the extension. In more philosophical terms, state of the universe at a given and the laws of nature are compatible with multiple different possible futures. Indeed, uncountably many distinct extensions (distinct from the analytic one of section 5.2.1) can in principle be constructed. See Earman (1995) for a classic exposition, and Doboszewski (2019b), Doboszewski (2019a), as well as Smeenk and Wüthrich (2021), for a recent discussion of these and related issues in the context of the Laplacian notion of determinism.

Recall that a spacetime $\langle M, g_{ab} \rangle$ has a Malament-Hogarth property iff there is a point p and a future complete timelike curve γ such that $\gamma \subseteq I^-(p)$. Notably, the Kerr spacetime satisfies this property for any future complete timelike curve fully contained entirely in region I and any point p in the acausal region III; this is illustrated in Figure 5.1. Because our usual notion of a computation (involves only finitely many steps, Malament-Hogarth spacetimes provide unexpected epistemic boons. In such spacetimes one might contemplate, for instance, whether placing computers on curves such as γ could tell us something profound about limits of computability. Such a computer could perform infinitely many steps (for instance, check all of countably many arithmetical statements), whereas an observer at p would have to only wait finite amount of time to learn, for example, whether Peano arithmetics is consistent, which is a undecidable problem in a standard sense of computation using finitely many steps. See Pitowsky (1990) and Earman and Norton (1993) for a classic discussion and Andréka et al. (2018), for a recent overview by some of the pioneers of such analysis. Just as with time machines (on which more in a moment) one might ask whether there is a sense in which laws of physics could enforce the Malament-Hogarth property; see Manchak (2020b) for a discussion of these issues.[4]

For the issue of time machines, the crucial observation is that region III contains CTCs. CTCs are a relativistic form of time travel to the past, and as such raise a range of questions, logical and philosophical as well as physical. On the conceptual side, one should point out that the

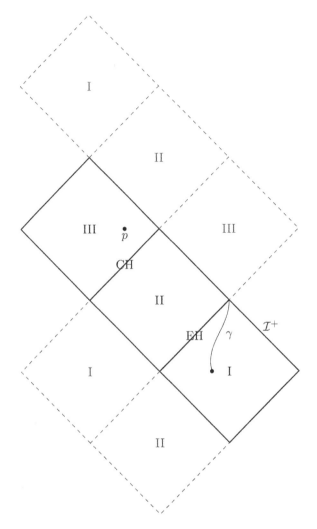

Figure 5.1 Penrose diagram showing the causal structure of a subset of subextremal Kerr spacetime. Region I represents the universe outside the black hole, region II the black hole interior, region III the deep interior. The maximal analytic extension continues infinitely in the manner shown on shaded copies of regions I, II, III. Acausal features of region III are not shown on this diagram. The event horizon is marked with EH, the Cauchy horizon with CH, point *p* and curve *γ* show the Malament-Hogarth property.

alleged paradoxes of time travel can, upon closer inspection, be dissolved (Lewis, 1976); see Wasserman (2017) and Effingham (2020) for a recent discussion. A physical analysis is then much needed. The main line of questioning focuses on general features of CTCs and aims to prove that under plausible conditions CTC do not occur. This led to the no-go theorems of Tipler (1976) and Hawking (1992), and their subsequent generalizations (Krasnikov, 2002; Manchak, 2009); see also Krasnikov (2018) and Luminet (2021), for a recent physics perspective, and Earman et al. (2009); Earman and Manchak (2016); Wüthrich (2021), for a philosophical analysis of these results.

But it also worthwhile spending some time looking at something less general, namely particular exact solutions: arguably the empirical success of general relativity is closely connected to particular solutions (including, as I gestured at in section 5.2.2, the Kerr spacetime). Accordingly, in the next section I will explore the extent to which Kerr spacetime can be thought of as a time machine spacetime. But what might it even mean that a region of spacetime functions as a machine for producing some features?

5.3 "Machines"

An important clarification of the sense in which laws of nature could be used by, say, an advanced civilization to produce closed timelike curves has been made by Smeenk, Earman and Wüthrich (SEW) in Earman et al. (2009). Under the SEW notion, a reliably operating time machine space-time satisfies, at minimum, the following conditions:

C1 There is a partial Cauchy surface Σ such that there are no CTCs to the past of Σ. In Hawking's words, this would ensure that the universe before the time machine begins its operation is safe for historians;

C2 "The conditions that encode the instructions for the operation of the time machine can be set on Σ; to assure that the instructions are obeyed, the region of spacetime *TM* corresponding to the operation of the time machine should belong to the future domain of dependence $D + (\Sigma)$," (*ibid.*, p. 101)

C3 "The time machine is supposed to be confined to a finite region of space and to operate for a finite amount of time, implying that *TM* should have compact closure," (*ibid.*, p. 102)

C4 The chronology violating region *V* is contained in the causal future $J^+ (TM)$ of a time machine,

C5 The time machine functions reliably, i.e., any extension of the time machine contains chronology violating region *V* (Potency Condition).

Are Kerr black holes time machines in the SEW sense? So far it might seems so: all that the hypothetical advanced civilization would have to do is to collapse some stellar object to a Kerr black hole. Jumping into the deep interior would amount to entering for a ride on a time machine. But some problems arise.

The first is that Kerr spacetime is asymptotically flat, so the Cauchy horizon is not compactly generated. On one hand this seems to evade various no go results against time machines (again, see Earman et al., 2009, for a discussion of those); but on the same grounds it also violates the assumption C3 of the SEW definition. So if asymptotic flatness of Kerr is taken literally, rotating black holes cannot be time machines in the SEW sense. But the situation is subtle. In physics practice, asymptotic flatness is not taken literally, but rather is interpreted as a form of idealization: astrophysical sources are only approximately flat, and our observations are not made from the perspective of anything like actual null infinity (see Ellis, 2002, for a discussion of the size of an "effective infinity" for various types of systems). Such a carefully interpreted property makes it amenable to the SEW treatment—and to the no go theorems.

The second problem concerns the Potency Condition. This condition demands that one considers only extensions which are "as large as possible," which is formally cashed out as a "no holes" kind of condition.[5] The idea is that unless some such condition is posited, one could construct many other spacetime extensions by (i) starting with some extension with chronology violating region *V*, and (ii) removing subregions of *V* in such a way that a new extension is chronological. For instance, one could remove a closed set of points *C* such that every CTC intersects with some element of *C* from region III of maximal extension of the Kerr spacetime. If that were allowed, time machine would be unreliable, because it might not produce chronology violations after all. The problem is that, arguably, there is no single condition adequately capturing the sense in which a spacetime is "as large as possible," and many of the known conditions give conflicting assessments in simple black holes spacetimes, including Kerr (Doboszewski, 2022).

In any case, if the Final State Conjecture turns out to be true but the Strong Cosmic Censorship turns out to be false (so Cauchy horizons in the interiors might occur generically within some parameter range), then Kerr black holes would be quite promising candidates for time machines. In that hypothetical situation, gravitational collapse (ending with Kerr spacetime) and/or merging and accretion would be a robust and physically plausible mechanism for producing time machines in a SEW sense.[6] Naturally, there are other types of systems and other mechanisms which might produce time machines; so that if it turns out that Strong Cosmic Censorship is true around Kerr, time machines are restricted, but not fully ruled out. Nevertheless, the Kerr spacetime is of particular physical importance, and with the strongest observational support, so keeping track of new results is of high importance.

Stakes are high when it comes to Strong Cosmic Censorship for the Kerr spacetime, and so the third crucial problem for a time machine advocate is whether SCC is true. And it is not only time machines in the SEW sense, but all three features of philosophical interest from section 5.2.3 which rely on the presence of Cauchy horizons in the interior of Kerr black holes. In this sense the issue of stable occurrence of Cauchy horizons is crucial for understanding whether these philosophically interesting properties might arise as models of particular physical systems in general relativity. Conditions under which that is possible would then be "machines" for producing these features. Accordingly, in what follows I will briefly survey some of the recent classical (section 5.4) and quantum (section 5.5) constraints on the occurrence of Cauchy horizons in spacetimes similar to the Kerr spacetime.

I would like to stress that we find ourselves in a very fortunate situation: interesting philosophical questions can be partially answered by exploiting a solution to a well-defined problem in mathematical physics. To some extent a philosophical problem reduces to a mathematical one. True, it is a hard problem, and the final outcome is not yet clear; but it is also true that a lot of progress has been recently made in that area.

5.4 Cosmic censorship in general relativity

We are now in a following situation: the Kerr spacetime has Cauchy horizons, and some of its extensions contain CTCs. Should we expect that this occurs generically, or is just a feature of the highly symmetric Kerr spacetime? In other words, what is known about whether strong cosmic censorship hypothesis is true?

Unfortunately, this quickly leads to a cornucopia of open problems. Even though mathematical relativity community agrees on the basic statement: development of a class of initial data is not globally hyperbolic only in non-generic situations,[7] SCC might be seen as a set of related conjectures. These are parametrized by choices concerning the matter model, type of perturbations, asymptotic structure, linearity, numerical code implementation, standard of distance (such as a topology over a set of spacetimes), and regularity of the resulting spacetime. Moreover, the situation can also change with the dimensionality of spacetime (an example of that is given in the next section). All of these choices express idealizations which these problems amenable to numerical or analytical treatment, and so are also natural points at which proponent of time machine might raise worries and objections.

The basic heuristics driving a lot of work on the SCC is blueshift instability argument suggested by Penrose (1979). Signals coming in from the exterior build up to arbitrarily high frequency at the Cauchy horizon inside the black hole, so perturbations of the initial data in the exterior could turn them into curvature singularities. For a scalar field model and asymptotically flat black holes, recent numerical results of Chesler et al. (2019) for Reissner-Nordström black holes[8] (see references therein for a long list of previous results) and Chesler et al. (2019) for Kerr suggest that for the late times the singularity structure inside rotating black holes indeed becomes spacelike. This is bad news for proponents of time machines.

In the presence of a cosmological horizon, heuristic blueshift instability argument may be countered by noting presence of a redshift factor outside the black hole. Which of these effects is stronger? Cardoso et al. (2018) have argued that de Sitter asymptotics may prevent curvature blowup at the Cauchy horizon for near-extremal Reissner-Nordström black holes, making them C^2-extendible. In a similar vein, Dias et al. (2018) considered linearized coupled gravitational and electromagnetic perturbations for near-extremal Reissner-Nordström black hole, and concluded that arbitrarily high differentiable SCC fails (for highly charged and sufficiently large black holes).[9] However, Dias et al. (2018) argue that in the Kerr case the decay rates of quasi-normal modes in the perturbation respect cosmic censorship. Moreover, Dafermos and Shlapentokh-Rothman (2018) claim that the Reissner-Nordström result is an artifact of the choice of the class of initial data sets. And, recently, Luna et al. (2021) provided even more *caveats*, by concluding their analysis of non-linearities in these results with the conclusion that "numerical codes are insufficient to draw conclusions about the potential failure of SCC for near-extremal RNdS black hole spacetimes." (*ibid.*, p. 104043–2) It seems fair to say that even in simple cases the situation with positive cosmological constant is not yet fully conclusive.

Playing for a moment the devil's advocate, let me voice three concerns a proponent of time machines might raise at this stage. First, how confident should one be that these results will carry over to other matter models, more realistic than (mostly massless) scalar fields? After all, the only known arbitrarily high differentiable way scalar field in nature, the Higgs field, is not even a classical field. Second, which of those results will generalize to the Kerr-Newman case (*pace* endnote 1)? And finally what is the physical significance of the choice of sets of initial data (smooth or rough) considered to be close to Kerr, and standards of stability used in these results? Non-uniqueness and unwelcome features of such topologies make this a hard problem (see Fletcher, 2016, and Curiel, 2014b, for a discussion). Having a clear picture of relations between these issues would be very useful in future assessments of the fate of time machines and other interesting physical possibilities general relativity presents us with.

Despite this, however, one might get a feeling that prospects for the occurrence of Cauchy horizons in generic near-Kerr spacetimes begin to look rather bleak. Many models and methods of calculations do seem point in the consistent direction (at least in the asymptotically flat case). If these results will continue to improve, proponents of time machines may be forced to look elsewhere for more robust (but also more exotic) examples, most likely losing the advantage of empirical significance of the Kerr spacetime.

5.5 Cosmic censorship beyond general relativity

Classical general relativity is not all there is, and perhaps lack of consensus about the SCC there is not that problematic. Hollands et al. (2020) have recently argued that a form of cosmic censorship holds in quantum field theory in curved spacetime. Namely, they showed that for any Hadamard state of a quantized massless real scalar field, the expectation value of that state diverges at a Cauchy horizon (in Reissner-Nordström-de-Sitter spacetime, so problematic in the preceding section). In particular, this is taken to imply that, if that quantity was backreacted through the semi-classical Einstein's field equations, the resulting spacetime would be inextendible (even as a weak C^0 solution).

On the other hand, Dias et al. (2019) showed that for a variety of matter models in 2+1 dimensional BTZ black holes, an analogous quantity remains bounded at the Cauchy horizon. They then propose a similar argument to Dafermos and Shlapentokh-Rothman (2018) concerning rough initial data to enforce cosmic censorship in the BTZ case.

This suggests that quantum effects might, but do not have to enforce cosmic censorship, and that at least in some cases what enforces strong cosmic censorship is not quantum effects on their own, but these effects together with other assumptions (such as enlargement of the set of admissible initial data—which brings many of the same issues concerning stability and physical significance). Whether such arguments are fully convincing depends also on the trust one places in semi-classical gravity; but they seem to provide a further constraint on the occurrence of Cauchy horizons, even if those turned out to be classically viable.

The situation is even more complicated in candidates for a more fundamental theory of quantum gravity. Space constraints do not allow for a complete discussion (see Wüthrich, 2021 for an overview), but let me briefly point out one delicate issue. In all forms of strong cosmic censorship considered so far Cauchy horizon is converted into some stronger form of spacetime singularity; a more benign singularity (in the form of curvature blowup) replaces a more worrisome one. However, it is widely expected that a full theory of quantum gravity will provide a form of singularity resolution for curvature singularities (see Crowther and De Haro in this volume for an in-depth discussion). If that indeed happens, what are the grounds for believing that Cauchy-horizon-turned-singular-subsequently-resolved will be well-controlled, rather than (so to speak) come back with vengeance, leading to acausal or otherwise highly non-unique behavior?

5.6 The second law of thermodynamics and time travel

At this stage the situation seems to be the following: conceptual reasons (*pace* Lewis, 1976) are not sufficient for ruling out time travel, and in the particular case of Kerr spacetime, physical reasons within general relativity and its immediate extensions are not *yet* sufficient to rule them out.

We could then ask ourselves whether time machines are compatible with other physical theories or principles, ones which, unlike the results discussed in the previous sections, may not be directly concerned with the structure of spacetime itself. One natural place to look at is thermodynamics. In this context Rovelli (2019) has recently argued that, although it may seem that time travel to the past is (in principle) possible due to there being general relativistic spacetimes with closed timelike curves,[10] "examin[ing] with care the full thermodynamical and statistical physics along the closed timelike curves" (*ibid.*, p. 1) leads to another conclusion: that time travel to the past is incompatible with the second law of thermodynamics (SL).

Rovelli begins with the observation that any mechanical clock dissipates energy, which makes it subject to the laws of thermodynamics, in particular the second law. Now, the argument goes,

> [L]et $S(\tau)$ be a measure of local entropy along the closed loop γ. For the clock to work all along γ, registering the number of its oscillations, we need $dS(\tau)/d\tau > 0$ everywhere going around the loop. But since γ is a loop, $S(\tau)$ cannot grow monotonically as we go round. Therefore it is impossible for a clock to count its own oscillations along a closed timelike curve.
>
> (*ibid.*, p. 2)

This claim is then extended to systems with memory, such as the human nervous system, and other irreversible phenomena. Since "(a) long a closed timelike loop γ the only possibility of having $dS/d\tau \geq 0$ everywhere is having $dS/d\tau = 0$," (*ibid.*, p. 2) any process happening on γ is reversible and cannot have memory, which makes time travel to the past thermodynamically impossible.

How convincing is this argument? Uncontroversially, it shows that (assuming that we have an in principle unrestricted epistemic access to our causal past) we are not currently living on a CTC. It also suggests that living on a CTC would (insofar as the SL in this form indeed applies to situations such as CTCs) be quite different from our everyday life.

It is also quite clear that for a simple time machine spacetime, such as the Misner spacetime (see Earman et al., 2009, for a discussion of this example), the chronology violating region has to saturate the entropy bound in the SL.[11] But closed timelike curves can arise in a myriad of different ways. Is this argument sufficient to rule out all of them at once? I do not think so.

First, inference from the claim that on a closed loop any process is reversible to the point that there can be no clocks in a spacetime with CTCs is too fast. To see that, consider the global structure of what, as a matter of fact, is the first solution with closed timelike curves that has been discovered: the Gödel spacetime. It is a rotating, non-expanding, geodesically complete, exact solution of the Einstein's field equations with a dust source and cosmological constant. This is not a time machine spacetime in the sense of the SEW definition of section 5.3, because there are CTCs through any two points in spacetime (so, contra the SEW notion, there is no region which is safe for historians). But it is well-known that although there are CTCs in the Gödel spacetime, there are are no closed timelike geodesics: some acceleration is needed in other to enter a CTC in this spacetime.[12] So it seems that there could be, to use Rovelli's phrase, physical ways for a clock to count its oscillations in a spacetime with CTCs: a local measure of entropy can grow monotonically if a clock is placed on a geodesic in the Gödel spacetime. Perhaps the thermodynamical argument could be somehow patched up in order to show that such a form of time travel would not be very useful. Plausibly the dissipated entropy should satisfy some additional self-consistency constraints (an idea going back to Friedman et al., 1990), potentially limiting physical processes exploiting travel along a CTC as a resource (for example for computational purposes). But uselessness is very different from impossibility, and so the Gödel spacetime example should make us wary: there could be irreversible processes in spacetimes with CTCs.

There is a second, perhaps stronger, objection to the thermodynamical argument. Rovelli assumed the second law of thermodynamics in the form $dS(\tau)/d\tau \geq 0$. But the second law in this form applies only to isolated systems. Can the interior of a Kerr black hole be thought of as an isolated system?[13] In particular, since the argument of Rovelli (2019) begins with considering $S(\tau)$ along a CTC and derives a contradiction with the Second Law, can interiors of these black holes be thought of as systems isolated from their exterior? Again, this depends on what physics we consider to be of relevance. Ignoring for the time being semi-classical effects, in particular black hole evaporation, I think the answer should be negative. Quite a few reasons for denying that the standard form of the second law is the appropriate one arise:

- In the asymptotically flat case, the deep interior of the black hole cannot be thought of as isolated from the exterior, in the sense that for any point q in the region III, there are infinitely many points p in the exterior region I such that $p \ll q$. This means that matter, energy, and a lot of entropy can flow in from the exterior region I to region III through both horizons.
- Furthermore, if we consider as a serious possibility the existence of other copies of the regions I and II, then the excess entropy could also leave region III, providing us with a natural entropy sink.
- Another sink into which excess entropy could be thrown into and destroyed is the black hole curvature singularity.[14]

- All of those issues arise before we consider the plausible thermodynamical nature of the black hole itself (see Curiel, 2014a and Prunkl and Timpson, 2019, as well as Wallace, 2018, and Wallace, 2019, for a recent introduction of these issues), or the thorny issue of gravitational entropy (Wallace, 2010); both of which, it seems, should be accounted for in a full discussion of the second law along closed timelike curves in the black hole interior.[15]

In the light of these worries, I do not think that the standard form of the SL is the appropriate equation to be used in the argument, at least in the contexts such as Kerr spacetime. The interior of Kerr black holes is not an isolated system—rather it is an open system[16], and so the right equation expressing the second law should be some entropy balance equation. But the choice of the appropriate terms in that equation seems to depend on a prior decision about which regions of Kerr are physically significant; for example, formulating the balance equation seems to demand that we first make a decision about how many entropy sinks are available. If so, setting up a thermodynamical problem of time travel *prima facie* seems to require prior choices about physical significance of various features of Kerr, making such generalizations of Rovelli (2019) argument liable to a *petitio principi* objection.

I do not claim that a more convincing argument against time machines could not be made on a similar basis. Perhaps the thermodynamical argument could be patched up or generalized. But the standard form of the the second law seems unsuitable for arguing against the possibility of time travel to the past in its full generality. This does not imply that time travel is compatible with thermodynamics—indeed, even more careful analysis of full physics of chronology violating regions would be needed to make a positive case for it. A clear understanding of the generalized second law and black hole horizon thermodynamics under an open system view would be a welcome step in shedding more light on these issues.

Acknowledgements

I would like to thank audiences at the Black Hole Initiative Colloquium, History and Philosophy of Physics Research Seminar of the Lichtenberg Group at the University of Bonn, and Logic, Relativity and Beyond 2021 conference for their feedback on previous versions of this paper; Erik Curiel, Harvey Brown, and Dennis Lehmkuhl for conversations on this topic; an anonymous referee for their suggestions; and Antonio Vassallo for his boundless patience. Funding acknowledgement: I would like to thank the Volkswagen Foundation for its support in providing the funds to create the Lichtenberg Group for History and Philosophy of Physics at the University of Bonn.

Notes

1. Strictly speaking: subextremal part of the family, as discussed in section 5.2. Partly to simplify the overall discussion, and partly because many results focus on a simpler yet still very hard case of Kerr, I will also tentatively follow the common working astrophysicist practice and assume that black holes have approximately net zero electromagnetic charge, thus ignoring the possibility that astrophysical black holes are a part of a more general Kerr-Newman family of solutions. See, however, Zajaček and Tursunov (2019) for a recent critical discussion of this assumption.

2. Spacetime is called globally hyperbolic if it has a Cauchy surface: $n-1$ dimensional (where n is the dimensionality of the spacetime) hypersurface Σ intersected exactly one by every inextendible causal curve. Equivalently, there is an achronal closed subset S whose domain of dependence (the set of points such that every inextendible causal curve passing through that point intersects S) is the entire spacetime. Here, the whole of Kerr is not globally hyperbolic, but there are partial Cauchy surfaces in various subregions. See Manchak (2020a) for an introduction to these notions.

3. Which is defined as the boundary of the domain of dependence. A physical situation at a partial Cauchy surface determines physical situation everywhere in its domain of dependence, but not necessarily everywhere, because its domain of dependence may be further extended if a Cauchy horizon forms. Here the Cauchy horizon is a boundary of the region I+II; sometimes it is also called an inner horizon. Note that typically extensions through Cauchy horizons are highly non-unique. In the Kerr spacetime there is a unique extension in region III only under the assumption that the extension is analytic and has the same symmetries as regions I+II.

4. Note—following Romero (2021)—that if a black hole was not asymptotically flat but rather asymptotically de Sitter, one still might have failure of determinism and a form of a time machine, but the Malament-Hogarth property is lost, in the sense that any computer placed on a geodesic in the exterior will end up causally separated from the interior, and hence cannot be used as a hypercomputational resource. So although Kerr black holes can be both time machines and Malament-Hogarth machines, they seem to be more robust as the former. (However, a discussion of the Potency Condition in the very next section complicates this nice picture.) Change of asymptotics will also be of relevance in section 5.4.

5. These "holes" refer to methods of constructing new spacetimes by removing closed subsets from a given spacetime, and not to the (in)famous Hole Argument, or black holes, which are characterized by event horizons (or their generalizations; see Curiel, 2019).

6. This, by the way, would have certain advantage over many other forms of "machines" (such as those discussed by Manchak (2019) or Manchak (2020b), or Deutsch-Polizer spacetimes), which seem to heavily rely on cut-and-paste constructions: properties of philosophical interest would be realized in a much more natural spacetime, and moreover one with all of the empirical support discussed in section 5.2.2.

7. One additional *caveat* should be mentioned here. My discussion so far assumed something that might be called a global, or 4-dimensional, or geometric view about general relativity. In particular, I have assumed that one can contemplate not only spacetimes which are not globally hyperbolic, but also make statements about their properties (such as causal properties of the extension). An alternative approach, sometimes called the Partial Differential Equation (PDE) view due to main mathematical tools it relies on,

focuses on dynamics and solving initial or characteristic initial value problem for the Einstein's field equations. This view can be summarized by two demands: that all sensible assumptions are about initial data, and all sensible questions concern their domains of dependence. (See Landsman, 2021 for an exposition of this view.) Under the PDE view only failure of determinism of section 5.2.3 is a genuine problem, since both the Malament-Hogarth property and the occurrence of time machines are phrased in terms of properties of non-globally hyperbolic extensions. But the interpretation of the PDE dictum is far from clear. Should it be understood as a fundamental statement about general relativity being a dynamical theory? Or is it rather a temporary mathematically convenient restriction (since there is very little control over wave equations outside their domains of dependence, solving those equations in such a situation is a rather hopeless task), and as such could be abandoned if a deeper understanding of the behavior of matter systems on non-globally hyperbolic background emerged? In any case, the PDE view also calls into question the significance of additional assumptions (analyticity and rotational symmetry) that ensure uniqueness of the extension of the Kerr spacetime and its features, such as CTCs. (Since these spacetime regions are outside of the domain of dependence of the initial data, why believe that these symmetries will continue to hold? Perhaps one could elevate the symmetries or a Potency Condition of section 5.3 to a metaphysical principle, and try to say something on that basis, but to me this like a rather risky move. The upshot is that philosophers interested in time machines in general relativity may have to somehow adjudicate between the global view and the PDE view.

8. Reissner-Nordström's (charged, spherically symmetric) black holes have similar causal structure to Kerr, so they are often used as a toy model for first tests of conjectures about Kerr black holes.

9. Here we encounter an interesting and as of yet unexplored by philosophers issue: a distinction is often made between Strong Cosmic Censorship: that domains of dependence of typical initial are C^2-inextendible, a Very Strong Cosmic Censorship: that said domains are C^0-inextendible, and intermediate statements, for example a demand that the spacetime is inextendible in such a way the the Christoffel's symbols are twice-differentiable. A discussion of these versions can be found in Landsman (2021). The novel philosophical problem concerns physical significance of the choice of regularity class of an extension for low regularity. Is there a single appropriate k to be demanded in C^k-extendibility of spacetime, and if so, what is it? Note also that although what is currently known (Ling, 2020) about low regularity causality theory has been developed with the aim of proving C^0-inextendibility of spacetime, it would be interesting to see whether it could be a promising way of expressing statements about causal structure of low regularity spacetimes, and potentially salvaging some claims about time machines in case it turned out that C^2 SCC holds, but C^0 SCC fails.

10. Rovelli considers such spacetimes problematic on the basis of paradoxes of time travel, but as I have gestured at in section 5.2.3, one might claim that on purely conceptual grounds they are not so problematic.

11. Interestingly, the second law could be satisfied in a number of other, causally better behaved, i.e., chronological, extensions of the Misner spacetime, ones which presumably do not satisfy SEW Potency Condition. So it seems that, when faced with non-globally hyperbolic spacetime extensions, demanding validity of the second law of thermodynamics could be a physical principle responsible for producing what Manchak (2014) time and Manchak (2020b) called a "hole machine." To see that, the reader is invited to compare chronological extension of Misner spacetime in Manchak (2019) (that clearly can satisfy the second law) with basically the same example

used by Earman et al. (2009) as a motivation for the Potency Condition (where the extension favored by the Potency Condition clearly can satisfy the second law only in the trivial sense, as an equality). In short, validity of the second law would produce holes by excluding non-chronological extensions from the set of all extensions of that spacetime. Note that two other remarkable features we have seen, lack of determinism and the Malament-Hogarth property, are compatible with the SL (because there are CTC-free Malament-Hogarth extensions).

12. The minimum total integrated acceleration required for entering a CTC has been investigated; see Malament (1985, 1987); Manchak (2011); Natàrio (2012); see also Buser et al. (2013) for a visualization of this spacetime. As a matter of fact, this is also true for CTCs in the Kerr spacetime; see Wüthrich (1999) for a calculation.

13. Of course in a general relativistic setting the answer has to be positive, because the notion of a gravitationally isolated system is spelled out by the global property of asymptotic flatness. But the issue is whether the black hole is an isolated system in the same sense as isolated thermodynamical systems. Asymptotic flatness does not seem to do that: it only provides a sense in which some system is isolated from gravitational interactions. Nothing is said about other interactions, and if their gravitational effect is negligible, there is plenty of room for them in the exterior regions. In addition, in the context of the thermodynamical argument against time travel to the past, such an identification would not be fortuitous, as one would like to apply the argument to asymptotically flat Kerr as well as Kerr-de Sitter black holes. So in this context asymptotic flatness should not be treated as equivalent to being thermodynamically isolated.

14. Note that this argument would not help with the Misner spacetime, where these features do not occur, so cannot serve as entropy sinks.

15. An interesting result important in this context is a no go result for time machines using a Generalized Second Law of (horizon) thermodynamics due to Wall (2013). This is a generalized. This is a generalization of Hawking (1992) result which replaces an energy condition with a Generalized Second Law. As such, it only applies to the Kerr case modulo issues concerning compactness of section 5.3.

16. See Cuffaro and Hartmann (2021) for a recent appreciation of the open system view.

Bibliography

Abbott, B., R. Abbott, T. Abbott, M. Abernathy, F. Acernese, K. Ackley, . . ., and Virgo Collaboration (2016). Observation of gravitational waves from a binary black hole merger. *Physical Review Letters* 116 (6), 061102.

Abuter, R., A. Amorim, N. Anugu, M. Bauböck, M. Benisty, J. Berger, . . ., and G. Zins (2018a). Detection of the gravitational redshift in the orbit of the star S2 near the galactic centre massive black hole. *Astronomy & Astrophysics* 615, L15.

Abuter, R., A. Amorim, M. Bauböck, J. Berger, H. Bonnet, W. Brandner, . . ., and S. Yazici (2018b). Detection of orbital motions near the last stable circular orbit of the massive black hole SgrA. *Astronomy & Astrophysics* 618, L10.

Andréka, H., J. X. Madarász, I. Németi, P. Németi, and G. Székely (2018). Relativistic computation. In M. E. Cuffaro and S. C. Fletcher (Eds.),

Physical Perspectives on Computation, Computational Perspectives on Physics, pp. 195–216. Cambridge University Press.

Buser, M., E. Kajari, and W. P. Schleich (2013). Visualization of the Gödel universe. *New Journal of Physics* 15 (1), 013063.

Cardoso, V., J. L. Costa, K. Destounis, P. Hintz, and A. Jansen (2018). Quasinormal modes and strong cosmic censorship. *Physical Review Letters* 120 (3), 031103.

Cardoso, V. and P. Pani (2019). Testing the nature of dark compact objects: A status report. *Living Reviews in Relativity* 22 (1), 1–104.

Chesler, P. M., E. Curiel, and R. Narayan (2019). Numerical evolution of shocks in the interior of Kerr black holes. *Physical Review D* 99 (8), 084033.

Chesler, P. M., R. Narayan, and E. Curiel (2019). Singularities in Reissner–Nordström black holes. *Classical and Quantum Gravity* 37 (2), 025009.

Cuffaro, M. E. and S. Hartmann (2021). The open systems view. *arXiv:2112. 11095* [physics.hist-ph].

Curiel, E. (2014a). Classical black holes are hot. *arXiv:1408.3691* [gr-qc].

Curiel, E. (2014b). Measure, topology and probabilistic reasoning in cosmology. *arXiv:1509.01878* [gr-qc].

Curiel, E. (2019). The many definitions of a black hole. *Nature Astronomy* 3 (1), 27.

Dafermos, M. and Y. Shlapentokh-Rothman (2018). Rough initial data and the strength of the blue-shift instability on cosmological black holes with $\Lambda > 0$. *Classical and Quantum Gravity* 35 (19), 195010.

Dias, O. J., F. C. Eperon, H. S. Reall, and J. E. Santos (2018). Strong cosmic censorship in de sitter space. *Physical Review D* 97 (10), 104060.

Dias, O. J., H. S. Reall, and J. E. Santos (2018). Strong cosmic censorship: Taking the rough with the smooth. *Journal of High Energy Physics* 2018 (10), 1–54.

Dias, O. J., H. S. Reall, and J. E. Santos (2019). The BTZ black hole violates strong cosmic censorship. *Journal of High Energy Physics* 2019 (12), 1–59.

Doboszewski, J. (2019a). Interpreting cosmic no hair theorems: Is fatalism about the far future of expanding cosmological models unavoidable? *Studies in History and Philosophy of Modern Physics* 66, 170–179.

Doboszewski, J. (2019b). Relativistic spacetimes and definitions of determinism. *European Journal for Philosophy of Science* 9 (2).

Doboszewski, J. (2022). *When is a Black Hole Spacetime as "Large as It Can Be"*? Under Preparation.

Earman, J. (1995). *Bangs, Crunches, Whimpers, and Shrieks: Singularities and Acausalities in Relativistic Spacetimes*. Oxford University Press.

Earman, J. and J. Manchak (2016). Time machines. In E. N. Zalta (Ed.), *The Stanford Encyclopedia of Philosophy* (Winter 2016 ed.). Metaphysics Research Lab, Stanford University. https://plato.stanford.edu/archives/ win2016/entries/ time-machine/.

Earman, J. and J. D. Norton (1993). Forever is a day: Supertasks in Pitowsky and Malament-Hogarth spacetimes. *Philosophy of Science* 60 (1), 22–42.

Earman, J., C. Smeenk, and C. Wüthrich (2009). Do the laws of physics forbid the operation of time machines? *Synthese* 169 (1), 91–124.

Eckart, A., A. Hüttemann, C. Kiefer, S. Britzen, M. Zajaček, C. Lämmerzahl, M. Stöckler, M. Valencia-S, V. Karas, and M. García-Marín (2017). The Milky

Way's supermassive black hole: How good a case is it? *Foundations of Physics* 47 (5), 553–624.

Effingham, N. (2020). *Time Travel: Probability and Impossibility*. Oxford University Press.

Ellis, G. F. (2002). Cosmology and local physics. *New Astronomy Reviews* 46 (11), 645–657.

The Event Horizon Telescope Collaboration, K. Akiyama, A. Alberdi, W. Alef, K. Asada, R. Azulay, . . ., and P. Yamaguchi (2019b). First M87 Event Horizon Telescope results. VI. The shadow and mass of the central black hole. *The Astrophysical Journal Letters* 875 (1), L6.

The Event Horizon Telescope Collaboration, K. Akiyama, A. Alberdi, W. Alef, K. Asada, R. Azulay, . . ., and L. Ziurys (2019a). First M87 Event Horizon Telescope results. I. The shadow of the supermassive black hole. *The Astrophysical Journal Letters* 875 (1), L1.

Fletcher, S. (2016). Similarity, topology, and physical significance in relativity theory. *The British Journal for the Philosophy of Science* 67 (2), 365–389.

Friedman, J., M. S. Morris, I. D. Novikov, F. Echeverria, G. Klinkhammer, K. S. Thorne, and U. Yurtsever (1990). Cauchy problem in spacetimes with closed timelike curves. *Physical Review D* 42 (6), 1915.

Hawking, S. W. (1992). Chronology protection conjecture. *Physical Review D* 46 (2), 603.

Hollands, S., R. M. Wald, and J. Zahn (2020). Quantum instability of the Cauchy horizon in Reissner–Nordström–deSitter spacetime. *Classical and Quantum Gravity* 37 (11), 115009.

Isi, M., M. Giesler, W. M. Farr, M. A. Scheel, and S. A. Teukolsky (2019, September). Testing the no-hair theorem with GW150914. *Physical Review Letters* 123, 111102.

Krasnikov, S. (2002). No time machines in classical general relativity. *Classical and Quantum Gravity* 19 (15), 4109.

Krasnikov, S. (2018). *Back-in-time and Faster-than-light Travel in General Relativity*. Springer.

Landsman, K. (2021). Singularities, black holes, and cosmic censorship: A tribute to Roger Penrose. *Foundations of Physics* 51 (2), 1–38.

Lewis, D. (1976). The paradoxes of time travel. *American Philosophical Quarterly* 13 (2), 145–152.

Ling, E. (2020). Aspects of C^0 causal theory. *General Relativity and Gravitation* 52, 1–40.

Luminet, J.-P. (2021). Closed timelike curves, singularities and causality: A survey from Gödel to chronological protection. *Universe* 7 (1).

Luna, R., M. Zilhão, V. Cardoso, J. L. Costa, and J. Natário (2021). Addendum to 'strong cosmic censorship: The nonlinear story'. *Physical Review D* 103 (10), 104043.

Malament, D. B. (1985). Minimal acceleration requirements for 'time travel' in Gödel space-time. *Journal of Mathematical Physics* 26 (4), 774–777.

Malament, D. B. (1987). A note about closed timelike curves in Gödel spacetime. *Journal of Mathematical Physics* 28 (10), 2427–2430.

Manchak, J. B. (2009). On the existence of time machines in general relativity. *Philosophy of Science* 76 (5), 1020–1026.

Manchak, J. B. (2011). On efficient 'time travel' in Gödel spacetime. *General Relativity and Gravitation* 43 (1), 51–60.

Manchak, J. B. (2014). Time (hole?) machines. *Studies in History and Philosophy of Modern Physics* 48, 124–127.

Manchak, J. B. (2019). A remark on 'time machines' in honor of Howard Stein. *Studies in History and Philosophy of Modern Physics* 67.

Manchak, J. B. (2020a). *Global Spacetime Structure*. Cambridge University Press.

Manchak, J. B. (2020b). Malament–Hogarth machines. *The British Journal for the Philosophy of Science* 71 (3), 1143–1153.

Natário, J. (2012). Optimal time travel in the Gödel universe. *General Relativity and Gravitation* 44 (4), 855–874.

O'Neill, B. (2014). *The Geometry of Kerr Black Holes*. Dover Publications.

Penrose, R. (1979). Singularities and time-asymmetry. In S. Hawking and W. Israel (Eds.), *General Relativity: An Einstein Centenary Survey*, pp. 581–638. Cambridge University Press.

Pitowsky, I. (1990). The physical church thesis and physical computational complexity. *Iyyun: The Jerusalem Philosophical Quarterly*, 81–99.

Prunkl, C. E. and C. G. Timpson (2019). Black hole entropy is thermodynamic entropy. *arXiv:1903.06276* [physics.hist-ph].

Romero, G. E. (2021). Black hole philosophy. Crítica. *Revista Hispanoamericana de Filosofía* 53 (159), 73–132.

Rovelli, C. (2019). Can we travel to the past? Irreversible physics along closed timelike curves. *arXiv:1912.04702* [gr-qc].

Smeenk, C. and C. Wüthrich (2021). Determinism and general relativity. *Philosophy of Science* 88 (4), 638–664.

Tipler, F. J. (1976). Causality violation in asymptotically flat space-times. *Physical Review Letters* 37 (14), 879.

Wall, A. C. (2013). The generalized second law implies a quantum singularity theorem. *Classical and Quantum Gravity* 30 (16), 165003.

Wallace, D. (2010). Gravity, entropy, and cosmology: In search of clarity. *The British Journal for the Philosophy of Science* 61 (3), 513–540.

Wallace, D. (2018). The case for black hole thermodynamics part i: Phenomenological thermodynamics. *Studies in History and Philosophy of Modern Physics* 64, 52–67.

Wallace, D. (2019). The case for black hole thermodynamics part ii: Statistical mechanics. *Studies in History and Philosophy of Modern Physics* 66, 103–117.

Wasserman, R. (2017). *Paradoxes of Time Travel*. Oxford University Press.

Wiltshire, D. L., M. Visser, and S. M. Scott (2009). *The Kerr Spacetime: Rotating Black Holes in General Relativity*. Cambridge University Press.

Wüthrich, C. (1999). *On Time Machines in Kerr-Newman Spacetime*. Unpublished MSc thesis, University of Bern.

Wüthrich, C. (2021). Time travelling in emergent spacetime. In J. Madarász and G. Székely (Eds.), *Hajnal Andréka and István Németi on Unity of Science*, pp. 453–474. Springer.

Zajaček, M. and A. Tursunov (2019). The electric charge of black holes: Is it really always negligible? *The Observatory* 139, 231–236.

Part II
Quantum gravity

6 Spacetime quietism in quantum gravity

Sam Baron and Baptiste Le Bihan

6.1 Introduction

Recent philosophical work on quantum gravity has focused on the notion of spacetime emergence: the idea that spacetime, as found in our most fundamental and successful theories at the moment—*general relativity* and *quantum field theory*—arises at some non-fundamental level from a more fundamental, non-spatiotemporal, reality (see e.g., Butterfield and Isham, 2001, Crowther, 2018, Huggett and Wüthrich, forthcoming). The fundamental level is expected to be the purview of a theory of quantum gravity, such as *string theory, loop quantum gravity* or *causal set theory*, which have all been claimed to question the very existence of a fundamental spacetime.

The focus on spacetime emergence has had an invigorating effect on existing philosophical disputes concerning the nature of spacetime. In order to say that spacetime arises at some non-fundamental level, and in order to identify some fundamental level as spacetime-free, it seems we first need an account of what spacetime is. Accordingly, the question 'what is spacetime?' has been tackled as one aspect of the broader quest to understand the many approaches to quantum gravity currently being developed within physics.

We believe however that the focus on *spacetime* emergence in this context, while understandable, is misguided. It would be better to avoid entirely the question of what spacetime is. In order to make this case, we develop two tranches of argument. First, we show that there is substantial disagreement about what spacetime is, and not much hope of sorting it out. Second, we argue that the philosophical issues arising from the quest to understand quantum gravity that seem to suggest the need for an account of what spacetime is can be addressed without one.

The chapter is structured as follows. We begin by situating the various accounts of spacetime that have been offered to date within a broad framework (§6.2–6.3). We then discuss what to do in light of the

DOI: 10.4324/9781003219019-9

many accounts of spacetime that there are, and conclude that a form of pluralism is the best bet (§6.4). However, we then show that in order to answer one of the most discussed issues in the philosophical literature on quantum gravity—namely, the problem of empirical incoherence—we don't need an understanding of the nature of spacetime that goes beyond this pluralistic outlook. We can get by, instead, with an account of how various features commonly associated with, but potentially independent of, spacetime arise (§6.5). We close with concluding remarks on related issues which have been raised in the literature (§6.6).

In the end, we propose a philosophical approach to quantum gravity that is quite close to the spacetime functionalist positions we review in §6.2. However, rather than focusing on the identification of a functional role for spacetime (and then checking to see whether that role is filled in the context of a more fundamental theory) we recommend identifying theoretical functions or explanatory roles that must be performed if we are to have a viable physics. The right approach to quantum gravity emphasizes the functional roles of observer and observable, rather than functionalism as a metaphysical thesis about the nature of spacetime.

6.2   Two approaches to spacetime

Current discussion of spacetime seems to centre around two approaches to understanding spacetime, what we call the 'narrow' approach and the 'broad' approach respectively. To a first approximation, the narrow approach to spacetime anchors spacetime to a particular theory, or theoretical framework: most obviously special relativity and general relativity, but also other theories with a special relation to relativistic physics, for instance *Newton-Cartan theory* or *string theory*. The broad approach to spacetime, by contrast, does not. Rather, spacetime is conceived of in a more or less theory-independent manner. These different approaches serve to identify a specific concept that we can then subject to further analysis. The approaches themselves do not constitute an analysis of that concept, however, and nor do they constitute a complete account of what spacetime is. We can sharpen up this initial characterization as follows:

The Narrow Approach: Provide an account of spacetime, where spacetime is something that is only found within the ontology of general relativity, and within theories that approximate general relativity sufficiently well.

The Broad Approach: Provide an account of spacetime, where spacetime is something that is found within the ontology of general relativity and within theories that approximate general relativity sufficiently well, but not only: it is found in the ontology of theories that don't approximate general relativity.[1]

To give this distinction some substance, it is important to say a bit about what it means for something to appear within the ontology of a theory, as well as what it takes to approximate a theory. With respect to the former, we take appearance in the ontology of a theory to be a matter of commitment. Spacetime is one of the ontological commitments of general relativity, which is to say that if one adopts a realist attitude toward that theory (whatever that might mean in the end) then one should accept the existence of spacetime. The narrow conception thus dictates that belief in the existence of spacetime is only warranted for a specific cluster of theories: namely general relativity, and those theories that can approximate it. The broad conception, by contrast, allows that one may be warranted in believing spacetime exists even if one does not adopt a realist attitude toward general relativity (but instead, adopts a realist attitude toward some other theory that brings with it an ontological commitment to spacetime, in a broad sense).

Approximation is understood in the sense used by physicists (as characterised by Butterfield and Isham, 1999, 2001): a theory T approximates a theory T^* when, by neglecting certain degrees of freedom or certain quantities in T, T and T^* numerically agree with only a very small margin of error. So, for instance, general relativity will numerically agree with Newtonian mechanics for a group of particles moving slowly (relative to the speed of light), in a weak, unchanging gravitational field. In this sense general relativity approximates a Newtonian theory. Similarly, a theory may approximate general relativity, when certain relativistic degrees of freedom are factored out in an analogous manner.

We take approximation to be a matter of degree: one theory T can be a better or worse approximation to a theory T^* compared to a third theory T'. In some sense, just about any two theories can stand in an approximation relation so long as we are willing to neglect enough degrees of freedom, or factor out specific quantities. That is why, for the narrow conception of spacetime, a certain threshold for approximation must be passed by a theory T with respect to general relativity before we can say that something within T qualifies as spacetime. We can understand this threshold in terms of empirical equivalence: a theory T approximates general relativity well enough for the ontology of T to satisfy the narrow conception of spacetime when every confirmed empirical prediction of general relativity is a confirmed empirical prediction of T as well.

The distinction between the broad and narrow approaches can be further developed by comparing two accounts of what spacetime is that fall on either side of the divide. Consider, first, the *functionalist conception of spacetime* offered by Yates (2021) and endorsed by Chalmers

(2021). According to this conception, spacetime is any *n*-tuple in any theory that satisfies the Ramsey sentence for general relativity either exactly, or well-enough. A Ramsey sentence for general relativity is, presumably, produced in the usual way. We start by treating 'spacetime' as a theoretical term. We then formulate a sentence that defines spacetime via a range of predicates within the context of general relativity. The Ramsey sentence is then produced by stripping out any mention of spacetime, and replacing it with a variable that is bound within the scope of one or more quantifiers.

In order to work out whether spacetime appears in another theory, we check to see whether something satisfies the Ramsey sentence. If something does satisfy it, well and good. If nothing satisfies the Ramsey sentence exactly, but the theory nonetheless manages to approximate general relativity in the manner described earlier (or perhaps, in some other way if a different approach to approximation is favoured), then the sentence is satisfied well enough. For theories that don't feature anything that satisfies the Ramsey sentence well-enough, spacetime does not make an appearance. In this way spacetime is narrowly construed since it is unlikely to feature in any theories that don't have a great deal in common with general relativity. Newtonian theories, for instance, are unlikely to feature spacetime in the narrow sense.[2]

Contrast this Ramsey-style functionalism with Knox's *inertial frame functionalism*. Knox (2019) builds on Brown's dynamical view of Minkowski spacetime. Brown's view supports two quite natural metaphysical pictures: a new form of relationism about spacetime, and a form of spacetime functionalism. Knox (2019) takes the second road and views spacetime as the theoretical entity that plays a certain role in defining inertial frames within any theory. These inertial frames are defined in turn in terms of preferred coordinates that maximally simplify, locally, the laws associated to all the non-gravitational interactions (Brown, 2005, p. 169). More intuitively, what this means is that the distinction between inertial and accelerated bodies is not a matter of relations between bodies and a background structure, or even a matter of relations between the bodies themselves, but rather of symmetries in the laws that dictate how the bodies move.

For Knox, then, any theory that features something within its ontology that satisfies the minimal condition she specifies will feature spacetime. This condition is satisfied within general relativity but, importantly, it is also satisfied within many theories that neither feature something that satisfies the Ramsey sentence for general relativity, nor approximate general relativity more generally. Knox's conception of spacetime is thus quite likely to be satisfied in Newtonian theories. Indeed, it is likely to be satisfied in a very wide range of theories.

We have used the contrast between two versions of spacetime functionalism to illustrate the distinction between the broad and narrow approaches to spacetime. This, however, should not be read as implying that the distinction itself is a distinction internal to a functionalist approach to spacetime. Rather, the question of whether spacetime is a functional kind (or not) cross-cuts the broad/narrow distinction. All that the broad/narrow distinction aims to capture is a difference in the theoretical anchorage for spacetime. Various different debates about what spacetime is can then be formulated on the basis of either the broad or the narrow conception.

Consider, for instance, the debate between relationism and substantivalism about spacetime, that can be traced back to discussions of Newtonian mechanics by Leibniz, Newton and Clark. Substantivalism is the view that spacetime is a substance on its own, which is not essentially connected to material objects. Relationism is the view that spacetime is a network of relations between material entities that could not exist in the absence of material entities and hence, is not fundamental. Of course, what this means exactly will depend on what one takes a substance to be and it is hard to find a consensus on how the two positions should exactly be characterised.

What we want to emphasize here is that the debate over whether relationism or substantivalism is true can be had in two ways. The first—and we take it, standard—way of having the debate involves thinking of spacetime along the lines of the narrow approach. Thus, the goal is to determine whether spacetime in general relativity is to be identified with a network of relations between material objects located in space and time or whether—if we take seriously its mathematical representation—an ontology of three structures: a *manifold* of points, the *metric field* and the *matter fields*. Without going into the details (cf. Bigaj's chapter in this volume for a more detailed discussion), some have suggested identifying spacetime to the manifold of points (a view sometimes called *manifold substantivalism* or *sophisticated substantivalism* in its more recent and refined version, see e.g., Teitel, 2022), others to the metric field, a view naturally dubbed *metric field substantivalism* (Hoefer, 1996).

The second way of conducting the relationist/substantivalist debate focuses on the notion of spacetime at issue in the broad approach. In this case, it may pay to work out what spacetime in general relativity is (since that may help to determine the nature of spacetime more generally), but that is certainly not the only way, and may be potentially misleading insofar as one ends up focusing on features that are peculiar to the relativistic context. Under a broad conception of spacetime, the question is whether the notion of spacetime focused on within the broad approach is relationist or substantivalist. Thus, if one thinks that the ontology of a Newtonian theory counts as spacetime, then one can discuss relationist and substantivalist approaches to spacetime in Newtonian mechanics and relativistic physics in the same breath.

Other metaphysical approaches to spacetime cross-cut the broad/ narrow divide in the same way. For instance, there is the mereological approach to spacetime (originally advocated by Paul, 2012, and more recently by Le Bihan, 2018a, 2018b and Baron and Le Bihan, 2022 in the context of quantum gravity). On this view, spacetime—whatever it is—is mereologically composed of parts. One way of developing this view is to treat spacetime as something that can be composed of non-spatiotemporal parts (Paul, 2012, pp. 242–242, for instance, proposes to build spacetime from bundles of properties, including location properties, which need not be spatiotemporal locations). This view can be held about spacetime in the broad sense, in which case anything that counts as spacetime broadly (perhaps by virtue of providing a structure for inertial frames) may be subject to a mereological analysis. Equally, however, the mereological approach can be held about spacetime in the narrow sense. Thus, one might hold that it is only spacetime anchored to general relativity that is to be treated mereologically, leaving it open as to whether other structures that might be spacetime in a broader sense are to receive a mereological treatment.

No doubt there are other accounts that one might offer concerning the nature of spacetime. It should be clear, however, from what we've said that such accounts are likely to work for broad or narrow conceptions of spacetime, and thus that within each way of thinking about what spacetime is, there is a substantive debate to be had about its nature.

6.3   Sources of disagreement

So far we have focused on the distinction between the broad and narrow approaches to spacetime, and on cross-cutting accounts of what spacetime is given this distinction. We want to now draw attention to two sources of disagreement. First, one might disagree about whether the broad approach to spacetime, or the narrow approach is the "correct" approach to understanding spacetime. Thus, one might argue that the broad approach is too broad, or the narrow approach too narrow. Or one might argue that one or other of these approaches has pride of place in developing a metaphysical or physical understanding of spacetime, and the other is to be discarded. Call the question of which approach to understanding spacetime is to be preferred (if any) the *indexing question*, since what's at stake is whether we should prefer an approach that treats spacetime as something that is essentially indexed to relativistic physics.

Second, one can also disagree about what the "correct" account of spacetime might be assuming either the broad or narrow approaches. Call this the *nature question*, since it is really the question of what the nature of spacetime might be. With respect to the narrow approach, one might debate about what spacetime within general relativity is.

The relationist/substantivalist debate is one example of a debate along these lines, as is the question of whether the causal theory of spacetime is true (which, arguably, is a question internal to a relationist program).

Similarly, one might debate about the correct account of spacetime under the broad approach. Knox (2019), Read and Menon (2021) and Baker (2021) all seem to engage in a debate along these lines. None of them are interested in a narrow approach to spacetime. Instead, they argue that spacetime in some sense that goes beyond general relativity is a structure of inertial frames (Knox); that which is measured by rods and clocks (Read and Menon) or a cluster of different notions, including a fundamentality constraint to the effect that the most fundamental things are the best candidates to be spacetimes (Baker).

In the process of providing an account of spacetime emergence in the context of quantum gravity, philosophers end up answering both the indexing and nature questions. For instance, Lam and Wüthrich (2018, 2021) advocate for a functionalist approach to spacetime as a philosophical approach to a number of programs in quantum gravity. In so doing, they seem to take a stand on both the indexing and nature questions. They tie their analysis closely to general relativity, and so seem to adopt the narrow approach. Their stated goal is to show how realism about spacetime in the context of general relativity is compatible with the emergence of spacetime from a fundamental, non-spatiotemporal structure. They deliberately set aside Knox's inertial frame functionalism, because it is a departure from the narrow conception. Additionally, they offer a functional analysis of what spacetime in the narrow sense is, thereby providing an answer to the nature question.

We are not particularly interested in criticizing Lam and Wüthrich's approach and others like it. Nor do we think that Lam and Wüthrich have erred in answering both the indexing and nature questions. Given their starting assumption, namely that spacetime is an emergent entity, it does seem that an answer to both questions is required, on pain of rendering the emergence of spacetime metaphysically obscure. Rather, what we are interested in is whether the starting assumption is a good jumping-off point for thinking about quantum gravity. Should we work within a framework of spacetime emergence at all?

6.4 Spacetime scepticism

Our goal in this section is to defend a sceptical approach toward questions of spacetime emergence. Simply put, we don't think that the prospects for answering the indexing or nature questions are very good. Since answering these questions would seem to be a necessary step toward giving an account of spacetime emergence, spacetime emergence is thus shown to be a troubled notion.

Let's start with the indexing question: the question of whether a broad or a narrow approach to understanding spacetime is the correct approach. In order to show that one of these approaches is "correct," one would need to argue that the concept of spacetime is, in fact, essentially indexed to relativistic physics, or that it is not. It is, however, very difficult to see how one might make the case that the concept of spacetime has any specific features. It is, of course, possible that there is some other way to defend an answer to the indexing question, and we invite suggestions for arguments along these lines.

First, one might take a historical approach. The idea here is that "spacetime" was introduced in a very specific context, namely as a way to interpret relativistic physics. Accordingly, the way spacetime ought to be understood is anchored to its initial introduction into the world of physics. A view along these lines is suggested by Chalmers, who writes:

> Spacetime as understood here is an essentially theoretical concept, one that emerges especially from the general theory of relativity.
> (Chalmers, 2021, p. 172)

Although Chalmers does not argue for the broad or narrow approaches, what he says can be formed into a historical argument in this direction. The basic idea is that the way the notion was historically introduced fully justifies a narrow conception of spacetime. In so far as there might be a notion of spacetime that is detachable from relativistic physics, that notion is, strictly speaking, a perversion of the core concept.

We don't think this historical argument carries much weight. Perhaps spacetime was introduced in the context of relativity originally, but concepts shift and change during their lives within science. The concept of mass, for instance, has changed in important ways, since its original introduction with Newtonian mechanics. We see no reason to suppose that this couldn't be true for spacetime as well.

A second approach to the indexing question appeals instead to contemporary physics. If we look into current physics, one might argue, we will find conclusive evidence for exactly one of the broad or narrow approaches to understanding spacetime being the correct approach. What would this look like exactly? Well, when we look into physics as currently practised, we will see that the notion of spacetime being used is, in every case, one that is linked to general relativity in the manner demanded by the narrow conception. Alternatively, we may look and see that the notion of spacetime is used much more liberally than that, thereby signalling that the broad conception of spacetime is correct. Either way, the indexing question gets its answer.

Full disclosure: we have not conducted a full and detailed analysis of the way in which spacetime is discussed within physics. By the same token, we know of no such analysis. That limits the extent to which we can evaluate this way of answering the external question. But this limitation cuts both ways: without a detailed analysis of how physicists think of spacetime, one can't argue in favour of any particular monistic answer to the external question either.

We also think it would be shocking if all physicists think about spacetime in the same way. We suspect that some physicists will see spacetime as anchored to general relativity, as per the narrow conception, while others will take a more liberal approach, as per the broad conception. Variation in how physicists talk about spacetime underdetermines whether they all adopt the broad conception, or whether only some do. There would, presumably, be a very similar pattern of responses either way. In short, there is little reason to suppose that the disagreement around what spacetime is, as captured by the distinction between the narrow and broad approaches to understanding spacetime is localized to philosophy only. Indeed, philosophers are, in part, responding to the way that physicists are conceiving of, and using, spacetime. So we don't see why the disagreement over the indexing question wouldn't find a home inside physics as well. The difference being, perhaps, that physicists don't really care about the philosophical issues at stake, and so are happy enough to live with the disagreement unresolved.

A third way to answer the indexing question is to argue that either the broad or narrow approaches to understanding spacetime line up with some "pre-theoretical" notion. Thus, rather than focusing on what physicists think about spacetime, we might instead focus on what the "folk" think. However, we take it to be fairly clear that the way the folk think about spacetime is a poor guide to what spacetime is, or how we should think about it. Absent some influence from the physics, the folk are unlikely to have any concept of spacetime that differs from a simple conjunction of the concepts of space and time. Thus, assessing folk semantic intuitions about spacetime will require preliminary work to introduce some concept of spacetime—which certainly counts as a form of theoretical contamination undermining any hope to gain substantive insight this way. Thus, even though the folk might be a good guide to other concepts of philosophical interest that do not essentially depend on scientific activity (like pain or colour), they will be poor guides to theoretical concepts like spacetime.

A fourth answer to the indexing question might appeal to background metaphysical facts. Spacetime, one might argue, has certain essential features. The correct account of spacetime is the one that attributes all and only these features to spacetime. This might seem obviously circular in a bad way: in order to say what the features of spacetime are, we first need an answer to the nature question. However, answering the nature question presupposes that we first answer the indexing question, since one cannot settle on an account of the nature of spacetime without first settling the question of whether spacetime is essentially indexed to relativity. Perhaps, however, the circularity can be avoided, if we combine the current approach with something like the historical approach discussed earlier.

Here's the idea. The notion of spacetime has its roots in the work of Minkowski, who put forward spacetime as a natural geometrical interpretation of special relativity (Minkowski, 1908).[3] This spacetime is four-dimensional, meaning that four numbers are required to mathematically ascribe any position within the manifold. This four-dimensionalist picture differs from a 3+1 ontology that would take the four-dimensional manifold to be simply the addition of two manifolds: a 1-dimensional time and a 3-dimensional space. Central to the notion of spacetime thus, is the view that spacetime cannot, ontologically, be decomposed into two distinct structures: space *and* time. It is fairly standard to take special relativity to be essentially tied to Minkowski spacetime (but some have resisted the view and tried to develop an ontology of relative facts between three-dimensional frames of reference). Core to the notion of spacetime as Minkowski thought of it, then, is the entanglement of space and time into a single unified object.

Now, consider once again the point made earlier about conceptual change. It is standard for notions within science to shift over time. This, in turn, undermines any strong reason for taking the way spacetime is understood now to be determined by the historical introduction of that notion into physics. One might argue, however, that this is a bit quick. Sure, concepts change, but while concepts change they must also stay the same in order for us to be able to say that we are dealing with the same concept. So it could be argued that while the current notion of spacetime does not need to be the same as the notion that Minkowski introduced, it should be continuous with it in the following minimal sense: the two concepts should overlap in some way. Given that Minkowski took the weaving of space and time together to be a deep feature of spacetime, we might suppose that it is this feature that should survive, above all else, through conceptual shifts in our understanding of spacetime.

If that's right, then we can, at the very least, require that a reasonable account of spacetime should be one according to which space and time are woven together in a deep sense. In this way, one can provide a metaphysical demand on what a viable notion of spacetime should require, in a non-circular fashion. This is fine as far as it goes, but it doesn't go far enough. Even if the weaving of space and time together is a necessary feature of spacetime, that is still not enough information to answer the indexing question. For it is compatible with both the narrow and broad approaches to understanding spacetime, that spacetime features a woven aspect.

At this point we have run out of ideas concerning how the indexing question might be answered. None of the options seem very promising.

Suppose, however, that we do manage to settle on an answer to the indexing question. This at least gives us the capacity to begin answering the nature question, since we know whether spacetime is essentially linked to relativity or not.

The prospects for answering the nature question are better, but even here there are problems. The prospects are perhaps best for answering the nature question if the narrow approach to understanding spacetime is the correct approach. For in this case, one has the theoretical structure of general relativity to use as a basis for determining what spacetime is. Indeed, we hold out hope that the nature question for the narrow conception is answerable, and thus that we can (for instance) determine whether spacetime in general relativity is best captured as relationalist, substantival or something else, and whether it is to be understood in causal terms or not (and so on). Indeed, we don't see any problems for answering the nature question under a narrow approach to understanding spacetime that aren't just problems with doing the metaphysics of a specific physical theory more generally.

Matters are a bit less certain when it comes to the nature question for the broad conception of spacetime. Here we don't have the structure of general relativity to use as a guide, at least not in the same way. But then it becomes unclear how we might determine what the right notion of spacetime might be. Take the debate between Knox, on the one hand, and Read and Menon, on the other. To over-simplify the debate somewhat, Knox argues that spacetime is a structure of inertial frames. Read and Menon demur because, in some sense, inertial structure won't capture causal structure in the right way. But what, exactly, are the rules of the game here? How do we determine whether causal structure is an important feature of spacetime or not? Without a background theory like general relativity to constrain our metaphysical theorizing, it is unclear what considerations might speak for or against the relevant proposals.

Here, again, the community of physicists might play a role. Perhaps there is some use to which spacetime is put in physics such that capturing causal structure is necessary for capturing that use. Or perhaps there is some way that physicists talk about spacetime that would strongly suggest that inertial structure is both necessary and sufficient for spatiotemporal structure. However, the problems already identified for appealing to physics in this way when it comes to the indexing question are likely to re-emerge. What physicists say or do when it comes to spacetime may not be much of a guide to the nature of spacetime.

In the section thus far we have been arguing that the prospects for answering the indexing and nature questions are not good. It is thus difficult to see how one might settle the question of what the correct account of spacetime is. An obvious response to the difficulties we have highlighted so far, however, is to challenge the conceptual foundations of our argument. We have been assuming, rightly or wrongly, that there is just one correct conception of spacetime, and thus there is pressure to determine what that conception is in order to give substance to the notion of spacetime emergence. It is only given this assumption that a choice between the narrow and broad approaches is forced. For it is only if there is a single concept of spacetime that one of these approaches must be selected over the other. Similarly, it is only if there is a single concept of spacetime that we must settle on a particular account of the nature of spacetime.

In short, everything we've said so far assumes conceptual monism about spacetime. One can thus take our arguments as evidence against conceptual monism. There isn't just one concept of spacetime, there are multiple concepts. This enables one to simply reject the indexing question outright. There is no need to determine whether the broad or narrow approach is "correct," they are both correct approaches to distinction spacetime concepts, one that is essentially indexed spacetime and one that is not. One can also reject the nature question: we have lots of different accounts of what spacetime is, and they are all equally good. We don't need to privilege one as the correct account. In short, the pluralist simply says: let a thousand spatiotemporal flowers bloom.[4]

Ultimately, we don't think the shift to pluralism automatically acquits one of the need to say what spacetime is, and so the difficulties we have identified will arise anew. To see this, it is useful to separate pluralism into two forms: weak and strong. Both the strong and weak pluralist maintain that there are many different, viable notions of spacetime. The strong pluralist, however, adds to this that all notions of spacetime have something substantive in common such that they count as viable notions of spacetime in the first place. The weak pluralist, by contrast, denies this: it is not the case that all notions of spacetime have something substantive in common. Note that by "substantive" we mean: more than the label. Thus a non-substantive commonality between the various notions of spacetime is just that we use the term "spacetime" for each. A substantive commonality is some feature that is common to all notions of spacetime that qualify them as such.

Strong pluralism requires an account of what the feature might be that is held in common between differing conceptions of spacetime. Specifying what this feature is looks about as difficult as answering the indexing and nature questions. For it is unclear what considerations might be brought to bear that would allow us to determine even a single common feature. Indeed, it would seem that the kinds of strategies discussed that might be used to answer the indexing or nature questions are the very strategies one would have available, were one to try and settle on a core feature to be used as the basis for strong pluralism. For this reason, we don't see strong pluralism as offering much in terms of benefit over the monistic alternative.

The same issue does not arise for weak pluralism. According to the weak pluralist, the only thing that different conceptions of spacetime have in common is the name. Moreover, the way the label is used is just an accident, based on various historical contingencies concerning the development of physics, and perhaps philosophy and mathematics too. Weak pluralism is a view that does indeed avoid the need to answer the indexing and nature questions, and in a way that avoids having to specify a basis for strong pluralism.

Given how difficult the indexing and nature questions are to answer, we recommend abandoning them. One then has the option of adopting weak pluralism about spacetime, or something a bit more radical. One could, instead, seek to eradicate the term "spacetime" replacing it with more specific notions. We don't have a view on what the best option might be, and so we leave the matter open.

If we abandon the indexing and nature questions, then we are giving up on the quest to say what spacetime is (or what it is to be a spacetime, as under strong pluralism). This, in turn, renders questions about whether spacetime exists effectively unanswerable. We can't really say whether spacetime exists or not, because we don't have a stable conception of what it is. Giving up on the indexing and nature questions might thus seem like a mistake. In the context of quantum gravity, the existence of spacetime is linked to the viability of specific theories. Schematically, the idea is that there are certain features that a theory of quantum gravity must possess in order for that theory to yield a viable physics. Spacetime delivers those features in virtue of what it is. Thus, by demonstrating the emergence of spacetime from an underlying theory of quantum gravity, one is thereby able to show that the theory possesses the right features for viability.

Something of a dilemma is starting to form: either we find a way to answer the indexing and nature questions, or we abandon them. The first option, as we've argued, is difficult to make work; the second option threatens to leave us without a way to establish the viability of various approaches to quantum gravity. In the last section, we embrace the second horn of this dilemma and show that the viability of an approach to quantum gravity can be established without demonstrating the existence of spacetime. Once we see this we can also see that focusing too much on spacetime emergence in quantum gravity may in fact be counterproductive.

6.5 No need for spacetime

In this section our goal is to show that the existence of spacetime is not needed to establish the viability of a given approach to quantum gravity. Note that we are not arguing for the stronger view that spacetime does not exist. A view along those lines has been introduced by Le Bihan

(2018b) and defended by Baron (forthcoming).[5] Indeed, given that we think the internal and external questions about spacetime should be abandoned, spacetime eliminativism is the kind of view that is unavailable to us (given that it involves saying that spacetime does not exist and thus agreeing beforehand on what exactly reality is denied to). Rather, we are arguing that the viability of an approach to quantum gravity can be upheld on the basis of a kind of *quietism about spacetime*, whereby one refuses to say one way or another whether spacetime exists.

To show this, we will focus, in the first instance, on the so-called *problem of empirical incoherence* (Huggett and Wüthrich, 2013). We focus on this problem because we see it as being the main philosophical threat to the viability of a number of approaches to quantum gravity. The problem is typically posed as follows. A number of approaches to quantum gravity seem to lack spatiotemporal structure. However, spatiotemporal structure, one might argue, is necessary for empirical confirmation to occur. That's because empirical confirmation is always a matter of gathering observations of local beables. A local beable, however, is just an object that is localized in spacetime. If spacetime does not exist, then it seems nothing can be localized in the manner require for observation. From there it seems to follow that if a theory lacks spacetime, then it cannot support observation. Thus if a theory of quantum gravity that lacks spacetime is true, then the truth of that theory would seem to undermine any prospect for empirically confirming it. Conversely, any empirical evidence gathered for such a theory would demonstrate that the theory is false. As Barrett puts it, such a theory would be empirically incoherent (Barrett, 2001, pp. 116–117).

A theory of quantum gravity is thus viable when it is empirically coherent. A number of philosophers maintain that viability in this sense is secured with the existence of spacetime. For, one might argue, we know that the existence of spacetime would support the observation of local beables (indeed the problem is set up this way, more on that in a moment). Accordingly, if the existence of spacetime can be demonstrated, then it would follow that a theory of quantum gravity is empirically coherent and thus viable. The perhaps most straightforward way to understand the existence of spacetime in this context is to claim that spacetime does not exist fundamentally but it exists in a derivative way—just as table and chairs can be viewed as being less fundamental than the matter they are made of (Wüthrich, 2017, p. 298). In other words, one way to solve the problem of empirical incoherence is to develop a rich ontology with different levels of reality, the level of empirical confirmation of a theory being less fundamental than the level at which lives the ontology of the theory.

Clearly, we can't rely on the existence of spacetime in the same way. But we also don't think that there is any need to give up our quietist take on spacetime to solve the problem. The way in which the problem of empirical coherence is often set up can make it seem like this is not so. As noted, the problem is often set up in spatiotemporal terms, which can make it seem as though a spatiotemporal answer is required. But the appeal to spacetime in the statement of the problem is tendentious. To see this, we need to take a step back and consider what would in fact be sufficient for observation. Of course, exactly how observation works is a vexing issue, and not one that we can hope to fully address here. However, one thing seems fairly clear. If space and time exist then that seems sufficient for entities to be observable.

A similar point has been made in the context of a particular approach to quantum mechanics by Ney (2015). One interpretative issue with quantum mechanics is that the wave function used to describe physical systems is not defined on the ordinary three-dimensional space, but on a so-called configuration space from which the ordinary space is taken, one way or another, to emerge. Ney defends *wave function monism*, the view that reality is a universal wave function located in a physical configuration space, raising the question of how to interpret philosophically the emergence of space from the configuration space. In this context, Ney argues that what's at stake to ensure the coherence of our theory is not space, but the local observers and observables. Although our proposal belongs to the same family's as Ney approach, there are at less three important differences worth emphasizing.

First, the context of quantum mechanics differs greatly from the context of quantum gravity by focusing on space instead of spacetime. Wave function monism is formulated in the context of non-relativistic quantum mechanics, where time is considered classical, and not brought into the picture. This has consequences for the problem of empirical coherence, as the issue is not only to situate observers and observations in space, but also in time, with enough structure to account for the empirical process of theory confirmation via well-connected sequences of observations.

Second, Ney, like Huggett and Wüthrich (2013), denies the need to analyze evidence in terms of local beables. They point out that our theories should tell us what the evidence is; and this may not include local beables in the light of our best physics. If our best physical theories do not contain local beables, then we should not argue that evidence is constituted of local beables. We present a different picture: the existence of local beables can be preserved, in non-spatiotemporal contexts, at the cost of a slight tweak of the notion of local beable. This revision concerns the concept of locality: "local" can be redefined as a broader notion than "spatiotemporally local." Indeed, the idea is that the concept of locality can be analyzed functionally—or grounded—in non-spatiotemporal terms.

A third major difference between our proposal and Ney's is that her account denies that we should place pre-theoretical constraints on our theory of scientific confirmation. Our scientific theories describe to us how confirmation works, and this sometimes means that pre-theoretical notions involved in our naive theory of confirmation—such as space or the observer—have to be discarded. Our view instead allows a role for pre-theoretical constraints. In our view, we start with pre-theoretical constraints and move back and forth between our theory and our pre-theoretical notion of confirmation to bring the two into a reflexive equilibrium. So our scientific theories can change the way we conceptualize confirmation. But they cannot lead to a complete abandonment of all pre-theoretical constraints on the nature of confirmation. In our view, there must be such constraints, otherwise we would never be able to embark on the scientific enterprise. If we rely on theorising to tell us how confirmation works, and if we allow that in general, then we will never be able to theorize, because we will not be able to agree on a picture of confirmation before we do science.

This presents a straightforward approach to demonstrating the viability of a theory of quantum gravity without having to say anything about spacetime at all. Rather than focusing on spacetime we can, instead, focus on space and time and show how these aspects emerge from an underlying physics of quantum gravity. Granted, showing that our fundamental theories of quantum gravity support these notions is a non-trivial task. The point, however, is that we can abandon the internal and external questions about spacetime, and abandon entirely the question of whether spacetime exists and still secure the viability of our physical theories.

One might respond that the very reasons that drove us toward quietism about spacetime may drive us toward quietism about these other space and time as well. Take time, for instance. It is unclear what the "correct" account of time might be. So too for space, one might argue. If not being able to settle on the correct account of spacetime is enough to abandon questions about what spacetime is, then surely the same should be true for space and time as well. But if we abandon the questions of whether space and time exist, we won't be able to demonstrate the viability of theories of quantum gravity in the manner described here.

It may well be that we can't settle on the correct account of space and time either. And perhaps this does license a similar quietist attitude towards them. But even if we can't determine what the correct account of space or time might be, we can still solve the problem of empirical coherence. We can do this by taking a functionalist approach to observables and observers. The idea is that being an observable—being the kind of thing that can be observed—is to play a certain functional role. Similarly, being an observer—being the kind of thing that can gather an observations—is to play a distinct functional role. Together, observer and observable enter into a relationship, characterized by the interplay between the two functional specifications.

Having specified the functional roles of observable and observer, we can then look into a theory of quantum gravity and see whether it features any entities that satisfy the relevant functional roles. In this way we avoid the question of whether the underlying theories have space, time or, indeed, spacetime in them, replacing those questions with questions about whether the functional demands of observation are met. Of course, it might turn out that the functional roles for observable and observer have aspects that some are inclined to call spatial, temporal or spatiotemporal. Nothing we say is supposed to constitute a blanket ban on appealing to features that go by those labels. The point is just that we don't need to settle any deep questions about the nature of space, time or spacetime in order to establish the viability of a theory. We just need to look for structure enough for the demands of observation to be met, where those demands are given by the functional characterization of observer and observable.

An important question thus emerges: namely, what are the functional roles of observation and observable? Answering that question would take us too far afield. However, we will note that the answer is unlikely to come from science alone. That's because the two functional roles—of observer and observable—are roles that should, ideally, be specified independently of any particular scientific theory. For if the functional roles are specified in terms of a specific theory, then it will be unclear how to apply the notion of observation to that very theory in a noncircular fashion. This doesn't mean, of course, that science is irrelevant to the specification of the roles. Indeed, it is likely to be highly relevant: it is the preconditions for successful science that likely determine the functional roles for observable and observer. However, determining what those preconditions are involves a deep consideration of the foundations of science itself and, we suspect, that is a job for philosophers.

So far in this section we have argued that the problem of empirical coherence can be addressed in a manner that upholds quietism about spacetime. This, it will be recalled, is important, since empirical coherence is needed to establish the viability of a number of approaches to quantum gravity. One might argue, however, that empirical coherence is not the *only* dimension of viability to be taken into account.

Any theory of quantum gravity aims at accounting for the predictive success of our best physical theories at the moment. General relativity is especially important as it is our best theory of spacetime. It is likely that accounting for the predictive success of general relativity requires approximating general relativity in the manner described in §6.2. Is there not, then, some connection between the narrow conception of spacetime and the viability of quantum gravity? And if so, doesn't that demonstrate that the viability of a theory of quantum gravity can't be had without settling on a particular account of spacetime after all?

We don't see why that should be so. The approximation of general relativity generally involves finding a mathematical connection between a more fundamental theory of quantum gravity and general relativity. That mathematical connection can be found without settling either the internal or the external questions. That is, one need not take any view regarding what spacetime is in order for the approximation relation to be in good standing. So while it is true that approximating general relativity is a constraint on the viability of various approaches to quantum gravity, this provides no pressure at all to give up our quietist attitude toward the nature and existence of spacetime.

6.6 Conclusion

By way of concluding, we will briefly sketch a couple of advantages of the quietist approach to spacetime that we are recommending. First, it dissolves what has been called the ontological problem of spacetime (Le Bihan, 2021). The ontological problem consists in asking about the ontology behind spacetime emergence: is it a monist ontology, or a dualist one with the fundamental non-spatiotemporal entities on the one hand, and the non-fundamental spatiotemporal entities on the other hand, the latter being less fundamental than the former? The ontological problem is premised on the view that the non-fundamental ontology is spatiotemporal, a problematic fact to be explained further. If we set aside spacetime emergence, however, and approach quantum gravity without it, then there is no ontological problem to answer. The central ontological question worth asking is whether a theory contains enough in its ontology to support structures that play the functional roles of observable and observer.

Second, our approach dissolves the explanatory gap or hard problem of spacetime that has been introduced in analogy with the hard problem of consciousness (Le Bihan, 2021): there is something psychologically puzzling in the difference between spatiotemporal and non-spatiotemporal entities such that the notion of spacetime emergence from a non-spatiotemporal reality triggers a form of cognitive dissonance. This dissonance is analogous to the dissonance experienced when trying to bridge the explanatory gap between body and mind, though may not have quite the same philosophical implications.

By doing away with spacetime emergence entirely, the cognitive dissonance fades away and, with it, any threat of an explanatory gap. That being said, we may be left with other potential explanatory gap issues that have to do with the features needed for observation. For instance, it may be that specific spatial and temporal features are needed for something to play the beable role. If those features are not in the fundamental ontology of a theory of quantum gravity, then some account of their emergence may be needed. Thus, the explanatory gap issue could be reframed as an issue about how to relate space and time to a non-spatial and non-temporal ontology.

Furthermore, our approach also finds support from the difficulty inherent to answering the internal and external questions. In the context of quantum gravity, at least, we've shown that there's no need to answer those questions, so we can avoid them. That's good: we don't need to argue about what spacetime is in any deep sense, since it doesn't matter. Nothing much seems to be lost by simply conceding that the term "spacetime" has escaped semantic control, and now picks out a hodgepodge of different things, which don't bear much, if anything, in common.

Acknowledgements

We are grateful to the reviewer for very helpful comments. The work of Sam Baron was supported by the Australian Research Council through a Discovery Early Career Researcher Award (DE180100414). The work of Baptiste Le Bihan was supported by the Swiss National Science Foundation via his Ambizione project "Composing the world out of nowhere."

Notes

1. We use the notion of "theory" in a fairly liberal sense as meaning a collection of pointedly grouped propositions: propositions that are grouped together for a specific purpose (cf. Vickers, 2013). Importantly, theory should not be restricted to physical theories in this context. What we are aiming at is a difference between the concept of spacetime as anchored in general relativity, and the concept of spacetime as it can be found elsewhere in views that do not approximate general relativity—other scientific theories but also potentially folk theories.
2. *Newton-Cartan theory*—a geometrical reformulation and generalization of Newtonian mechanics as a field theory (Bain, 2004) that features spacetime—might be viewed as conflicting with this claim. As before, this depends on what "approximation" means. Newton-Cartan theory won't approximate general relativity in the sense described earlier (involving empirical recapture) but it might on some other conception of approximation.
3. Minkowski introduces spacetime *points* that are then used to define spacetime *vectors*, *lines* and *filaments*. The concept of spacetime is not explicitly introduced but it's natural to think of it as the totality of spacetime lines.
4. Note that the broad answer to the indexing question simply implies that the concept of spacetime must be broad enough to apply to non-relativistic contexts. It is then another question whether or not the spacetime concepts involved in these non-relativistic contexts share a common minimal core with the relativistic context. Only if these different concepts do not share a common core will the broad answer involve a form of pluralism.
5. See also Ismael (2021) for discussion.

Bibliography

Bain, J. (2004). Theories of Newtonian gravity and empirical indistinguishability. *Studies in History and Philosophy of Modern Physics* 35 (3), 345–376.

Baker, D. J. (2021). Knox's inertial spacetime functionalism. *Synthese* 199, 277–298.

Baron, S. (forthcoming). *Eliminating Spacetime*. Erkenntnis.

Baron, S. and B. Le Bihan (2022). Composing spacetime. *Journal of Philosophy* 119 (1), 33–54.

Barrett, J. A. (2001). *The Quantum Mechanics of Minds and Worlds*. Oxford University Press.

Brown, H. R. (2005). *Physical Relativity: Space-time Structure From a Dynamical Perspective*. Oxford University Press.

Butterfield, J. and C. Isham (1999). On the emergence of time in quantum gravity. In J. Butterfield (Ed.), *The Arguments of Time*, pp. 111–168. Oxford University Press.

Butterfield, J. and C. Isham (2001). Spacetime and the philosophical challenge of quantum gravity. In C. Callender and N. Huggett (Eds.), *Physics Meets Philosophy at the Planck Scale: Contemporary Theories in Quantum Gravity*, pp. 33–89. Cambridge University Press.

Chalmers, D. (2021). Finding space in a non-spatial world. In C. Wüthrich, B. Le Bihan, and N. Huggett (Eds.), *Philosophy Beyond Spacetime: Implications from Quantum Gravity*, pp. 154–181. Oxford University Press.

Crowther, K. (2018). *Effective Spacetime: Understanding Emergence in Effective Field Theory and Quantum Gravity*. Springer.

Hoefer, C. (1996). The metaphysics of space-time substantivalism. *Journal of Philosophy* 93 (1), 5–27.

Huggett, N. and C. Wüthrich (2013). Emergent spacetime and empirical (in)coherence. *Studies in History and Philosophy of Modern Physics* 44 (3), 276–285.

Huggett, N. and C. Wüthrich (forthcoming). *Out of Nowhere: The Emergence of Spacetime in Quantum Theories of Gravity*. Oxford University Press.

Ismael, J. (2021). Visible and tangible reality of space (without space). In C. Wüthrich, B. Le Bihan, and N. Huggett (Eds.), *Philosophy Beyond Spacetime: Implications from Quantum Gravity*, pp. 199–221. Oxford University Press.

Knox, E. (2019). Physical relativity from a functionalist perspective. *Studies in History and Philosophy of Modern Physics* 67, 118–124.

Lam, V. and C. Wüthrich (2018). Spacetime is as spacetime does. *Studies in History and Philosophy of Modern Physics* 64, 39–51.

Lam, V. and C. Wüthrich (2021). Spacetime functionalism from a realist perspective. *Synthese* 199, 335–353.

Le Bihan, B. (2018a). Priority monism beyond spacetime. *Metaphysica* 19 (1), 95–111.

Le Bihan, B. (2018b). Space emergence in contemporary physics: Why we do not need fundamentality, layers of reality and emergence. *Disputatio* 49 (10), 71–85.

Le Bihan, B. (2021). Spacetime emergence in quantum gravity: Functionalism and the hard problem. *Synthese* 199, 371–393.

Minkowski, H. (1908). Die Grundgleichungen für die elektromagnetischen Vorgänge in bewegten Körpern. *Nachrichten von der Gesellschaft der Wissenschaften zu Göttingen, Mathematisch-Physikalische Klasse* 1908 (1), 53–111.

Ney, A. (2015). Fundamental physical ontologies and the constraint of empirical coherence: A defense of wave function realism. *Synthese* 192 (10), 3105–3124.

Paul, L. A. (2012). Building the world from its fundamental constituents. *Philosophical Studies* 158 (2), 221–256.

Read, J. and T. Menon (2021). The limitations of inertial frame spacetime functionalism. *Synthese* 199, 229–251.

Teitel, T. (2022). How to be a spacetime substantivalist. *Journal of Philosophy* 119 (5), 233–278.

Vickers, P. (2013). *Inconsistency in Science*. Oxford University Press.

Wüthrich, C. (2017). Raiders of the lost spacetime. In D. Lehmkuhl, G. Schiemann, and E. Scholz (Eds.), *Towards a Theory of Spacetime Theories*, pp. 297–335. Springer.

Yates, D. (2021). Thinking about spacetime. In C. Wüthrich, B. Le Bihan, and N. Huggett (Eds.), *Philosophy Beyond Spacetime: Implications from Quantum Gravity*, pp. 129–153. Oxford University Press.

7 Space the many substances

Vera Matarese

7.1 Introduction

What is space? Metaphysicians and philosophers of physics alike have pondered this question, but they have tackled it by adopting, broadly speaking, two different approaches. The former usually develop different metaphysical conceptions of space and inquire into their coherence with a priori, conceptual reasoning, and only at a later stage they may become concerned with their compatibility with actual physics.[1] The latter usually begin with the most fundamental physical theories of space and infer different characterizations and models of space from them. They are interested in interpreting the formalism of the theory in a meaningful way, and in defending their particular interpretation against other possible ones.

This chapter will bring the two approaches in fruitful dialogue in the context of one of the most important metaphysical debates concerning space, which is the one contrasting the continuous and the discrete views of space. Such a debate has hitherto found no conclusive resolution, as each of the two positions encounters conceptual difficulties that are hard to dispel. In particular, in this chapter, I focus on the discrete view of space, which for decades has been considered only a speculative theory for two different reasons. The first is that there has not been a yet fully worked out theory that really supports it. Our only fully worked out fundamental physical theory for space is General Relativity, which clearly portrays a continuous space. The second reason is that such a view is vulnerable to Weyl's argument, which has been found to be "devastating" (Van Bendegem, 2020).

However, Weyl's argument works only as long as it relies on some specific metaphysical assumptions about how the discrete model of space has to be constructed. More specifically, it relies on certain metaphysical assumptions on how length should be measured and on the metric of the discrete space. These assumptions are reasonable and reflect our pretheoretical conception of space, but they are not *sine qua non* conditions for models of discrete space. Breaking these assumptions is possible and for this reason, Weyl's argument cannot be considered decisive. This,

DOI: 10.4324/9781003219019-10

however, is not enough to fully rehabilitate the discrete view of space. While it is true that Weyl's argument does not apply to *any* discrete model of space and so does not determine the impossibility of the discrete view, simply looking for violations of the metaphysical assumptions implicit in Weyl's argument would lead to an arbitrary proliferation of many different models, since these violations are motivated only for the sake of escaping the challenge of Weyl's argument. Moreover, once it is established which models, by breaking some of the standard assumptions, are compatible with actual physics, none of these can be seriously preferable over the others unless on the basis of highly debatable theoretical virtues. What we lack is a physical justification for taking the discrete view of space seriously, and a physical ground for opting for some discrete models instead of others.

In this chapter, I argue that the violation of these assumptions finds not only conceptual but also physical justification if we consider cutting-edge models of space drawn from quantum gravity. In particular, I will develop a model from Loop Quantum Gravity, analogous to the one presented in Vassallo and Esfeld (2014), which gives a novel characterization of discrete space that has not been considered before by metaphysicians working on space. According to such a model, space consists of different substances, "the atoms," which are different in shapes and number of adjacency relations.

The plan of the chapter is the following. First of all, in section 7.2, I will review the atomistic discrete view of space, which is normally considered to be vulnerable to Weyl's argument. After this, in sections 7.3 and 7.4, I will present Loop Quantum Gravity and a possible model for the Loop Quantum Gravity discrete space, which is very similar to the one developed in Vassallo and Esfeld (2014). Section 7.5 will be devoted to the discussion of the metaphysical implications of such a model of discrete space. Some final considerations will follow.

7.2 State of the art: The atomistic discrete view of space and Weyl's tile argument

One of the most important debates in the metaphysics of space concerns the structure of space, whether it is continuous or discrete.[2] On the one hand, the continuous view, also called the "standard" view of space, holds that space is gap-free, dense, completely filled with an infinite number of extensionless points. There is no limit on the number of points contained in a portion of space, and for this reason, there is also no limit on how far we can divide space: space is infinitely or indefinitely divisible. On the other hand, we have the discrete view, according to which space is finitely divisible into atomic regions of space. If those regions are individuals and extended, without any parts, the space is said to be atomic.

Discrete Space: Space is discrete if and only if every finite extended region of space is composed of finitely many regions of space, where regions of space can be individuals or sets of points, and individuals and points are finite.

Atomic Discrete Space: Discrete space is atomic if and only if the regions of space are individuals rather than sets of points, and such regions are extended and have no proper parts, i.e., they are mereological atoms.

Since it is often the case that the discrete models of space discussed in the literature are also atomic, from now onwards, for the sake of simplicity, I will use "discrete space" to denote "atomic discrete space."[3]

As mentioned in the introduction, the discrete view was considered only a metaphysical speculation, since General Relativity, which is our only completely worked out fundamental theory of space, presents us with a continuous space. Moreover, it has been normally regarded as impossible, being vulnerable to the famous and allegedly devastating Weyl's argument, which aims to establish the unfeasibility of discrete models with conceptual reasoning (Van Bendegem, 2020).[4] According to Weyl's argument, discrete space can be modelled as a plane with a grid, where each component, a "tile," represents one atomic region of space and has exactly the same size and shape as the others. Within this model, Weyl notices the following problem:

If a square is built up of miniature tiles, then there are as many tiles along the diagonal as there are along the sides; thus, the diagonal should be equal in length to the side.

(Weyl, 1949, p. 43).

Consider an isosceles rectangular triangle *ABC*. If the length of a side is measured by counting the atoms, the diagonal turns out to have exactly the same length as each of the other two sides. This result clearly violates the Pythagorean theorem, according to which the diagonal should be longer than the two sides (Figure 7.1).

From this, it can be concluded that discrete space, whose metric violates the Pythagorean theorem, is not able to approximate Euclidean space, and so it cannot exist.

P1: If discrete space exists, it must approximate Euclidean space.
P2: Discrete space does not approximate Euclidean space.
C: Discrete space does not exist.

If one stipulates that what is metaphysically untenable cannot be empirical, it is also possible to establish the "non-empiricality" of discrete space even without any empirical work.

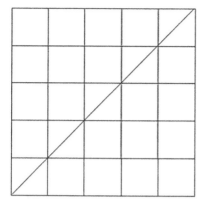

Figure 7.1 Weyl's example.

P3: If the discrete model of space is "metaphysically" untenable, then it cannot be empirical.
P4: The discrete model is "metaphysically" untenable.
C: The discrete model cannot be empirical.

As Salmon (1980) remarks, Weyl's argument goes through even if we go more microscopically by dividing the square in much finer grained level:

> No matter how small we make the squares, the hypotenuse remains equal to the length of the other two sides. This is one case in which the transition to very small atoms does not help at all to produce the needed approximation to the obvious features of macroscopic space.
>
> (*ibid.*, p. 66)

But is Weyl's argument really a fatal blow for the discrete view of space? In order to evaluate this, it is important to identify the assumptions of the model presupposed by the argument and check whether they are all necessary. Weyl's argument presupposes the so-called "tile model," within which atoms are arranged like the regular square tiling and where length is measured by counting the atoms: 1. Atomic regions of space have the same square shape and size; 2. The metric is such that each tile has the same number of adjacent tiles; 3. The length between two atoms is measured in terms of the number of atoms along the path connecting the two atoms. Moreover, a consequence of the particular model considered by Weyl, which is the standard square model, is such that the arrangement of its atoms gives rise to distinguished or preferred directions (horizontal, vertical, diagonal).

The question is whether all three assumptions of the tile models are necessary for a discrete space. The answer is, of course, in the negative. The discrete view of space simply dictates that there is a finite number of building block regions of space. It does not say anything about the particular tile arrangement, about their shape, adjacency relations, such as the ones assumed in the tiling model. Would the violation of one of the assumptions escape the challenge of Weyl's tile argument?

At first, it seems that violation of 1 and 2 would bring some improvement but no definite positive result. Among the different proposals that have been examined in the literature, there is the hexagonal tile spice model, within which the distance relations approximate Euclidean geometry much better than the square tiling; however, they do not do so sufficiently well. The only way to escape Weyl's argument seems to opt for revolutionary models such as the aperiodic model by Penrose, which has not even room for the notion of a straight line (and so for the notion of triangle as well), or such as those that feature a highly irregular and non-uniform arrangement of atoms.

What about assumption number 3? At first, one may feel sceptical about changing it. Indeed, the way we measure length in an atomistic space seems to be rather straightforward. How can we measure the length between two atoms if not by counting how many atoms are in between? An initial proposal was a model by Van Bendegem (1987), where the three sides of the triangle have width and so their length is determined by the number of squares it covers/contains. However, as Dainton (2009) and Chen (2021) notice, this model requires a continuous Euclidean geometry to be worked out in all the details, violating the tenets of the discrete view. While 3 is definitely a reasonable assumption, Chen (2021) has recently investigated whether it has any kind of conceptual necessity, and found that it does not. Her reasoning is the following. She starts by pointing out that within a discrete space, length should be understood not in terms of the notion of line, which implies continuity and so would be at odds with the discreteness of space, but in terms of the notion of path. We can therefore define a path as (*ibid*, p. 7536):

Path: A path from atom a_1 to atom a_n is a sequence of atoms $a_1, a_2, \ldots, a_k, \ldots, a_n$, such that for every k, a_k and $a_k + 1$ are adjacent $(1 \leq k \leq n{-}1)$.

Assuming now that the unit of length is the length of a path containing one atom, Weyl's argument can be taken as relying on the following assumption (*ibid*, p. 7536):

Length-by-counting: The length of a path is equal to the number of atoms it contains.

As Chen points out, the length-by-counting assumption is normally accepted because it allegedly relies on size-by-counting assumption (*ibid*, p. 7536):

> **Size-by-counting:** The measure or size of a region is equal to the number of atoms it contains.

However, the size-by-counting assumption generally does not imply the length-by-counting assumption.[5] Indeed, suppose that we have a three-dimensional space where the atoms have a particular shape that is given by primitive lengths along the three different directions (*ibid.*, p. 7537). Then, we should not infer the length-by-counting assumption from the size-by-counting assumption. This means that even if we accept the latter, we are not forced to accept the former.

At this point, one option would be to replace assumption 3 with an intrinsic account of distance McDaniel (2007), according to which the distance between two atoms does not depend on the path between them, but is intrinsic to their fusion, that is, if we duplicate the fusion of the atoms only, we obtain the very same distance. This account however is not comparable with general relativity (Chen, 2021). To solve this problem, Chen (2021) develops a mixed account model where she replaces length-by-counting assumption with an intrinsic definition of length, plus a definition of path-dependence distance. Within her model, two adjacent atoms do not need to be represented by two square tiles that are directly next to each other but can also be represented by tiles that are far apart.[6]

> **Intrinsic Local Distances:** Some atoms bear primitive distances.

> **Path-dependent Distances Distance between two atoms** = the least sum of primitive distances from one to the other.

Examining and evaluating all the solutions that have been proposed to solve Weyl's argument will not be my task here, since other works have already embarked on such enterprise.[7] However, overall, it can be fairly claimed that models that violate assumptions 1 or 2 and feature a simple and regular metric do not recover Euclidean space, at least not sufficiently well. On the contrary, models that are extremely complicated and irregular can escape Weyl's challenge. Another option, as we saw, would be to abandon tiling models, and to abandon assumption 3, which is based on the length-by-counting assumption. In the end, in one way or in another, the atomistic discrete view of space seems to be able to turn out to be metaphysically coherent, and so at least physically possible. Should we be content with this conclusion reached by metaphysicians of science?

My view is that more work should be done to rehabilitate the discrete view of space. Indeed, to my mind, the proliferation of all these atomistic discrete models looks rather arbitrary, in the sense that the reason for their birth is to escape the challenge of Weyl's argument by breaking one of its assumptions. In this regard, the only justification for breaking one of the assumptions of the tile model is to vindicate the metaphysical possibility of the discrete atomic space view at all costs. But is there any other justification for endorsing discrete models, and for breaking such assumptions, other than escaping Weyl's tile argument at all costs? More importantly, once we identify a family of models that is conceptually coherent and that is not vulnerable to Weyl's tile argument, is there any reason for actually endorsing one instead of another, other than an evaluation of metaphysical virtues? Of course, we can further ask which of these models are compatible with general relativity; however, this kind of assessment can help us only select a family of models compatible with physics but does not inform us of whether physics is actually corroborating discrete models rather than the continuous ones, and which discrete model we should opt for. In sum, the literature in metaphysics of science has recently shown that the atomistic discrete view of space is a possible candidate but cannot tell us whether there is any physical justification for taking the discrete view as a serious candidate, and what the real model for a discrete view of space is, if any.

In this regard, I advocate that turning to the most cutting-edge advancements in physics, in particular quantum gravity, can 1. vindicate the physical salience of discrete models of space; 2. justify the violation of the assumptions implicit in the tile models; 3. inform us of which violation physics allegedly breaks, if any; 4. shed light on some concrete possible models it would be interesting to examine in metaphysics. My intention is not to turn to quantum gravity models of space because they can teach us how to resolve Weyl's challenge (this might be regarded as a misleading approach, see endnotes 8 and 16). My goal is to provide a naturalistic ground for the discreteness of space. In particular, following this approach, in this chapter I turn to Loop Quantum Gravity, one of the most promising theories of quantum gravity, which predicts an atomistic discrete model of space.[8]

One may object to my proposal by pointing out that Loop Quantum Gravity is just a tentative approach, as it is still not possible to ascertain its consistency, and no direct empirical corroboration of the theory is available yet (Rovelli and Vidotto, 2015, p. 3). These are important aspects that certainly prevent us from establishing the *truth* of discrete view of space, but they do not undermine the claim that Loop Quantum Gravity can at least vindicate the physical salience of discrete models of space. Indeed, as long as Loop Quantum Gravity, which is one of the two most successful candidates for a theory of quantum gravity, and which is rooted in well-established theories such as General Relativity and Quantum Mechanics, presents us with discrete models of space, it is unfair to regard discrete views of space as just speculative armchair metaphysics.

7.3   Taking a peek at Loop Quantum Gravity

Loop Quantum Gravity (hereafter LQG) is a program that aims to develop a non-perturbative and background independent theory of quantum gravity. Two versions of LQG have been proposed, the canonical one and the covariant one. Even though neither of them has been fully developed, they are expected to be physically equivalent and they both feature a discrete model of space. Here I will focus only on the former.[9]

The canonical version of LQG starts with a canonical Hamiltonian formulation of General Relativity, based on Ashtekar connection variables,[10] which are meant to replace the spacetime metrics, conjoined with loop representation. According to the canonical formulation of GR, the 4-dimensional spacetime manifold M is interpreted as being artificially split into a 3-dimensional spatial manifold S and a one-dimensional temporal manifold R, breaking in this way the symmetry between space and time. Since LQG works with Ashtekar variables, this splitting is between the $3D$ spatial connections and a time parameter.

Both the Ashtekar variables and the loop representation are used to ensure smooth calculations in the quantization of the theory, which follows the standard Dirac procedure. In particular, in the context of geometric quantization, the Dirac procedure aims at identifying states of Hilbert space that satisfy the constraints of the system, and the quantum operators. The starting point is the identification of the unconstrained Hilbert space, which is defined by the normal Hamiltonian of the theory. Such a space, however, contains redundant information, such as identical or even physically impossible states. In order to restrict the unconstrained Hilbert space to a subspace that faithfully represents the state space of the theory and which determines the meaningful degrees of freedom of the system, one proceeds with implementing the constraints of the theory. By solving the constraints, we can identify first the kinematical Hilbert space, and later the physical Hilbert space of the theory:

$$H \xrightarrow{SU(2)} H_0 \xrightarrow{Diff} H_{Diff} \xrightarrow{\hat{H}} H_{phys}.$$

The constraints that need to be implemented are: i) the $SU(2)$ Gauss (internal) gauge constraint, which requires the theory to be invariant under internal gauge transformation; ii) the vector (or spatial) diffeomorphism constraint, which requires the theory to be invariant under diffeomorphism; iii) the scalar or Hamiltonian constraint, which requires the theory to be invariant under reparametrization of time coordinates (Norton, 2020, p. 14).

Within this process, the loop representation is particularly useful for the implementation of the $SU(2)$ and the diffeomorphism constraints. With this representation, indeed, in order to solve the Gauss constraint, we just need to identify a graph of links (oriented curves) and nodes (the end-points of the oriented curves) (Rovelli, 2004; Gambini and Pullin, 2011; Norton, 2020) embedded in the mathematical manifold M, which is the General Relativity manifold stripped of its metrical structure. The central aspect of this graph consisting of links and the nodes is that both these two elements are colored by a representation of the $SU(2)$ group. Such a colored network is called "spin-network" and can be expressed as the following construction:

$$\text{Spin-network} \rightarrow |\Gamma(\vec{x}, j_n, i_m)\rangle = |S(\cdot)\rangle.$$

More specifically, each link is associated with a half-integer number j_n, which is the 'spin' of the irreducible $SU(2)$ representation, and each node, represented by i_m, is specified by "intertwiners," which are other $SU(2)$ quantum numbers. At this point, the Hilbert space of our theory can be identified by associating to each embedded spin network a spin network state $|S\rangle$. Indeed, each of these states constitutes a basis vector of the gauge-invariant Hilbert space of the theory (Rovelli and Smolin, 1995b).

The second step consists in making sure that our states are diffeomorphically invariant. Such invariance is crucial since it is a fundamental feature of general relativity, and we want to preserve it in the quantum domain. It requires that our physical states be invariant under smooth stretching and shifting of spin-networks around the manifold (Norton, 2020, p. 17); in other words, they must be independent of the manifold where spin-networks are embedded. For this reason, in order to implement diffeomorphism constraints, we disembed our spin network states from the manifold M, by constructing several equivalent classes of diffeomorphically related states and mapping each of them to a single state in the Hilbert space. This way, our mathematical states should be tied to equivalence classes under diffeomorphisms of such networks. These are called 'abstract spin network states' or 's-knots' (Norton, 2020; Rovelli, 2004). A complete specification of the theory could be achieved only by solving the Hamiltonian constraint mentioned earlier. The problem is that there is not yet a fully worked out strategy to do so. Different attempts have been explored, but the most promising ones, like using spin-foams, fall within the LQG covariant version of the theory (Rovelli and Vidotto, 2015).

The arguably most important discovery in LQG was made by Rovelli and Smolin (1995a), who found out that spin-networks are eigenvectors of area and volume observables, and that their respective geometrical operators have a discrete spectrum, and so discrete eigenvalues. More specifically, while the area operator depends on the spin associated with the link of the network, the volume depends on the intertwiner associated with the node of the network. This important insight was followed by the discovery of the length operator, which also has a discrete spectrum. While the area operator measures the area of the surfaces pierced by the links of a spin-network and the volume operator measures the volume around a four-valent (or higher-valent) node, the length operator measures the length of a curve defined in the following intrinsic way (Bianchi, 2009). Imagine a spin network state featuring a node and two links originating at that node. Then, the two surfaces dual to the two links intersect at a curve. The elementary length operator measures the length of this curve.[11]

What is important to stress is that the area of surfaces, the volume of regions, and the length of curves come in discrete Planck-sized packages. This feature can be explained by the fact that there is a minimal unit of length, area, and volume on the order of Planck-scale and below which the LQG spacetime breaks down. Moreover, the discreteness is also a consequence of the fact that the graph of a network can be modified only by adding or subtracting whole numbered links and nodes to it. This implies that the process of quantization involves the loss of the continuous differentiable manifold and of the metric as described by General Relativity, which together are standardly thought to represent spacetime.

This however does not mean that we cannot recover the smooth space of General Relativity. Indeed, it is possible, and the eigenstates of areas, volumes and lengths that approximate flat space (or other smooth geometries) at large scales are called "weave" states. The physical procedure for recovering our smooth 3D space is, however, not entirely developed yet.[12] Moreover, it is not straightforward because of the problem of non-locality, also called disordered locality, discussed in Markopoulou and Smolin (2007), and addressed in philosophy works by Wüthrich (2017) and Huggett and Wüthrich (2013). This problem consists in the fact that parts of the spin network that are connected by a link and thus are fundamentally adjacent may end up giving rise to parts of emergent spacetime that are spatially very distant from each other. In a nutshell, adjacent atoms can be mapped to very distant events in our emergent spacetime. This phenomenon, examined in the simplest cases of weave states, which are designed to match the classical metric, clearly shows a mismatch between micro-locality and macro-locality, and is possible because disordered locality may still satisfy the conditions of correspondence to a classical metric. States can be semiclassical and nonetheless feature disordered locality. Such states are thought to be common in the LQG kinematical Hilbert space as well as among solutions to the quantum constraints.[13] Another point to remark when talking about the connection between micro-locality and macro-locality is the following (Markopoulou and Smolin, 2007). Since the state of space is not a spin network but a linear superposition of spin networks, we expect macro-locality to emerge from superpositions of many spin-network states (and not from single spin-networks).

7.4   A realist interpretation of space in Loop Quantum Gravity: space the many substances

Summing up the previous section, we can conclude that the geometry of LQG space is represented by a graph of nodes and links. Each node provides information of the volume of a particular chunk of space associated to it, while the links intersecting at the node provide information on the area associated to that chunk of space. Each chunk is not in space, but constitutes space; better, it is a particular localization of space.

While associating volume and area to the elements of the graph is a standardly accepted,[14] the specific ontological characterization of space that results from LQG is still controversial. As Vidotto correctly notices,[15] spin-networks are just mathematical auxiliary artefacts that, at the present stage, are useful tools to represent the LQG space. But what kind of ontology should we derive from this mathematical representation? How should we imagine the LQG space? What are these 'chunks' of space whose volume and area are specified by the nodes and links of the graphs?

Answering these questions by developing an ontological interpretation for LQG is important, and not only for LQG itself. It can also vindicate the physical significance of the metaphysical view of discrete space and can indicate which models of discrete space are "physically grounded." This was the goal that the chapter set up from the very beginning. If LQG presents us with a discrete model of space, it provides us with a physical justification for endorsing the discrete view, and also, depending on whether and how it violates Weyl's tile model assumptions, it gives us physically grounded reasons for endorsing one particular discrete model of space over the others.[16] For all these reasons, in this section I dig into the ontological interpretations of LQG hitherto proposed in the literature. And here is where other problems come, as interpreting the formalism is a difficult and controversial process, especially in a theory such as LQG, which features an extremely complicated mathematical apparatus and at the same time lacks empirical "graspability."

When it comes to interpreting LQG, two opposite roads have been taken. One road is the one trodden by Rovelli and Vidotto (2015), who are very cautious in interpreting the theory and extrapolating an ontology from it. Even though they use the word 'chunks of space' while describing the LQG space, they warn against a literal interpretation. According to them, indeed, these chunks should not be thought as concrete atoms of space that constitute the space mereologically; rather, they should be understood as modes of interaction:

> The quantum discreteness is due to the discrete spectrum of area and volume. The quanta of space described by the spin network states should not be thought of as quanta moving in space. Quantum numbers include the intrinsic physical size of the quanta themselves (area and volume) as well as the graph G that codes the adjacency relations between these quanta. . . . Finally, a warning: the quanta of space of loop quantum gravity should not be taken too naively as actual entities, but rather as modes of interaction. . . . Trying to think too literally in terms of concrete "chunks" forming the quanta of space can be misleading, as always in quantum mechanics. A pendulum is not "made of" its quanta. A basis in the Hilbert space should not be mistaken for a list of "things."
>
> (*ibid*, p. 141)

This is also because, as said before, a generic LQG state is a superposition of weave states:

> [S]pace is a quantum superposition of weave states. Therefore, the picture of physical space suggested by LQG is not truly that of a small scale lattice. . . . Rather, it is a quantum probabilistic cloud of lattices.
>
> (Rovelli, 2004, p. 271)

The picture that arises from this interpretation is very different from tile models. Rovelli and Vidotto's model, indeed, does not provide a picture of a concrete arrangement of atoms, in the same way the tile models do; it rather features, at the fundamental level, quantum probabilities. Moreover, within Rovelli and Vidotto's interpretation, the discreteness of space results from the fact that there is a minimum Planck-size unit of length, area, and volume and the fact that since their operators have discrete spectrum, their observables assume only discrete values. This is not the kind of discreteness that is under consideration and discussed in metaphysics, which is rather due to the granularity of concrete, extended atoms of space, which cannot be arbitrarily small. The picture proposed by Rovelli and Vidotto (2015) is therefore extremely interesting but cannot vindicate the discrete view of space that we find in the metaphysics debate. It is unhelpful with regard to the justification and legitimization of the discrete views of space, as it simply does not fit with the metaphysical debate on the existence of discrete space. Rather, it offers new, interesting, revolutionary ways of thinking about discreteness, which evade the definition of discreteness assumed in metaphysics debates on the nature of space. For this reason, in this chapter, I will not investigate it further.

The opposite road has been taken by Vassallo and Esfeld's interpretation of LQG within a Bohmian framework (Vassallo and Esfeld, 2014). Contrary to the warning by Rovelli and Vidotto (2015), they interpret the links and nodes of the spin-networks as real, concrete chunks of area and of volume constituting the space. That Vassallo and Esfeld (2014) interpret the chunks of space as concrete atoms may induce one to believe that their intention is to interpret literally the formalism of the theory and to read off the ontology from it. Such an inference could not be more incorrect. Esfeld and Vassallo strongly oppose any literal interpretation and reading off of ontology from formalism, in exactly the same way Rovelli and Vidotto do.[17] They do not interpret the chunks as concrete atoms because they endorse a literal interpretation of the mathematical formalism, but because they champion the primitive ontology approach, which advocates a clear, parsimonious, discrete ontology of "building blocks."

Within Vassallo and Esfeld's model, the discreteness of space is therefore given by these concrete chunks of space, also called the atoms, which mereologically compose the space. Space is divisible in a finite number of extended atomic regions of space that are mereological atoms, having no proper parts. In this regard, the kind of discreteness that is at play within this interpretation is exactly the same kind of discreteness inherent in traditional metaphysical view of discrete space. For this reason, such a model can play the role of vindicating the discrete view of space by providing a concrete example of a discrete model and by showing in what regards Weyl's tile models differ from such a model. In what follows, I will elaborate and discuss a model of space that is analogous to the characterization of space within Bohmian quantum gravity, presented in Vassallo and Esfeld (2014).

First of all, according to this model, there is one particular physical configuration of concrete atoms of space at the fundamental level of our space ontology, and there is a fact of the matter as to what shapes the concrete atoms at the fundamental level instantiate. Since the volume and surface area of these atoms are determined by the quantum numbers associated to the nodes and links of the spin-networks representing the atoms, these atoms come with different volumes and areas of surfaces, depending on their colors. In a nutshell, at the fundamental level, there is a variety of atoms, all different from one another in shape and size: there are "many substances" of space. It is a fundamental fact of the matter, for instance, that a spin-network has an intertwiner at its node such that its volume is eight cubic Planck lengths, while the spins associated to their links are such that each of its surfaces has four square Planck lengths. As Ashtekar and Pullin (2017) remark, each atom has its own particular "decoration," that is, its own geometrical properties determined by the quantum numbers associated to its links and node. That LQG presents us with a quantum space that is mereologically constituted by "many substances," all different one from another with regard to their geometrical properties, revolutionizes our most basic concepts of atoms. Indeed, the variety of atoms might sound perplexing, as metaphysics debates on discrete space have traditionally presented us with atoms that are all alike in size and shape.[18] Even concerning the development of discrete space models that could escape Weyl's challenge, none of them present a variety of atoms that differ with regard to their shape, volume and area.[19] LQG gives us precise information on the value of the specific Planck units of volume and area, and we would expect that our atoms are the smallest possible, as they must be further indivisible.[20] These would certainly be legitimate expectations; however, two distinctions should be considered.

The first concerns theoretical indivisibility and what is actually there, at the fundamental level. It is certainly true that one could conceptually divide non-minimal chunks of space into smaller chunks, until one reaches the minimal size of the atom, given by the minimal volume and the minimal area of the minimal possible number of surfaces. As long as the atoms are extended and feature volumes or areas that are not "minimal quanta," i.e., the Planck unit of volume and area, they can always be theoretically divisible. However, this does not imply that, actually, at the fundamental physical level, all our atoms feature Planck unit of volume and Planck unit of area. Even though, hypothetically, we can divide the atoms until the minimum Planck size is reached, the actual size of our atoms is decided by the theory, in particular by the quantum numbers attributed to the networks of which the chunks of space are the duals.

The second distinction concerns the units, or "quanta," of volume and area, and the smallest size of the atom itself. Given that only certain combinations of values of volume and values of area are possible, it may well be the case that the atom cannot feature the minimum "quantum" of volume, the minimum "quantum" of area, and the minimum "quantum" of length. There are specific ways to combine the values of the three observables. This means, for instance, that some atoms will have the minimum "quantum" of volume but not the minimum "quantum" of area, for some other atoms the opposite may obtain. Again, the possible combinations are restricted by the theory. Overall, it seems completely reasonable to accept this first important feature of our model, which is that at the fundamental level there are extended atoms, with different sizes and shapes. They are mereological atoms and yet they do not necessarily all feature one quantum of area and one quantum of volume.

Secondly, these atoms of space are related by adjacency relations, which are symmetric and irreflexive. The adjacency relations play the important role of determining locality at the microscale. If two atoms, identified by their two respective nodes, are both endpoints of the same link, they can be said to be adjacent. Contrary to Weyl's tile model, the adjacency relation is not determined by some preferred directions of space (such as horizontal, vertical, diagonal lines), but it is established by the links, which, as mentioned before, can reach nodes that are even "farther away." While within the spin-network representation it is possible that two nodes that are very far away are adjacent in virtue of sharing a link, within the physical interpretation I provided (in the dual picture), two atoms are adjacent only if at least one surface of an atom is "glued" with the surface of another atom, both of which must have identical area, given that they share the same link. According to my interpretation, therefore, that two nodes are "far away" in a graph does not carry any physical significance once we move in the dual picture of chunks of space, as in this space the adjacency relation is determined by the "gluing" of one surface of a chunk with the surface of another chunk. Even if two nodes that are adjacent in virtue of sharing the same link are "far away" in the spin-network graph, they will always represent two chunks of space that share at least one of their surfaces in the dual picture. Another important aspect to consider about the adjacency of the atoms is that the number of adjacency relations of each atom depends on the number of its links. For this reason, atoms differ not only concerning volumes and areas, but also concerning the number of their adjacent atoms. This constitutes another striking difference from the standard tiling model of metaphysics debates, in which all atoms are usually featured as having the same number of adjacent atoms.

In this respect, let me briefly discuss one option to account for the individuality of the chunks of space. Apart from being distinguished and individuated by their different quantum numbers, and so, ontologically, by their different sizes and shapes, one may also argue that they possess their adjacency relations essentially. This perspective is analogously endorsed within metrical essentialism (Maudlin, 1989; Hoefer, 1996; Bigaj, 2015), which is a form of spacetime substantivalism applied to General Relativity, and so to a continuous spacetime. While within metrical essentialism, we identify spacetime with a particular metrical field, whose points are individuated by their metrical relations, which are essential to them, within the model of LQG that we are adopting, each atom possesses its adjacency relations essentially.[21] This means that an atom cannot exist if its adjacency relations change. It is essential for each of them to have the adjacency relations it has. Of course, this is an option that is not needed. A strong reason for adopting metrical essentialism for General Relativity spacetime was to account for the distinguishability of points. In our models, on the contrary, atoms of space are already distinguishable. Moreover, I suspect that it may generate some problems at the dynamical level of the theory. However, it is an option that should be taken into account for a complete metaphysical discussion of this discrete model of space.

The last important aspect that is worth discussing is how the LQG model treats length. One of the most important problems of our metaphysical models of discrete space was the notion of distance, which, within the standard Weyl's tile model, is based on the length-by-counting assumption, according to which we should count the number of atoms. Other definitions have been proposed, among which the intrinsic account of distance and the mixed account of distance. However, it seems that no perfect notion of distance has been developed yet, and this issue remains controversial. Something analogous is happening within the LQG model of space. For years, the length operator was not understood at all, and even now, it is not sufficiently well understood. Two aspects, however, are usually undisputed within the LQG community, and are worth mentioning here. The first aspect is that, as we mentioned earlier, the length operator has a discrete spectrum exactly like the volume and the area operator. This means that the length operator has a discrete spectrum of eigenvalue and that there is a minimum Planck unit of length. This kind of discreteness is different from the kind of discreteness of distance or length found in metaphysical models, in which length was equal to the numbers of the atom along the shortest path. In the LQG model length does not depend on the number of the atoms, rather on the quantum numbers associated to their edges. The operator of length corresponds to the length of a curve at which the surfaces of two adjacent atoms intersect. This differs from how length is conceived within the standard Weyl's tile models. The second aspect that is worth considering is the marginal role of length in the LQG model, with respect to the more central role of volume and area. Rovelli (2008) briefly discusses whether this is simply a technical difficulty or it reflects some deep fact, but finds no conclusive answer. It is therefore an open possibility that discrete spaces do not have a proper or natural notion of length, which however emerges as a novel concept in continuous spaces.

7.5   Conclusion

Is space infinitely divisible or there are indivisible, mereological atoms at its fundamental level? The atomistic discrete view of space has been traditionally regarded as conceptually flawed, because of Weyl's argument, and as physically ungrounded, given that General Relativity presents us with a continuous space. In the last three decades, however, there was a surge of interest in the atomistic discrete view, with a proliferation of attempts, more or less successful, at constructing atomistic models that are not vulnerable to Weyl's argument, by violating one of its implicit assumptions. Even though the success of some of these models has established the possibility of the discrete view of space, more work needs to be done to fully rehabilitate the view. Indeed, all these models were created just for the sake of escaping Weyl's threat, and we still lack a physically grounded motivation for taking any of these models seriously. In this regard, turning to our most promising and fundamental theories of space may give us insight into whether the atomistic discrete view is naturalistic or just a conceptual possibility, and may indicate what kind of atomistic discrete space should be taken as a real, physical possibility that is worth of metaphysical investigation. My contribution has hopefully, even if only partially, filled this lacuna. I presented a physical model of discrete space, which derives from a particular interpretation of Loop Quantum Gravity and which, to my mind, should be taken as a starting point for developing a metaphysics of discrete space.

My intention was not to present LQG model as a case that shows exactly how to resolve Weyl's challenge. There is a sense in which LQG proves Weyl's tile argument wrong because LQG is expected to recover the continuous space of General Relativity (and in turn the Euclidean space with its Pythagorean theorem) from an atomistic discrete space. However, one might object that, strictly speaking, the LQG presents a case that evades Weyl's argument by rendering it inapplicable.[22]

My intention was rather to provide a naturalistic ground for developing a metaphysical theory of atomistic discrete space. The LQG model that I have presented in this work provides some indications and suggestions on how we should metaphysically conceive of realistic, physically grounded, metaphysical models of discrete space. While in the past many different models of discrete space were built just in order to prove Weyl's argument wrong, my invitation is to turn to this model, which supposedly is able to evade Weyl's argument, and check what features of discrete space deserve particular attention. Starting from those, one can elaborate a naturalistic metaphysical theory of space.

Given that LQG is not fully developed yet, the full metaphysical characterization of its space is still under discussion. However, several points are already clear. The atomistic model I inferred from LQG differs from standard Weyl's tile models in significant ways. First of all, the resulting LQG model presents a variety of atoms, "the substances" of space, which differ one from another with regard to shape and size (in particular, volumes and areas of their surfaces). Given that their arrangement is not uniform, they also differ concerning the number of their adjacency relations. This revolutionizes our pre-theoretical intuitions of atoms, as the minimal entities that not only do not have actual parts, but are also theoretically further indivisible. Secondly, the LQG model confirms a conceptual difficulty that all discrete views of space are encountering in the metaphysics debate, which concerns a definition of distance. Far from equating distance with the number of atoms along a path, the LQG model of space seems to understand length as a measure of particular curves originated at the intersection of surfaces of chunks of space. The discreteness characterizing length, in the LQG model, is not given by the number of atoms, but is the result of the discreteness of the spectrum of the eigenvalues of the length operator. Moreover, its non-fundamental status and the fact that it remains ill-defined may suggest that it is a notion that is natural in continuous space, but unnatural in discrete contexts.

Acknowledgements

I would like to thank the editor, Antonio Vassallo, for his kind invitation to contribute to this volume and for his insightful comments on the manuscript. I also thank an anonymous referee and Casey McCoy for very helpful suggestions. I am grateful to the audience of my talk at the Warsaw Spacetime Colloquium Series organized by the Philosophy of Physics group at the Warsaw University of Technology for questions and feedback on the topic of this chapter.

Notes

1. For a discussion on how metaphysicians approach the nature of space, see Dainton (2009): "Metaphysics also has its own distinctive domains and methods of inquiry. While it may well be that to discover the answers to *some* questions about the space . . . we will have to listen to the physicist, there are other questions for which this is not so. . . . Our concepts of space and time are to some degree independent of particular scientific theories, and so we can investigate these concepts independently of any one scientific theory. . . . What must a world be like in order for it to be spatial and temporal? What are the different forms that space and time can take? In their attempts to answer these questions philosophers have introduced many novel and counterintuitive ideas, a good many of which are stranger than anything seriously entertained by scientists engaged in the struggle to ascertain the particular forms that space and time take in this world." (*ibid.*, pp. 5–6).
2. In what follows, I only provide an intuitive, non-technical characterization of the continuous and discrete view of space, as for instance presented in Dainton (2009) and in Bell (2017). In particular, in this chapter I assume that the difference between the two views concerns whether space is infinitely or finitely divisible. However, different definitions and differentiations between the two views are possible.
3. There are, however, views of discrete space that do not imply atomicity. For instance, see Forrest (1995) for a discrete view that posits points, rather than individual regions, as the constituents of discrete space.
4. I call this kind of reasoning "conceptual" to differentiate it from "empirical" reasoning.
5. For the argument, see Chen (2021).
6. Forrest's model (Forrest, 1995), which develops a model of space that is locally anisotropic and whose atoms have no shape, is a model in this vein, but because of these properties it does not count as an atomistic space as we are defining it.

7. See for instance, Dainton (2009); Chen (2021); Crouse and Skufka (2019).
8. Before proceeding, let me clarify again that the goals of the chapter are those spelled out in 1 – 4. My intention in the following sections is *not* to present the LQG model as a case that defeats Weyl's argument. This move, indeed, may be vulnerable to the objection that if LQG evades Weyl's challenge is because it presents a case very different from the one assumed in Weyl's tile argument. Weyl's tile argument presupposes a discrete structure *embedded* in a Euclidean space. LQG describes a fundamental, non-metrical description of atoms of space from which, supposedly, an Euclidean space should emerge. For an extension of Weyl's argument as a no-go theorem for the emergence of an isotropic space from a discrete structure within the context of classical physics, see Fritz (2013).
9. In the following, I provide a non-technical but faithful description of the theory, with particular attention on those aspects relevant to my later philosophical discussion. For detailed explanations of LQG, see Rovelli (1991, 2004, 2008); Crowther (2016); Gambini and Pullin (2011); Norton (2020). For a careful explanation of the covariant version, see Rovelli and Vidotto (2015).
10. The Ashtekar variables consist in a spatial $SU(2)$ connection variable and an orthonormal triad. This formulation is crucial, since it uses connection variables instead of metric ones. This allows the formulation of General Relativity to take a form similar to a Yang-Mills theory, which is naturally regarded as a theory of loops. This means that LQG is also cast as a theory of loops, from which it derives its name.
11. For a visualization of the curve measured by the length operator, see Bianchi (2009).
12. Substantial progress in this direction has been made in the covariant form of LQG. See Rovelli and Vidotto (2015). For a philosophical discussion on how spacetime emerges from canonical LQG, see Wüthrich (2017).
13. This however, should not be regarded as hindering LQG, according to Rovelli. According to him, indeed, these non-local connections could be possibly encoded as non-local wormholes in the emerging relativistic spacetime.
14. But see McCabe (manuscript) for a criticism regarding the discreteness of these observables.
15. She remarked this in her comments on Baptiste Le Bihan's talk "Quantum gravity and mereology: not so simple" at the Warsaw Spacetime Colloquium. She described the spin-networks as just useful or convenient auxiliary mathematical structure that could be eventually replaced by another mathematical structure (https://www.youtube.com/watch?v=J1aaXWVCjPI, 00:58:28—00:58:30 and 00:59:25—00:59:48).
16. As already flagged in endnote 8, the question I focus on is not whether the Loop Quantum Gravity model can approximately recover the Pythagorean theorem. Rovelli has recently admitted that, in the first years of his career, the discreteness of LQG made him concerned about the possibility that the LQG space was vulnerable to Weyl's argument (https://www.youtube.com/watch?v=eTadrU_Mhmc, 01:03:53 – 01:04:55). Indeed, as we said earlier, while LQG proposes a discrete model of space, Weyl's argument is usually considered as establishing the metaphysical and physical impossibility of discrete models of space. LQG establishes the possibility of discrete models of space, because LQG presents us with a discrete model of space from which GR and an Euclidean space can be recovered. So, in this sense, LQG does invalidate Weyl's conclusion that discrete space cannot exist. However, it is important to notice that Weyl seems to assume that the discreteness manifests itself at a classical level, and that it is embedded in

Euclidean space. Therefore, it may be objected that, strictly speaking, it does not apply to cases, such as the LQG one, where the discrete model of physics is at a quantum level and it is expected to approximate Euclidean space at the continuum level. See Fritz (2013, p. 1300) for the same point: "The tile argument does not address the question of whether there could be discrete models of physics for which the continuum limit corresponds to ordinary (non-relativistic) physics in Euclidean space." See, again, Fritz (2013) for an extension of Weyl's argument to cover such cases.

17. Indeed, it is exactly such a mismatch between mathematical formalism and ontology that justifies Vassallo and Esfeld to consider the superpositions of weave-states as just mathematical tools to compute measurement results and therefore as ontologically insignificant.

18. In case, of course, they belong to the same element or substance.

19. However, the possibility of having an irregular arrangement of identical atoms was presented when discussing possible ways to escape Weyl's argument (Fritz, 2013; Chen, 2021).

20. The minimum volume is 10^{-99} cubic centimeters, while the minimum area is 10^{-66} square centimeters.

21. There is also an interesting analogy between the discrete view of space described here and metrical essentialism. Supporters of metrical essentialism may identify points of spacetime as different substances, in the same way we identify chunks of space as different substances. For a good discussion of metrical substantivalism and the individuality of spacetime points, see Bigaj (2015), Hoefer (1996), as well as Bigaj's chapter in this collection.

22. See endnotes 8 and 16. Fritz (2013), for instance, claims while Weyl's argument aims to prove the impossibility of a discreteness embedded in a Euclidean framework where the Pythagorean theorem is expected to hold, the LQG model is clearly a model that describes a more fundamental space, a quantum arena, from which our Euclidean space is supposed to emerge at the continuum limit.

Bibliography

Ashtekar, A. and J. Pullin (Eds.). (2017). *Loop Quantum Gravity: The First 30 Years*, Volume 4. World Scientific.

Bell, J. (2017). Continuity and infinitesimals. In E. Zalta (Ed.), *The Stanford Encyclopedia of Philosophy* (Summer 2017 ed.). Metaphysics Research Lab, Stanford University. https://plato.stanford.edu/archives/sum2017/entries/continuity/.

Bianchi, E. (2009). The length operator in loop quantum gravity. *Nuclear Physics B* 807 (3), 591–624.

Bigaj, T. (2015). Essentialism and modern physics. In T. Bigaj and C. Wüthrich (Eds.), *Metaphysics in Contemporary Physics*, pp. 145–178. Brill/Rodopi.

Chen, L. (2021). Intrinsic local distances: A mixed solution to Weyl's tile argument. *Synthese* 198 (8), 7533–7552.

Crouse, D. and J. Skufka (2019). On the nature of discrete space-time. Part 1: The distance formula, relativistic time dilation and length contraction in discrete space-time. *Logique et Analyse* 246, 177–223.

Crowther, K. (2016). *Effective Spacetime: Understanding Emergence in Effective Field Theory and Quantum Gravity*. Springer.

Dainton, B. (2009). *Time and Space*. Routledge.

Forrest, P. (1995). Is space-time discrete or continuous?—an empirical question. *Synthese* 103 (3), 327–354.

Fritz, T. (2013). Velocity polytopes of periodic graphs and a no-go theorem for digital physics. *Discrete Mathematics* 313 (12), 1289–1301.

Gambini, R. and J. Pullin (2011). *A First Course in Loop Quantum Gravity*. Oxford University Press.

Hoefer, C. (1996). The metaphysics of space-time substantivalism. *The Journal of Philosophy* 93 (1), 5–27.

Huggett, N. and C. Wüthrich (2013). The emergence of spacetime in quantum theories of gravity. *Studies in History and Philosophy of Modern Physics* 44, 273–275.

Markopoulou, F. and L. Smolin (2007). Disordered locality in loop quantum gravity states. *Classical and Quantum Gravity* 24 (15), 3813.

Maudlin, T. (1989). The essence of space-time. *Proceedings of the Biennial Meeting of the Philosophy of Science Association* 1988 (2), 82–91.

McCabe, G. *Loop Quantum Gravity and Discrete Space-time*. http://philsci-archive.pitt.edu/14959/. Accessed 21 June 2021.

McDaniel, K. (2007). Distance and discrete space. *Synthese* 155 (1), 157–162.

Norton, J. (2020). Loop quantum ontology: Spacetime and spin-networks. *Studies in History and Philosophy of Modern Physics* 71, 14–25.

Rovelli, C. (1991). Ashtekar formulation of general relativity and loop-space nonperturbative quantum gravity: A report. *Classical and Quantum Gravity* 8 (9), 1613.

Rovelli, C. (2004). *Quantum Gravity*. Cambridge University Press.

Rovelli, C. (2008). Loop quantum gravity. *Living Reviews in Relativity* 11 (1), 5.

Rovelli, C. and L. Smolin (1995a). Discreteness of area and volume in quantum gravity. *Nuclear Physics B* 442, 593–622. Erratum: Ibid. 456:753–754, 1995.

Rovelli, C. and L. Smolin (1995b). Spin networks and quantum gravity. *Physical Review D* 52 (10), 5743–5759.

Rovelli, C. and F. Vidotto (2015). *Covariant Loop Quantum Gravity*. Cambridge University Press.

Salmon, W. (1980). *Space, Time, and Motion*. University of Minnesota Press.

Van Bendegem, J. (1987). Zeno's paradoxes and the Weyl tile argument. *Philosophy of Science* 54 (2), 295–302.

Van Bendegem, J. (2020). Finitism in geometry. In E. Zalta (Ed.), *The Stanford Encyclopedia of Philosophy* (Fall 2020 ed.). Metaphysics Research Lab, Stanford University. https://plato.stanford.edu/archives/fall2020/entries/geometry-finitism/.

Vassallo, A. and M. Esfeld (2014). A proposal for a Bohmian ontology of quantum gravity. *Foundations of Physics* 44 (1), 1–18.

Weyl, H. (1949). *Philosophy of Mathematics and Natural Sciences*. Princeton University Press.

Wüthrich, C. (2017). Raiders of the lost spacetime. In D. Lehmkuhl, G. Schiemann, and E. Scholz (Eds.), *Towards a Theory of Spacetime Theories*, pp. 297–335. Birkhäuser.

8 Quantum metaphysics and the foundations of spacetime

Vincent Lam, Laurie Letertre,
and Cristian Mariani

8.1 Introduction

In recent years we have witnessed a novel tendency in the research on the metaphysical implications of quantum theory. Given the difficulties in providing a shared ontological picture of *how the world is like* if quantum theory is true—in large part due to the many ways in which we could address the measurement problem—researchers have attempted to focus on those features of the theory that can be considered to some extent *interpretation-neutral*, to use Wallace (2019) expression. Phenomena such as entanglement and superposition, along with the mathematical features underpinning them, seem to be essential for how we define what a quantum theory is, and this is arguably true independently of one's preferred approach to the measurement problem. For instance, research in the metaphysics of quantum non-separability has shown that a certain *structuralist* (Lam, 2017) or *holistic* (Miller, 2016) attitude seem natural when it comes to understanding the phenomenon of entanglement. More recently, philosophers have suggested that the notion of *ontological indeterminacy* (a.k.a metaphysical indeterminacy) may provide an explanation of various features of quantum theory, and in particular it may explain the lack of value-definiteness affecting quantum systems in a state of superposition (Calosi and Wilson, 2021).

These metaphysical strategies are not in the business of providing novel solutions to the measurement problem. Rather, the idea behind them is to refine the overall metaphysical understanding of the theory, which could then be implemented by specifying the many ontological alternatives inspired by the various solutions to the measurement problem. And in effect, there are now several concrete proposals on how to implement these strategies within the context of specific interpretations of the theory,[1] and under the overall assumption of scientific realism towards physics.

Once we grant that the relevant quantum features used to derive such metaphysical conclusions are essential to any quantum theory, it is natural to expect this claim to be true also in the context of the novel research programs in quantum gravity (QG). Our main working

DOI: 10.4324/9781003219019-11

hypothesis is precisely that the aforementioned metaphysical strategies will prove insightful to better understand the world according to QG. And as a matter of fact, such an *interpretation-neutral* attitude is not just preferable in the context of QG, but may even seem necessary if we consider that no specific quantum interpretation is assumed or suggested at the current stage of research.

A common striking conceptual feature of many research programs in QG is that they suggest, in one way or another, that most (if not all) spatiotemporal structures are not fundamental (Huggett and Wüthrich, 2013). Consequently, one should expect that those metaphysical and foundational results in quantum theory that are already in tension with certain spatio-temporal features (or that already point towards the existence of certain non-spatiotemporal features) may turn out to be particularly relevant. In this chapter we are going to provide two such examples, and we will then consider how they may give us novel insights on the ontology of QG. The first of them, already well-known in the literature, concerns the metaphysical implications of quantum entanglement, whereas the second example pertains to some rather novel results in quantum foundations regarding the notion of causal non-separability. As we will show, each of these cases arguably forces us to reconsider some plausible and intuitive idea about the nature of spacetime, specifically as regards to the notion of locality, and of definiteness of the causal order among distinct events. If seen through the lenses of these recently proposed metaphysical views, the problematic implications of QG may perhaps seem more natural and acceptable, or so we will argue.

Road map. In §8.2, we focus on the metaphysical tools that have been developed to account for quantum entanglement and non-locality, as within the *structuralist* and the *holistic* metaphysical strategies. In §8.3, we first look at recent developments in quantum foundations about causally non-separable processes that do not assume a definite global causal structure. We then exploit the recently developed tools of *quantum indeterminacy* in order to provide an understanding of the notion of superposition of causal order. In §8.4 we show how the tools that we have introduced can help to articulate a coherent but not necessarily spatio-temporal worldview at the level of quantum gravity. In §8.5 we conclude.

8.2 The metaphysics of quantum entanglement

This section provides an overview of some of the metaphysical tools that have been articulated within the framework of standard quantum mechanics (and quantum field theory) to account for the key features of quantum entanglement and non-locality. These metaphysical tools do not aim to provide a full ontology for quantum mechanics (or quantum field theory), since this would require addressing the measure-ment problem. Rather, they aim to capture central features that, to some extent, cut across (most of) the various realist interpretations of quantum mechanics; indeed, they can often be further specified within these inter-pretations. Since we are interested in exploring to what extent these metaphysical tools can be relevant in the quantum gravity context, we highlight their links to spacetime structures.

8.2.1 Entanglement and non-Locality

In many ways, quantum entanglement and non-locality are central fea-tures of quantum mechanics—and, to some extent, of any quantum theory (such as quantum field theory). Within the standard quantum for-malism, entanglement is encoded in the ubiquitous entangled quantum states for composite systems. Quantum states can be represented by a vector in a Hilbert space, noted $|\psi\rangle$. A more general representation is provided by appeal to density matrices, noted ρ, which are linear opera-tors acting on the Hilbert space assigned to the system under consider-ation. A density matrix encodes either pure quantum states, which can be described with a vector in a Hilbert space, or mixed quantum states, which express situations in which a system is in a probabilistic mixture of pure states. The latter case expresses a situation in which we describe a system in a probabilistic mixture of pure states. In the remainder of this work, we will appeal to density matrices to represent a system's quantum state, not only because this mathematical object more naturally connects with the discussions of section 8.3, but also because it provides a more generalized framework.

Let a composite system, labelled 1–2, be composed of two subsystems, labelled 1 and 2. The quantum states of the subsystems 1 and 2 are said to be *nonseparable*, or *entangled*, if the global quantum state of system 1–2 cannot be expressed as follows:

$$\rho_{1-2} = \sum_i q_i \, \rho_1^i \otimes \rho_2^i \tag{8.1}$$

where the index i sums over classical probabilities (q_i) to have the subsystem x in the (pure or mixed) quantum state described by ρ_x^i. This notion of entangled states is purely formal at this stage, and needs to be interpreted to get assigned a meaning. If one adopts a realist approach towards quantum mechanics, the quantum state is considered as pointing towards objective features of nature. Yet, there is a debate regarding the exact nature of these objective features (see section 8.2.2). Most accounts see the wavefunction as the mathematical object representing the objective content of the quantum state, while density matrices are seen as encoding a mere epistemic information about the quantum state. This is however debated, and previous work emphasizes that there is no need for an episte-mic interpretation of density matrices (Aharonov et al., 1993). Several authors have defended a view called *density matrix realism*, in which it is the density matrix that represents the objective content of the quantum state (Chen, 2021).

Importantly, quantum entanglement can lead to empirically verified non-classical correlations violating Bell-type inequalities among space-like separated entangled subsystems (Hensen et al., 2015). A Bell inequality, as famously defined in Bell's theorem (Bell, 1964), is an algebraic inequality, the violation of which by any given probability dis-tribution is commonly understood in terms of the violation of the premise called "local causality." This premise encodes the fact that causes precede their effects, and that causal influences travel continu-ously through spacetime at subluminal speeds. Such quantum correla-tions violating a Bell inequality are said to be *non-local*, and are not determined by and do not supervene on the states of the entangled sub-systems[2] or by additional local variables not encoded in the entangled states, and are independent of the distance between the spacelike sepa-rated subsystems. In this context, quantum entanglement is naturally con-sidered as involving some form of non-locality.[3] Since Bell inequalities can be defined in a purely operational way—i.e., by appealing exclusively to notions such as inputs and outputs of quantum operations treated as black boxes—non-local correlations are said to be model-independent.[4] For this reason, Bell non-locality (i.e., the existence of non-local correla-tions) is naturally taken as reflecting some objective fact about the phys-ical world that any quantum theory has to account for.[5]

8.2.2   *Structuralism and holism*

Several metaphysical tools have been articulated to account for quantum entanglement and the related non-local correlations. Indeed, the fact that the modal connections that these quantum correlations exemplify cannot be understood in terms of intrinsic properties of the entangled subsys-tems (as encoded in their reduced density matrices or with the help of possible additional—"hidden"—variables)[6] provides a strong motiva-tion for a structuralist interpretation in the sense of ontic structural realism (OSR). Indeed, in the context of quantum entanglement, a natural structuralist understanding takes the novel, experimentally veri-fied non-local correlations among entangled subsystems as the manifes-tation of a new fundamental physical relation—often simply called "entanglement relation"—connecting the subsystems (whatever these latter precisely are according to the quantum theory under consideration and the preferred quantum ontology). The metaphysical details of the relationship between relations (or structure) and relata within OSR can be articulated in different (and sometimes controversial) ways and have been much discussed in the literature (e.g., see the references in Lam, 2017, §1). The *moderate* structuralist conception according to which the relations are on a par with their relata—forming together "structures"— seems especially appropriate for many situations in (fundamental) physics, including the entanglement case. Most importantly, in this structuralist perspective, the entanglement relation connects the entangled subsystems such that these latter have no independent existence. On this view, the existence of the entangled subsystems (ontologically) depends on the entanglement structures they are part of, that is, on there being entanglement relations, but also on there being other subsystems to which they are entangled to—these latter being conceived as (ontologically) interdependent on one another. This characterization of entanglement in terms of (symmetric) ontological interdependence or mutual dependence has been recently nicely discussed in Calosi and Morganti (2021) within their coherentist conception—which, to us, constitutes a refinement rather than a departure from the moderate structuralism considered here and further detailed elsewhere,[7] *contra* what Calosi & Morganti seem to suggest.

Leaving aside the metaphysical subtleties, what is important to highlight for our purpose in this chapter is that, within this structuralist conception, the entanglement structures encoding the interdependence among the entangled subsystems are conceived as physical structures on their own right, without being necessarily tied to spacetime.[8] As we will see in section 8.4, this metaphysical feature may well prove very useful when trying to interpret the quantum gravity context where spacetime may not play its usual roles.

Another main metaphysical strategy to account for quantum entanglement is quantum holism (recently defended in Ismael and Schaffer, 2020), which consists in arguing for the ontological priority of the quantum whole (that is, the total composite quantum system) over its entangled parts (the entangled subsystems). Such an interpretative move gets direct inspiration (and support) from the fact that the quantum state of the total composite system determines those of its entangled subsystems, while the converse fails. Beside this characterization in terms of ontological priority, various holistic aspects of quantum entanglement have been articulated for some time in the physics and philosophy literature (see Healey, 2016, for a review), some of which more or less explicitly encode structuralist elements (to some extent, certain types of holism can be considered as precursors of the recent structuralist conceptions in the quantum context). However, it is not the place to discuss the commonalities and the disanalogies between quantum holism and structuralism (see Calosi and Morganti, 2021, for a recent critical look).

What is especially interesting in our perspective here is the common ground argument recently put forward for quantum holism (Ismael and

Schaffer, 2020, §4). In many ways, the structure of this argument is similar to the familiar Reichenbach's common cause principle, which roughly states that two correlated events, where neither is the cause of the other, have a common cause that screens off the correlations between them. In view of the well-known difficulties of this principle in the quantum context (the issue is subtle though, see Hitchcock and Rédei, 2021, for a recent review), Ismael and Schaffer (2020) articulate a common ground account of quantum entanglement, which relies on the principle that if "non-identical entities a and b are modally connected, then either (i) a grounds b, or (ii) b grounds a, or (iii) a and b are joint results of some common ground c" (p. 4137)—where grounding is understood, in the way advocated by Schaffer (2009) as a (metaphysical) asymmetric dependence relation between more fundamental and less fundamental entities.[9] The application of this principle to entangled quantum subsystems then naturally leads to consider the total composite quantum system as their common ground, which clearly amounts to a form of holism since the whole (the total composite system) is then considered as ontologically prior to (more fundamental than) its parts (the entangled subsystems).

Similarly to the structuralist case, we want here to highlight the fact that this holistic common ground strategy need not be tied to spacetime in any way. Indeed, for instance, Ismael and Schaffer (2020) explicitly consider wavefunction realism as a case where ordinary 3-dimensional space and entities located in 3-dimensional space are grounded in (and emerge from) a holistic common ground that is not 3-dimensional, namely the quantum wavefunction on (high-dimensional) configuration space.[10] In an analogous way, as we will see in section 8.4, this holistic common ground strategy can be naturally extended to full 4-dimensional spacetime, thereby helping to make sense of the non-spatiotemporal features suggested in many QG research programs.

8.3   Causal non-separability and quantum Indeterminacy

The aim of this section is to introduce the recently discussed notion of *causally non-separable process*, and to discuss its metaphysical implications. We first introduce, in §8.3.1, the formal and technical details needed to understand causal non-separability, and we show how it is to be conceptually distinguished from quantum non-separability. In §8.3.2 we then suggest a way to understand the notion of causal superposition by exploiting the metaphysical tools developed within the debate on quantum indeterminacy.

8.3.1   *Causally non-separable processes*

In the previous sections, the discussion focused on the theory of quantum mechanics in its standard form, i.e., as expressed in the formalism of

Hilbert spaces. Yet, the theory has various formulations, on the basis of which different questions can be asked. More specifically, the field of quantum foundations also seeks to reformulate the theory within different formalisms in order to investigate the very structure of the theory, on the one hand, and inquiry about possible generalizations of the theory on the other hand. One of the formalisms used to reformulate quantum mechanics is that of operational theories. It offers an interesting framework, as it anchors the theory to a set of physical principles (from which the theory is recovered). Those principles select the probability distributions possibly correlating results of measurements performed on a given system (prepared in a specific way). Operational theories therefore widely appeal to the notions of preparation and measurement procedures, performed in isolated laboratories and in the context of given experimental setups. As discussed by Letertre (2021), the operational formalism, in spite of the instrumental connotation of its core concepts, does not need to be interpreted in an antirealist manner, and suits realist approaches as well.

There is one particular operational theory that not only recovers quantum mechanics, but generalises it by relaxing an important assumption according to which quantum events (which are, roughly, pairs of input and output systems connected via quantum operations) take place according to a definite (although possibly dynamical) causal structure. Such a theory, called the process matrix formalism, allows to investigate whether more general causal structures than those encountered in classical physics are allowed in the quantum realm. The connection between causal structures and spacetime geometries is tight, which makes this theory particularly relevant in order to study how quantum features can possibly impact the structure of spacetime (already at the non-relativistic level, and without taking gravity into account).

The process matrix formalism introduces the concept of quantum process, which is a function describing how n local quantum operations (noted A_j, with j being an integer ranging from 1 to n) are combined together to form a global operation. This notion is represented mathematically by a process matrix, noted W (see Figure 8.1). A process matrix then describes the causal order among local quantum operations, by encoding how their respective inputs and outputs are connected to each other. Importantly, this description makes no assumptions about the spatiotemporal locations of the various interacting parties performing the local operations.

In some analogy with the notion of quantum non-separability, a quantum process can be causally non-separable, which means that it is incompatible with any fixed (although possibly dynamical) causal structure among the local operations involved. By definition, in the bipartite case,[11] a causally separable process $W_{A,B}$ involving the two parties A and B is a process that can be expressed as a probabilistic mixture of

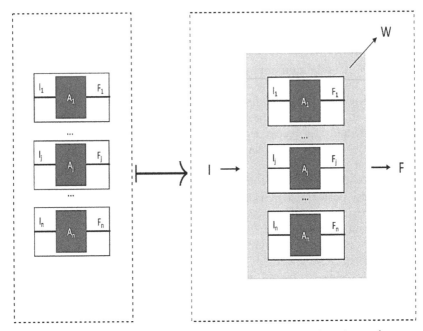

Figure 8.1 A generic quantum process. A quantum operation A_x sends some input quantum system described by a quantum state noted I_x on a final output quantum system described by a quantum state noted F_x. A quantum process W maps n such local quantum operations over a global operation noted A, of which the input is noted I and the output is noted F.

processes compatible with a given fixed causal structure (Oreshkov et al., 2012; Oreshkov and Giarmatzi, 2016):

$$W^{A,B} = q\,W^{A \preccurlyeq B} + (1 - q)\,W^{B \preccurlyeq A} \tag{8.2}$$

where q is a number between 0 and 1 and $W^{X \preccurlyeq Y}$ represents a process for which signalling is only possible from X to Y.

For now, this notion of causal non-separability is purely formal. In a realist perspective, it is however natural to investigate to what extent this generalized concept may encode (possibly novel) physical features. We now first focus on this question, addressing subsequent interpretative issues in the next section.

There are at least some causally non-separable processes that have a rather straightforward physical implementation (Wechs et al., 2021). Among those, a particular process (see Figure 8.2) called the *quantum switch* has been studied extensively these recent years (Chiribella et al., 2013; Oreshkov and Giarmatzi, 2016; Araújo et al., 2015). Two

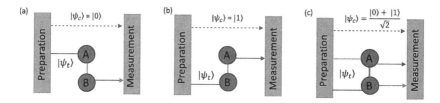

Figure 8.2 Diagram of the quantum switch. An initial preparation procedure prepares the initial quantum states of the target and
control systems. $|\psi_t\rangle$ represents the quantum state of the
control system. $|\psi_c\rangle$ represents the quantum state of target
system. A and B represent Alice and Bob's parties, respectively. A final measurement procedure is performed by third
party Fiona. The state of the control system is entangled to
the order between operations A and B. While scenarios (a)
and (b) represents definite causal orders, the scenario (c), in
which the control state is in a superposition of states, leads
to a superposition of causal orders between A and B. This corresponds to the quantum switch.

parties, Alice and Bob, perform an operation in their respective closed
local laboratory, on a shared system called the target system. This
system is entangled with a control qubit,[12] the state of which determines
the temporal order between Alice's and Bob's operations. More pre-
cisely, if the control qubit is in the state $|0\rangle$, the target system undergoes
Alice's operation (noted A) before undergoing Bob's one (noted B), and
vice versa if the qubit's state is in $|1\rangle$. A third party, Fiona, will perform
an operation on the control qubit after Alice and Bob made their operations on the target, erasing the information about the causal order
between Alice and Bob. If the control qubit is in a superposition of
states $|0\rangle$ and $|1\rangle$, then the process involves a superposition of causal
orders between Alice and Bob. It can be shown that the corresponding
process describing the causal order among those parties is indeed causally non-separable, i.e., it is incompatible with any fixed causal order.
In this case, we can say that the causal structure is indefinite.

The quantum switch has been experimentally implemented in a variety
of ways (Procopio et al., 2015; Rubino et al., 2017; Goswami et al.,
2018; Wei et al., 2019; Guo et al., 2020). Yet, objections have been
raised against the idea that those implementations are physical realizations of a genuine indefinite causal order. The indefinite character of
the causal order in the quantum switch indeed vanishes if we consider
the possibility of causal cycles between A and B (MacLean et al.,
2017). Yet, it is still possible, conceptually and experimentally, to
imagine implementations of the quantum switch in which such cycles

are not present. Another objection points out that only a gravitational quantum switch (i.e., involving an actual superposition of spacetime metrics) would constitute the only proper implementation of an indefinite causal order (Zych et al., 2019; Paunković and Vojinović, 2020). Indeed, it can be argued that the quantum switch, since it is defined within a classical non-relativistic spacetime, cannot guarantee a faithful description of spacetime at quantum scales. Yet, it can also be argued that the quantum switch, and more generally indefinite causal orders, point towards a tension between quantum features and classical spacetime. As such, their investigation can potentially lead to fruitful conceptual tools that could be of use in more advanced quantum theories of spacetime. Finally, some object that implementations of the quantum switch were not real processes, as each operation A and B were not performed once and only once (as is requested in the process matrix formalism). Oreshkov (2019) provides a technical argument against such worries, arguing that the formalism shows that operations within a process each takes place on specific input and output systems.

It seems therefore rather reasonable to consider that indefinite causal orders can possibly be found in physically implementable processes. The interesting question is now to articulate the interpretative implications of this generalized framework. Before doing so, and since the notion of causal non-separability has been defined in analogy with that of quantum non-separability, it is important to emphasize the extent and limits of this analogy in order to avoid potential unwarranted shortcuts or hasty conflations.

First, the notion of causal non-separability involves quantum processes, while quantum non-separability involves quantum states. These two notions are related, yet conceptually very distinct. A process links different operations, independently of the actual operations performed or of the actual systems undergoing those operations. A quantum state (mathematically represented by a density matrix) describes the state of a given physical system. In certain cases, a process can reduce to a density matrix, e.g., when a process describes separated operations (including the initial preparation procedure) performed on uncorrelated parts of a shared system, and of which we do not care about the outcomes. In that sense, processes generalize quantum states: a process can encode the same information as density matrices, but they usually allow much more, namely the encoding of correlations between different quantum operations. Because of these conceptually deep discrepancies, non-separability of quantum states and causal non-separability of quantum processes refer to different features. They characterize a certain kind of correlations between quantum states for the former, and between quantum events for the latter.[13] For this reason, causal non-separability can encode relations among relata that are timelike separated, contrary to quantum non-separability. Indeed, the various quantum events involved in a given quantum process, once embedded in spacetime, can be timelike separated, while the quantum states of entangled subsystems are (usually) spacelike separated. Yet, it is not the case that causal non-separability is a mere extension of quantum non-separability encoding correlations between quantum states at different times. Indeed, causal non-separability is about correlations among quantum operations, while the initial quantum state for these operations can be left unspecified. To summarise, we have two distinct notions. On the one hand, quantum non-separability expresses that for some composite non-separable quantum state, the quantum states of the subsystems is indefinite. As discussed in section 8.2.2, various metaphysical theories can be applied to non-separable states in order to assign them a precise meaning. On the other hand, causal non-separability, by contrast, expresses that for some quantum processes, the causal order among the operational events involved are indefinite. Assigning a meaning to this formal feature will constitute the task of the next section.

Since all the currently known causally non-separable processes that are physically implementable can be described within standard quantum mechanics, their characterization within the process matrix formalism can be regarded as purely formal—i.e., not capturing any novel physical features. However, it remains an open issue whether there exist causally non-separable processes that are instantiated in nature, without being describable within standard quantum mechanics. Because such a scenario may well bring new physical and metaphysical insights, it is the one that will be explored in the next section.

Before doing so, it is worth noting that, in the same way that quantum non-separability has a model-independent counterpart, namely non-local correlations, causal non-separability has such an equivalent too. Indeed, causal non-separability leads to correlations that may violate the causal equivalent of Bell inequalities, i.e., causal inequalities (Oreshkov et al., 2012; Branciard et al., 2015). Their violation necessarily indicates that the correlation is incompatible with any definite causal structure. Causal non-separability is necessary but non-sufficient for noncausal correlations (i.e., correlations violating a causal inequality) (Oreshkov et al., 2012; Wechs et al., 2019; Oreshkov and Giarmatzi, 2016; Araújo et al., 2015). No empirical protocol that generates noncausal correlations has been found so far (Wechs et al., 2021).

8.3.2 *The metaphysics of causal non-separability*

If we adopt a realist attitude towards causal non-separability, the nature of the related indefiniteness of causal orders will very much depend on the exact meaning assigned to the process matrices characterizing causal non-separability. In the case of metaphysically indefinite causal orders, the world itself is such that causal order among events is ontologically indefinite, independently of any observer. This claim needs to be further clarified, e.g., by relying on already existing metaphysical tools developed by previous work in the field of metaphysical indeterminacy (Wilson, 2013; Barnes and Williams, 2011), and in particular on its application to quantum theory (Calosi and Wilson, 2019; Calosi and Mariani, 2020; Calosi and Wilson, 2021). Quantum theory seems to violate a rather standard principle regarding the way in which properties are instantiated by physical systems. Contrary to what happens in classical physics, where every quantity gets assigned a definite value at all times, quantum theories are affected by what has been called *lack of value-definiteness* (LVD). In short, LVD indicates that quantum observables do not always possess precise values. LVD can be easily grasped by looking at the standard way of assigning values to physical systems given the quantum formalism, the so-called Eigenstate-Eigenvalue Link (EEL):

EEL. A physical system s has a definite value v of an observable $\mathcal{O}$ *iff* s is an eigenstate of $\mathcal{O}$.

Following Calosi and Wilson (2019), we can provide the following threefold classification of cases of LVD in quantum theory:

> *Superposition.* A linear combination $|\psi\rangle = q_1|\phi_1\rangle + q_2|\phi_2\rangle$ of different eigenstates $|\phi_1\rangle$ and $|\phi_2\rangle$ of an observable $\mathcal{O}$ is not always an eigenstate of $\mathcal{O}$. If a system S is in $|\psi\rangle$ it does not have a definite value of $\mathcal{O}$.
> *Incompatible Observables.* Consider two observables $\mathcal{O}_1$ and $\mathcal{O}_2$. The observables commute *iff* $[\mathcal{O}_1, \mathcal{O}_2] = \mathcal{O}_1\mathcal{O}_2 - \mathcal{O}_2\mathcal{O}_1 = 0$. If they do not, they are *incompatible*. If two observables are incompatible, they do not share all the same eigenstates. Thus if S is in one such non-shared eigenstate of $\mathcal{O}_1$ $(\mathcal{O}_2)$, it follows that it does not have a definite value for $\mathcal{O}_2$ $(\mathcal{O}_1)$.
> *Entanglement.* Consider an *entangled* system S_{12} composed by S_1 and S_2 with corresponding Hilbert space $\mathcal{H}_{12} = \mathcal{H}_1 \otimes \mathcal{H}_2$. S_{12} might be in an eigenstate $|\psi\rangle$ of $\mathcal{O}_{12} = \mathcal{O}_1 - \mathcal{O}_2$ that is neither an eigenstate of $\mathcal{O}_1$ nor an eigenstate of $\mathcal{O}_2$—with $\mathcal{O}_1$ and $\mathcal{O}_2$ defined on $\mathcal{H}_1$ and on $\mathcal{H}_2$ respectively. Both S_1 and S_2 will therefore lack a definite value for the corresponding observables.

Consider that, in each of these cases, applying the EEL entails that one or more observables do not always possess a definite a value. Such lack of definiteness has been taken at face value as to indicate the existence of an ontological kind of indeterminacy (a.k.a. *metaphysical indeterminacy*, henceforth MI), namely one that we cannot explain away as due to our ignorance or to semantic indecision. According to Calosi and Wilson (2019), MI is pervasive in quantum theory, and affects in one way or another every interpretation of quantum theory. We shall notice, however, that this claim is not so straightforward, and requires many details that we cannot enter here. For one, consider that the argument leading from LVD to the existence of quantum indeterminacy is essentially based on the EEL. The EEL, however, is rejected by any interpretation of quantum theory other than the Orthodox one. Therefore, in order to establish the existence of quantum indeterminacy in the context of other interpretations of the theory it seems that much more needs to be said. For our purposes, however, we simply notice that similar arguments have been put forward in recent years in many of the existing interpretations,[14] thus showing that, if not forced upon us, MI is at least to some extent suggested by quantum theory, and could then be a useful explanatory tool.[15]

One of the major problems with accepting the existence of quantum MI is that the very notion of ontological indeterminacy has been considered for many decades incoherent. Lewis (1986) was famously against this notion, claiming that "the only consistent account of [indeterminacy][16] is one that locates it in either our thought or our language." (p. 212) Moreover, Evans (1978) has influentially argued that indeterminate identity is contradictory, and many philosophers believed that indeterminate identity is required to make sense of MI. For all these reasons, philosophers have long been sceptical towards MI. This situation has started to change in the last decade, when several proposals have been developed as consistent ways to make sense of this idea. Two quite distinct families of approaches should be mentioned: the *meta-level* views, and the *object-level* views. Very roughly, the distinction between them is the following. According to the former, indeterminacy is understood as wordly unsettledness between fully precise alternatives. Therefore, on this view, there is MI when it is indeterminate which determinate state of affairs obtain. According to the latter, the *object-level* view, indeterminacy is understood as the obtainment of an indeterminate state of affairs. Thus, on this view, there is MI when an indeterminate state of affairs (determinately) obtains. Both approaches to MI have been exploited in order to make sense of the lack of value-definiteness in quantum theory, although it is fair to notice that the *object-level* view seems to be preferred (see Calosi and Mariani, 2020, for discussion).

Quantum MI has been so far discussed only in the context of non-relativistic quantum theory. Nonetheless, notice that the main arguments for its existence rely on certain features of the theory that we expect to find in any quantum theory, quantum gravity included. The major reason for such a pervasiveness of quantum MI is arguably that this view proposes an ontological account of superposition states. Consequently, it also seems quite natural to apply these tools to the case of a superposition of causal orders emerging in the PMF. To that end, we shall first require that *every* causal order expressed (or represented) by a given process matrix is to be taken as objective, i.e., independent from observers (in some standard sense). This assumption is largely justified if we consider that realism towards the PMF basically means, if anything, that this formalism gives us a language for expressing causal claims in the most generalized way (as input-output physical operations).

As we have shown, causally non-separable processes that are physically implementable display superpositions of indefinite causal orders which result from the entanglement of the causal structure with an ancillary system's state. Despite some technical differences between this case and that of indeterminacy in standard QM (differences that do not concern us here), this situation could be analyzed according to the same rationale found in Calosi and Mariani (2021), namely by interpreting such superpositions as instances of metaphysical indeterminacy. To recall, in the standard case a given observable may lack a *unique* determinate property, say *spin up* or *spin down*, while still instantiating the relevant determinable, say $spin_x$. In the case at hand, instead, we would take the determinable to be something like *event* E *being causally connected with event* F, and the corresponding determinates to be E *causing* F and F *causing* E. As we have shown, causally non-separable processes instantiate this determinable while they do not uniquely instantiate the corresponding determinates.[17] As a result, superposition of causal orders can *in principle* be interpreted as an instance of metaphysical indeterminacy in the same way as it is the superposition of values for standard quantum observables. The crucial assumption we are making for such an analogy to hold is obviously that there is a genuine *determinable-determinate* relation involved when we think of a causal relation between two distinct events. Notice however that, if the existence of a certain determinable property is to be taken seriously whenever there exists a physical observable that corresponds to it, our assumption looks justified since in the case of causally non-separable processes there is indeed an observable corresponding to the causal order.[18]

The main conceptual difference between indeterminacy of quantum states and that of causal orders is that in the standard case the indeterminacy affects the *state* of a given system and its properties. Here, instead, the indeterminacy affects the causal relations among systems. This difference is quite crucial for our purposes. Indeed, consider that metaphysical indeterminacy in the objective causal order between events seems to point towards indeterminacy in the spacetime structure itself (given some plausible assumptions concerning how causal structures relates to spacetime structures). Thus, the result of applying the tools of quantum MI to the PMF, along with the assumption that this formalism faithfully represents the objective causal structure of the world, seem to suggest that the spacetime structure is itself indeterminate. Of course, the most natural reaction to this is simply to see in these arguments a *modus tollens*, and so to reject one of the premises. However, as we will argue in the next section, the notion of *indeterminate spacetime* may be useful in the context of QG as well. If we are right, this would not only show that this notion should be taken seriously, but would also indirectly reinforce the argument just given by suggesting a certain continuity between PMF and QG.

8.4 Quantum metaphysics and quantum gravity

The aim of this section is to explore the relevance in the quantum gravity context of the quantum metaphysics tools and the recent work in quantum foundations we have discussed in sections 8.2 and 8.3. The main research programs in quantum gravity (QG) tend to suggest in one way or another that most (if not all) standard spacetime structures are not fundamental but emergent (Huggett and Wüthrich, 2013). This radical suggestion raises two main (sets of) philosophical issues (which, to some extent, run in parallel with the main physical and technical issues in QG). The first concerns the sense in which spacetime features can emerge from non-spatiotemporal structures, while the second is about the very characterization of a non-spatiotemporal physical ontology. A functionalist perspective has been recently articulated in order to elucidate the emergence issue: the idea is that spacetime functionally emerges from the non-spatiotemporal in the sense that under the right circumstances, the quantum gravitational structures can play the relevant spatio-temporal roles (Lam and Wüthrich, 2018).

It is interesting to note that, if functionalism has a wide range of applications in philosophy of science, this spacetime functionalism strategy that aims to provide a conceptual framework for the emergence of spacetime in QG is partly inspired by quantum metaphysics, more precisely by the functionalist understanding of standard 3-dimensional situations within wavefunction realism (see Lam and Wüthrich, 2018, p. 41, as well as references therein). To a certain extent, this case exemplifies a fruitful link between quantum metaphysics and QG.

Similarly, the metaphysical tools of structuralism, holism and indeterminacy we have discussed in section 8.2 can help to characterize a physical ontology in non-spatio-temporal terms.[19] Indeed, to the extent that approaches to QG aim to be quantum theories, it is largely expected that quantum entanglement and Bell-type correlations remain a central feature at the QG level. For instance, quantum entanglement plays a central role in loop quantum gravity (LQG) and group field theory, where there is a precise sense in which quantum entanglement constitutes a glueing relation among fundamental non-spatiotemporal degrees of freedom (e.g., see Baytaş et al., 2018, and Colafranceschi and Oriti, 2021). More broadly, quantum entanglement is also expected to play a role in the very emergence of spacetime, as suggested for instance in the context of the AdS/CFT string duality by the Ryu-Takayanagi relationship between entanglement entropy (in the boundary) and area (in the bulk).[20]

Now, to the extent that the structuralist and holistic accounts of quantum entanglement do not depend on spacetime, they can be articulated in the context of quantum gravity, and so may help to characterize its non-spatio-temporal ontology. Within the structuralist perspective, this latter is then understood in terms of entanglement structures, that is, in terms of networks of entanglement relations among (entangled) quantum gravitational subsystems, which are interdependent on one another. Since, in this context, quantum entanglement is the glueing relation holding things together-a kind of Lewisian worldmate relation that would be non-spatio-temporal (see the discussion in Darby, 2009; more recently, see Jaksland, 2021, who uses the notion of "world-making" relation)—quantum gravitational subsystems are entangled, and hence interdependent, in virtue of being part of the same world (or in virtue of giving rise to the same spacetime).[21] This is an interesting metaphysical implication of the structuralist perspective, suggesting some nontrivial insight into the nature of the physical world as described by some prominent current QG approaches.

The holistic account of quantum entanglement provides a slightly different (but not necessarily incompatible) perspective, emphasizing the ontological priority of the whole entangled quantum gravitational structure (over its entangled parts or subsystems). Following Ismael and Schaffer (2020)'s common ground strategy (see §8.2.2), it can then be argued that the quantum gravitational whole (e.g., the total, entangled spin network or spin foam structure in LQG) is best understood as the common ground of its quantum gravitational subsystems, some of which corresponding to spacetime regions. This naturally leads to consider the total, entangled quantum gravitational structure as the common ground for spacetime itself. This view does not necessarily clash with the functional emergence mentioned earlier, but can be seen as capturing an additional aspect of the relationship between the emergent spacetime and the underlying quantum gravitational structure.

The recent work in quantum foundations discussed in section 8.3 also connects to certain suggestions from research in QG. In order to articulate this potential link, we start by emphasizing the connection between spacetime and causal structures.

The link between causation and spacetime has been extensively discussed in several theories, both in metaphysics and in philosophy of science. To make but one prominent example, consider that in the standard Lewisian supervenience picture causation emerges from the mosaic of actual events plus spatiotemporal relations among them. In a way, Lewis starts from spacetime to get to causation, but in principle the converse can be done too. For instance, in the context of QG, causal set theory aims to recover spacetime, as described by (sectors of) general relativity, from fundamental, non (fully) spatio-temporal causal structures.[22] No matter what the priority relation between spacetime and causation is—or whether there is such a relation, or whether they are best conceived as being metaphysically on a par—nobody would deny that there is an intimate connection between spacetime and causation.

From this consideration, and under the assumption that the causal orders encoded in causally non-separable process matrices are objective, indefinite causal orders can be seen as reflecting indefiniteness in the spacetime structure itself. In this regard, as discussed in section 8.3.1, Zych et al. (2019) introduced a thought experiment called gravitational quantum switch, for which indefiniteness of causal order would arise from an actual superposition of spacetime regions in a quantum gravity context.[23] This clearly resonates with the idea, present in many approaches to QG, that spacetime itself could be in a quantum superposition. It also shows that, while indefinite causal orders point towards a tension between a classical background spacetime and quantum features such as superposition and entanglement, the implications of indefinite causal orders should be considered within the context of a fully gravitational theory.

The route leading from causal indefiniteness to spacetime indeterminacy is not straightforward, yet it may give us some initial motivations for exploring the possible role (and nature) of indeterminacy in approaches to QG. First, recall from §8.3.2 that one of the three instances of quantum indeterminacy individuated in the literature concerns the *non-commutativity* of the algebra of operators. This feature is an essential component for any quantum theory, QG included. A working hypothesis consists in individuating first the relevant operators in the theory that arguably describe spacetime properties. By doing so, and having shown that these do not commute, we would get an initial grasp on how superposition of spacetime regions could emerge. In a recent paper, Cinti et al. (2021) consider in detail two cases of non-commutative operators, one in string theory (ST) and one in LQG, that can both lead to spatiotemporal superpositions. In both cases, it can be shown that the operators describing geometric properties (the centre of mass position/momentum in ST, and the minimal length in the case of LQG) do not commute. By analogy with the standard argument given by Calosi and Wilson (2019), we could take these properties, along with the systems instantiating them, as instances of metaphysically indeterminate states of affairs. Of course, such a line of reasoning is to be developed further. For one, consider that the standard argument for quantum indeterminacy is based on the EEL (as shown in §8.3.2). Clearly though, nothing like the EEL is explicitly endorsed in the approaches to QG we have mentioned. Although this is an interesting issue *per se*, it has to be noticed that even in the case of QG, we would likely need to provide a way to ascribe values to physical properties. It may be expected that at least in some way of doing so, the argument based on the EEL could then be given.

Allowing for indeterminacy in the fundamental spacetime structure of QG may also provide a novel, rather provocative perspective on the most debated philosophical issue in the QG context. Indeed, as we mentioned, it is widely argued that spacetime is not fundamental in QG. This conclusion often relies on the assumption that quantum superpositions of (spacetime-like) structures at the fundamental level cannot be understood in any spatio-temporal sense. For instance, quantum superpositions of spin networks (or spin foams) are generic in LQG, and this is often taken as indicating that spacetime vanishes at this level, and so is not fundamental (see e.g., Rovelli, 2004, § 6.7.1, in the physics literature, and Huggett and Wüthrich, 2013, § 2.3, in the philosophy of physics literature). Up to now, and to the best of our knowledge, the interpretative and metaphysical strategy of considering (certain) spacetime structures as being indeterminate in QG (in some appropriate sense) has not been seriously investigated (at least in the current philosophy of physics literature). In certain cases, it seems that spacetime fundamentality is rejected because spacetime indeterminacy is rejected.[24] We notice that such a rejection of spacetime indeterminacy is only justified if we have good reasons to believe that this notion does not make sense, or even that it is inconsistent. However, there are now various proposals to make sense of metaphysical indeterminacy, and this notion has been argued to be explanatory useful already in the context of non-relativistic QM. This points to the intriguing suggestion that what is taken as the non-fundamentality of spacetime (or of certain spacetime features) in the QG context could also be understood in terms of some spacetime indeterminacy. Articulating this suggestion in details and its implications for the debates around the emergence of spacetime thus constitutes a worthwhile project that may shed an interesting new light on spacetime in QG.

8.5   Conclusion

A great deal of the recent philosophical discussions on non-relativistic quantum mechanics can be seen as an attempt to provide an *interpretation-neutral* understanding of the most crucial features of this theory, namely *entanglement* and *superposition*. The line of reasoning we developed in this chapter starts by considering how pervasive these features are, even once we look well beyond quantum mechanics and we start taking into account the recent approaches to QG. As we have shown, and despite QG being in its early stage of development, these key aspects of non-relativistic quantum mechanics, instead of being erased or overcome, will likely remain crucial for our understanding of the natural world. The main suggestion of this chapter is that we should look at the metaphysical views that have already been developed within the philosophy of quantum mechanics to provide a more systematic understanding of these features, and we apply similar approaches to QG. In that spirit, we first reviewed two stances, holism and structuralism, as strategies to interpret quantum *entanglement*. We argued that, to the extent that these views are already meant to describe non-spatiotemporal features of reality, they can be used as a conceptual basis for developing an ontology for QG. Second, we have shown how the notion of quantum indeterminacy has been articulated in the context of standard quantum mechanics as an explanation for *superposition* states. Similarly, this notion has proven useful, as we have argued in details, to make sense of the notion of *causal superposition* within the process matrix formalism. This can arguably lead, as we suggested, to a form of spacetime indeterminacy which may provide a novel perspective on the thorny issue of the emergence of spacetime in QG.

Acknowledgements

We wish to thank Antonio Vassallo for the invitation to contribute to this volume and for his valuable comments. We are also grateful to Cyril Branciard, Claudio Calosi, Hippolyte Dourdent and Daniele Oriti for helpful exchanges. L.L.'s work was supported by the Agence Nationale de la Recherche under the programme "Investissements d'avenir" (ANR-15-IDEX-02), and the Formal Epistemology Project funded by the Czech Academy of Science. V.L. acknowledges support from the Swiss National Science Foundation (PP00P1_170460). C.M.'s work was supported by the Foundational Questions Institute Fund (Grant number FQXi-IAF19-05 and FQXi-IAF19-01).

Notes

1. For instance, see Calosi and Mariani (2020) for indeterminacy within the main realist interpretations of QM, and Esfeld (2017) for structuralism.
2. This failure of supervenience is often referred to as a form of non-separability in the philosophical literature (see recently Ismael and Schaffer, 2020); in the physics literature, quantum non-separability often specifically denotes the non-factorizability of entangled quantum states, as it is reflected by equation (8.1).
3. Indeed, we here focus on entanglement leading to the violation of some Bell-type inequalities, and hence to some non-local correlations among the entangled subsystems—in particular, note that all pure entangled states lead to the violation of a Bell-type inequality, whereas mixed entangled states may not violate any Bell-type inequalities.
4. In this context, a model-independent notion is one that does not appeal to any specific machinery, tool or apparatus; the experimental setup is reduced to a black box fed with some inputs and returning some output.
5. By this, of course, we do not mean to argue that non-locality is unavoidable. As a matter of fact, even within a broadly realist approach, there are accounts that may escape this conclusion. Examples include the acceptance

of retrocausality (Price, 2012; Leifer and Pusey, 2017; Friederich and Evans, 2019), versions of superdeterminism ('t Hooft, 2016), and perhaps some versions of the many-worlds approach to QM (Vaidman, 2021). It is highly debated whether any of these strategies really help us avoiding non-locality—see Myrvold et al. (2021) for discussion.

6. Arguably, even within Bohmian mechanics, the non-local modal connections among Bohmian particles cannot be accounted for only in terms of intrinsic (and local) properties of the particles (Lam, 2016).

7. See for instance Lam (2013), who writes that the "interdependent existence of fundamental entangled quantum systems [is] one of the main morals of QM, which OSR aims to encode" (p. 65). As a matter of fact, while Calosi and Morganti (2021) critically analyze the standard version of OSR at length, they do not really discuss the *moderate* version in any depth. The relationships between *coherentism* and *structuralism* are obviously intricate, and yet we simply point out that in both views the main explanatory job is done by relations of *symmetric dependence*. Thus, we take it that the two views are not necessarily in conflict

8. We believe that the question of what, if not spacetime, makes such structures physical (Lam and Esfeld, 2013, §3) can be answered in functionalist terms (Lam and Wüthrich, 2018).

9. Note, however, that *grounding* is more commonly understood as an explanatory relation between facts (Fine, 2012; Correia and Schnieder, 2012). On Schaffer's view, instead, *grounding* applies to every kind of entity (not just facts), which is also why his notion of grounding closely resembles that of *ontological dependence* (Fine, 1994; Tahko and Lowe, 2020).

10. As a reviewer correctly points out, in order to get the holistic conclusion here we need the further claim according to which the 3D world is *part* of the wavefunction. This claim—recently defended at length by Ney (2021b)—is not however straightforward and can be resisted.

11. A generalization to the multipartite case can be found in Oreshkov and Giarmatzi (2016) and Wechs et al. (2019).

12. A qubit is a quantum system of which the quantum state pertains to a two-dimensional Hilbert space, and that can be in any quantum superposition of two independent states forming a complete basis of this Hilbert space.

13. We recall that process matrices do not encode quantum states *per se*. They only connect local operations to form a global one. As such, they correlate quantum events, which are distinct from the standard notion of events in relativity (namely spacetime points). Instead, in this context, a quantum event refers to a pair of input and output connected by a quantum operation.

14. In particular, notice that in the context of spontaneous collapse interpretations of QM, indeterminacy can arise even by revising the EEL. See, e.g., Lewis (2016); Albert and Loewer (1996); Mariani (2022). For a critique, see Glick (2017).

15. The Editor points out—we think correctly—that appealing to indeterminacy here does not provide a further explanation as to why the indeterminacy occurs. In a way, the existence of the indeterminate states of affairs is to be taken as a brute fact. However, what the accounts of metaphysical indeterminacy can do is to elucidate and explain what *it means* for a state of affairs to be indeterminate, and to to distinguish indeterminate from determinate states of affairs. It is in this sense that we contend the accounts can be useful tools. We thank the reviewer for inviting us to say more on this.

16. *Editor's note*: Lewis uses the word "vagueness" in the original source.

17. It seems safe to presume that the resulting indeterminacy would be of the *glutty* kind, with both determinates instantiated at the same time. The *gappy* version seems ruled out by the simple fact that, if neither E *causes* F nor F *causes* E are instantiated, the whole causal structure displayed by a causally non-separable process could not be empirically tested in the first place.

18. The order in which the transformations A and B are applied on the target system can be seen as a quantum property of the particular process that describes the quantum switch (for which, by definition, the initial quantum state of the control system is fixed in a superposition of states (in the $\{|0\rangle, |1\rangle\}$ basis)). We note that property q, such that $q = 1$ means "The target system has gone through laboratory A and then laboratory B," and $q = 2$ means "The target system has gone through laboratory B and then laboratory A." By construction of the QS, the observable corresponding to the property q can be measured by measuring the state of the control system (in the $\{|0\rangle, |1\rangle\}$ basis). The corresponding observables $\mathcal{O}$, $\mathcal{O}_{AB}$ and $\mathcal{O}_{BA}$ are mathematically expressed as follows:

$$\mathcal{O} = \mathcal{O}_{AB} + \mathcal{O}_{BA} = |0\rangle\langle0| + |1\rangle\langle1| = \mathbb{1} \tag{8.3}$$

with

$$\mathcal{O}_{AB} = |0\rangle\langle0| \tag{8.5}$$

$$\mathcal{O}_{BA} = |1\rangle\langle1| \tag{8.6}$$

The observables $\mathcal{O}_{AB}$ and $\mathcal{O}_{BA}$ can be applied on the process W_{QS} itself, acting as a measurement of property q. First, it can be showed that W_{QS} can be expressed as follows:

$$|W_{QS}\rangle = \frac{1}{\sqrt{2}}(|\psi\rangle^{A^I}|\mathbb{1}\rangle^{A^OB^I}|\mathbb{1}\rangle^{B^OF^I}|0\rangle^{F^c} + |\psi\rangle^{B^I}|\mathbb{1}\rangle^{B^OA^I}|\mathbb{1}\rangle^{A^OF^I}|1\rangle^{F^c})$$
$$\tag{8.7}$$

with $|\psi\rangle^{A^I}$ ($|\psi\rangle^{B^I}$) being the quantum state of the target system at the entrance of laboratory A (B), $|0\rangle^{F^c}$ and $|1\rangle^{F^c}$ are the quantum states of the control system at the entrance of the third party's laboratory (Fiona's), and $|\mathbb{1}\rangle^{XY}$ is the Choi vector of the identity channel sending the output quantum state of laboratory X on the input of laboratory Y.

If we apply, say the observable $\mathcal{O}_{AB}$ on W_{QS}, we indeed "project" the process W_{QS} on the process W_{QS}^{AB} which instantiate a definite causal order in which A causally precedes B:

$$\frac{\mathcal{O}_{AB}W_{QS}}{\sqrt{W_{QS}\mathcal{O}_{AB}\mathcal{O}_{AB}W_{QS}}} = |\psi\rangle^{A^I}|\mathbb{1}\rangle^{A^OB^I}|\mathbb{1}\rangle^{B^OF^I}|0\rangle^{F^c} = |W_{QS}^{AB}\rangle \tag{8.8}$$

Similarly, applying the observable $\mathcal{O}_{BA}$ on W_{QS} will project the process on W_{QS}^{BA} which instantiate a definite causal order in which B causally precedes A.

19. In the light of our discussion, it looks quite natural to consider in more details how the metaphysics of structuralism, holism, and coherentism interact with the hypothesis of quantum indeterminacy. This is an intricate issue, and although we do not aim to provide a systematic view in this chapter, we believe it is important to provide a few lines on our own perspective on this. What is common to the various metaphysical strategies for interpreting entanglement is the attempt to provide an *explanation* to some peculiar relations of dependence between distinct systems. It is in fact no surprise that the discussion on entanglement gets intimately connected to the issue of *fundamentality*, which is seen in contemporary philosophy as a way of regimenting the notion of explanation itself. On the other hand, it seems quite natural to interpret much of what goes on in the literature on quantum indeterminacy as an attempt to indicate an inherent lack of completeness in our explanations of certain physical facts, for instance the systems in a superposition state. But more importantly, if we consider the indeterminacy which affects entangled systems, it appears immediately clear that by adding (indeterminate) facts about the component systems we do not get to a full explanation of the dependence relations exhibited by the composite system. Therefore, one would expect that the lack of completeness in the explanation one finds with quantum indeterminacy could be supplemented by the specific metaphysics of entanglement. The details will then depend on the specific metaphysics one has in mind among the ones on the market. For instance, take structuralism, according to which (very roughly!) entanglement can be explained by positing fundamental relations on top of (or along with) the physical objects. By adding quantum indeterminacy to the picture, one could expect that the fundamental relations of entanglement (instantiated by the composite system) provides an explanation to the existence of the indeterminacy affecting the component systems. We thank an anonymous reviewer of this journal for a very insightful discussion and for inviting us to say more on this important issue.

20. See for instance Ryu and Takayanagi (2006) and Van Raamsdonk (2010). See Jaksland (2021) for a philosophical discussion of how distance can be recovered from quantum entanglement, and Ney (2021a) for a critical perspective. The Ryu-Takayanagi formula is expected to generalize beyond the AdS/CFT correspondence, such as within the framework of LQG and group field theory, see Chirco et al. (2018).

21. There may well be sectors of QG that do not give rise to any classical spacetime, and in particular that may not be entangled with anything else. This then raises the question of what makes them part of the same world.

22. A strong motivation for causal set theory is provided by a series of important results in general relativity basically showing that the geometry of spacetime is fixed by its causal structure (up to a conformal factor—and assuming the condition of past and future distinguishability; see Huggett and Wüthrich, forthcoming, ch. 2, for a recent philosophical discussion of causal set theory, as well as references therein).

23. Zych et al. (2019) emphasize that spacetime in the standard quantum switch is classical, in contrast to the gravitational quantum switch. Indeed, indefinite causal order in the quantum switch is the result of a temporal order entangled with some ancillary system, while the spacetime background is classical. This leads to several differences with the gravitational version, in

which one has an explicit superposition of spacetime regions, originating from a mass superpositions linked to spacetime via the principles of general relativity. In the standard quantum switch, only very specific causal orders escape to a classical description, and the time of the events depends on their temporal order. This is not the case in the gravitational quantum switch, in which an entire region of spacetime is non-classical (in a quantum superposition), and within this region, the time of events does not depend on their temporal order.

24. The situation is actually more subtle than that: as we have mentioned earlier, there may well be sectors of QG that do not give rise to any classical spacetime (e.g., in some limit), and so that are not spatio-temporal in a way that is not directly related to quantum indeterminacy (thanks to Daniele Oriti for highlighting this point to us).

Bibliography

Aharonov, Y., J. Anandan, and L. Vaidman (1993). Meaning of the wave function. *Physical Review A* 47 (6), 4616.

Albert, D. Z. and B. Loewer (1996). Tails of Schrödinger's cat. In R. Clifton (Ed.), *Perspectives on Quantum Reality*, pp. 81–92. Springer.

Araújo, M., C. Branciard, F. Costa, A. Feix, C. Giarmatzi, and C. Brukner (2015). Witnessing causal nonseparability. *New Journal of Physics* 17 (10), 102001.

Barnes, E. and J. R. G. Williams (2011). A theory of metaphysical indeterminacy. In D. W. Zimmerman and K. Bennett (Eds.), *Oxford Studies in Metaphysics*, Volume 6, pp. 103–148. Oxford University Press.

Baytaş, B., E. Bianchi, and N. Yokomizo (2018). Gluing polyhedra with entanglement in loop quantum gravity. *Physical Review D* 98, 026001.

Bell, J. (1964). On the Einstein-Podolsky-Rosen paradox. *Physics* 1 (3), 195–200.

Branciard, C., M. Araújo, A. Feix, F. Costa, and C. Brukner (2015). The simplest causal inequalities and their violation. *New Journal of Physics* 18 (1), 013008.

Calosi, C. and C. Mariani (2020). Quantum relational indeterminacy. *Studies in History and Philosophy of Modern Physics* 71, 158–169.

Calosi, C. and C. Mariani (2021). Quantum indeterminacy. *Philosophy Compass* 16 (4), e12731.

Calosi, C. and M. Morganti (2021). Interpreting quantum entanglement: Steps towards coherentist quantum mechanics. *The British Journal for the Philosophy of Science* 72 (3), 865–891.

Calosi, C. and J. Wilson (2019). Quantum metaphysical indeterminacy. *Philosophical Studies* 176 (10), 2599–2627.

Calosi, C. and J. Wilson (2021). Quantum indeterminacy and the double-slit experiment. *Philosophical Studies* 178, 3291–3317.

Chen, E. K. (2021). Quantum mechanics in a time-asymmetric universe: On the nature of the initial quantum state. *The British Journal for the Philosophy of Science* 72 (4), 1155–1183.

Chirco, G., D. Oriti, and M. Zhang (2018). Group field theory and tensor networks: Towards a Ryu–Takayanagi formula in full quantum gravity. *Classical and Quantum Gravity* 35, 115011.

Chiribella, G., G. M. D'Ariano, P. Perinotti, and B. Valiron (2013). Quantum computations without definite causal structure. *Physical Review A* 88 (2), 022318.

Cinti, E., C. Mariani, and M. Sanchioni (2021). Lack of value definiteness in quantum gravity. *arXiv:2109.10339* [physics.hist-ph].

Colafranceschi, E. and D. Oriti (2021). Quantum gravity states, entanglement graphs and second-quantized tensor networks. *Journal of High Energy Physics* 52.

Correia, F. and B. Schnieder (2012). *Metaphysical Grounding: Understanding the Structure of Reality*. Cambridge University Press.

Darby, G. (2009). Lewis's worldmate relation and the apparent failure of Humean supervenience. *Dialectica* 63 (2), 195–204.

Esfeld, M. (2017). How to account for quantum non-locality: Ontic structural realism and the primitive ontology of quantum physics. *Synthese* 194 (7), 2329–2344.

Evans, G. (1978). Can there be vague objects? *Analysis* 38 (4), 208.

Fine, K. (1994). Ontological dependence. *Proceedings of the Aristotelian Society* 95 (1), 269–290.

Fine, K. (2012). Guide to ground. In F. Correia and B. Schnieder (Eds.), *Metaphysical Grounding: Understanding the Structure of Reality*, pp. 37–80. Cambridge University Press.

Friederich, S. and P. W. Evans (2019). Retrocausality in quantum mechanics. In E. N. Zalta (Ed.), *The Stanford Encyclopedia of Philosophy* (Summer 2019 ed.). Metaphysics Research Lab, Stanford University. https://plato.stanford.edu/archives/sum2019/entries/qm-retrocausality/.

Glick, D. (2017). Against quantum indeterminacy. *Thought* 6 (3), 204–213.

Goswami, K., C. Giarmatzi, M. Kewming, F. Costa, C. Branciard, J. Romero, and A. White (2018). Indefinite causal order in a quantum switch. *Physical Review Letters* 121 (9), 090503.

Guo, Y., X.-M. Hu, Z.-B. Hou, H. Cao, J.-M. Cui, B.-H. Liu, Y.-F. Huang, C.-F. Li, G.-C. Guo, and G. Chiribella (2020). Experimental transmission of quantum information using a superposition of causal orders. *Physical Review Letters* 124 (3), 030502.

Healey, R. (2016). Holism and nonseparability in physics. In E. N. Zalta (Ed.), *The Stanford Encyclopedia of Philosophy* (Spring 2016 ed.). Metaphysics Research Lab, Stanford University. https://plato.stanford.edu/archives/spr2016/entries/physics-holism/.

Hensen, B., H. Bernien, A. E. Dréau, A. Reiserer, N. Kalb, M. S. Blok, J. Ruitenberg, R. F. Vermeulen, R. N. Schouten, C. Abellán, et al. (2015). Loopholefree Bell inequality violation using electron spins separated by 1.3 kilometres. *Nature* 526 (7575), 682–686.

Hitchcock, C. and M. Rédei (2021). Reichenbach's common cause principle. In E. N. Zalta (Ed.), *The Stanford Encyclopedia of Philosophy* (Summer 2021 ed.). Metaphysics Research Lab, Stanford University. https://plato.stanford.edu/archives/sum2021/entries/physics-Rpcc/.

Huggett, N. and C. Wüthrich (2013). Emergent spacetime and empirical (in)coherence. *Studies in History and Philosophy of Modern Physics* 44 (3), 276–285.

Huggett, N. and C. Wüthrich (forthcoming). *Out of Nowhere: The Emergence of Spacetime in Quantum Theories of Gravity*. Oxford University Press.

Ismael, J. and J. Schaffer (2020). Quantum holism: Nonseparability as common ground. *Synthese* 197, 4131–4160.

Jaksland, R. (2021). Entanglement as the world-making relation: Distance from entanglement. *Synthese* 198, 9661–9693.

Lam, V. (2013). The entanglement structure of quantum field systems. *International Studies in the Philosophy of Science* 27, 59–72.

Lam, V. (2016). Quantum structure and spacetime. In T. Bigaj and C. Wüthrich (Eds.), *Metaphysics in Contemporary Physics*, pp. 81–99. Brill/Rodopi.

Lam, V. (2017). Structuralism in the philosophy of physics. *Philosophy Compass* 12 (6), e12421.

Lam, V. and M. Esfeld (2013). A dilemma for the emergence of spacetime in canonical quantum gravity. *Studies in History and Philosophy of Modern Physics* 44, 286–293.

Lam, V. and C. Wüthrich (2018). Spacetime is as spacetime does. *Studies in History and Philosophy of Modern Physics* 64, 39–51.

Leifer, M. S. and M. F. Pusey (2017). Is a time symmetric interpretation of quantum theory possible without retrocausality? *Proceedings of the Royal Society A* 473 (2202), 20160607.

Letertre, L. (2021). The operational framework for quantum theories is both epistemologically and ontologically neutral. *Studies in History and Philosophy of Science* 89, 129–137.

Lewis, D. (1986). *On the Plurality of Worlds*. Blackwell.

Lewis, P. J. (2016). *Quantum Ontology. A Guide to the Metaphysics of Quantum Mechanics*. Oxford University Press.

MacLean, J.-P. W., K. Ried, R. W. Spekkens, and K. J. Resch (2017). Quantum-coherent mixtures of causal relations. *Nature Communications* 8 (1), 1–10.

Mariani, C. (2022). Non-accessible mass and the ontology of GRW. *Studies in History and Philosophy of Science* 91, 270–279.

Miller, E. (2016). Quantum holism. *Philosophy Compass* 11 (9), 507–514.

Myrvold, W., M. Genovese, and A. Shimony (2021). Bell's theorem. In E. N. Zalta (Ed.), *The Stanford Encyclopedia of Philosophy* (Fall 2021 ed.). Metaphysics Research Lab, Stanford University. https://plato.stanford.edu/archives/fall2021/entries/bell-theorem/.

Ney, A. (2021a). From quantum entanglement to spatiotemporal distance. In C. Wüthrich, B. Le Bihan, and N. Huggett (Eds.), *Philosophy Beyond Spacetime: Implications from Quantum Gravity*, pp. 78–102. Oxford University Press.

Ney, A. (2021b). *The World in the Wavefunction*. Oxford University Press.

Oreshkov, O. (2019). Time-delocalized quantum subsystems and operations: On the existence of processes with indefinite causal structure in quantum mechanics. *Quantum* 3, 206.

Oreshkov, O., F. Costa, and Č. Brukner (2012). Quantum correlations with no causal order. *Nature Communications* 3 (1), 1–8.

Oreshkov, O. and C. Giarmatzi (2016). Causal and causally separable processes. *New Journal of Physics* 18 (9), 093020.

Paunković, N. and M. Vojinović (2020). Causal orders, quantum circuits and spacetime: Distinguishing between definite and superposed causal orders. *Quantum* 4, 275.

Price, H. (2012). Does time-symmetry imply retrocausality? How the quantum world says 'maybe'? *Studies in History and Philosophy of Modern Physics* 43 (2), 75–83.

Procopio, L. M., A. Moqanaki, M. Araújo, F. Costa, I. A. Calafell, E. G. Dowd, D. R. Hamel, L. A. Rozema, Č. Brukner, and P. Walther (2015). Experimental superposition of orders of quantum gates. *Nature Communications* 6 (1), 1–6.

Rovelli, C. (2004). *Quantum Gravity*. Cambridge University Press.

Rubino, G., L. A. Rozema, A. Feix, M. Araújo, J. M. Zeuner, L. M. Procopio, C. Brukner, and P. Walther (2017). Experimental verification of an indefinite causal order. *Science Advances* 3 (3), e1602589.

Ryu, S. and T. Takayanagi (2006). Holographic derivation of entanglement entropy from the anti–de Sitter space/conformal field theory correspondence. *Physical Review Letters* 96, 181602.

Schaffer, J. (2009). On what grounds what. In D. J. Chalmers, D. Manley, and R. Wasserman (Eds.), *Metametaphysics*, pp. 347–283. Oxford University Press.

Tahko, T. E. and E. J. Lowe (2020). Ontological dependence. In E. Zalta (Ed.), *The Stanford Encyclopedia of Philosophy* (Fall 2020 ed.). Metaphysics Research Lab, Stanford University. https://plato.stanford.edu/archives/fall2020/entries/dependence-ontological/.

't Hooft, G. (2016). *The Cellular Automaton Interpretation of Quantum Mechanics*. Springer.

Vaidman, L. (2021). Many-worlds interpretation of quantum mechanics. In E. Zalta (Ed.), *The Stanford Encyclopedia of Philosophy* (Fall 2021 ed.). Metaphysics Research Lab, Stanford University. https://plato.stanford.edu/archives/fall2021/entries/qm-manyworlds/.

Van Raamsdonk, M. (2010). Building up spacetime with quantum entanglement. *General Relativity and Gravitation* 42, 2323–2329.

Wallace, D. (2019). What is orthodox quantum mechanics? In A. Cordero (Ed.), *Philosophers Look at Quantum Mechanics*, pp. 285–312. Springer.

Wechs, J., A. A. Abbott, and C. Branciard (2019). On the definition and characterisation of multipartite causal (non) separability. *New Journal of Physics* 21, 013027.

Wechs, J., H. Dourdent, A. A. Abbott, and C. Branciard (2021). Quantum circuits with classical versus quantum control of causal order. *PRX Quantum* 2(3), 030335.

Wei, K., N. Tischler, S.-R. Zhao, Y.-H. Li, J. M. Arrazola, Y. Liu, W. Zhang, H. Li, L. You, Z. Wang, et al. (2019). Experimental quantum switching for exponentially superior quantum communication complexity. *Physical Review Letters* 122 (12), 120504.

Wilson, J. M. (2013). A determinable-based account of metaphysical indeterminacy. *Inquiry* 56 (4), 359–385.

Zych, M., F. Costa, I. Pikovski, and Č. Brukner (2019). Bell's theorem for temporal order. *Nature Communications* 10 (1), 1–10.

9 Four attitudes towards singularities in the search for a theory of quantum gravity

Karen Crowther and
Sebastian De Haro

9.1 Introduction

Singularities feature prominently in the best theories of fundamental physics: *quantum field theory* (QFT), being the framework within which the Standard Model of particle physics is formulated, describing all fundamental particles and forces, and *general relativity* (GR), describing gravity as the curvature of spacetime. These singularities are of various types, prompting differing diagnoses in regards to what they suggest about the status of these theories and the development of future theories. At least some of them, however, are standardly interpreted as motivating the search for a more fundamental theory: *quantum gravity* (QG). Furthermore, the appearance of these singularities in GR and QFT is often taken to suggest some features of QG that would render the singularities in the less-fundamental theories unproblematic; i.e., there is an expectation that the new theory will *resolve*, or remove, particular singularities, and explain their appearance in current theories. Singularities are thus usually treated not only as motivation to look for the new theory, but also as providing valuable insights as to the form of this theory. This is of great importance in the search for QG, given the paucity of empirical motivations, guiding principles, and constraints available to aid its development.

Given their prominence and potential value, it is worthwhile to more thoroughly investigate the significance of singularities in GR and QFT in regards to what they may suggest for the search for QG. In particular, it is interesting to contrast the different attitudes towards the different singularities in these theories, and to ask if the conjectured implications for QG are well-motivated. This is the aim of the present essay.

We begin by considering two types of spacetime singularities in GR: geodesic incompleteness (§9.2.1) and curvature singularities (§9.2.2). Already there is disagreement between the prevailing attitude in physics compared to that in philosophy regarding the meaning of these singularities in GR. In physics, spacetime singularities are usually said to represent the "breakdown" of GR, and thus to point to the need

DOI: 10.4324/9781003219019-12

for QG. One detects the opposite attitude in philosophy, as some prominent literature tries to make precise the sense in which GR "breaks down," and finds no answer that warrants an accusation of incompleteness of the theory. We outline some arguments for why each of these types of singularities may be considered problematic, prompting the need for resolution. In particular, §9.2.3 presents an argument for how curvature singularities may be said to signal the "breakdown" of GR, which we believe has been under-appreciated in the philosophical literature.

We then consider two types of singularities in QFT: UV-divergences which are typically thought to stem from the use of perturbation theory (§9.3.1), and Landau poles, which are UV-divergences not typically thought to stem from the use of perturbation theory (§9.3.2). Following this (§9.3.3), we consider the divergences in GR when treated perturbatively in the framework of QFT (i.e., those associated with the non-renormalizability of the Einstein-Hilbert action), and a potential resolution proposed by the *asymptotic safety scenario*. We find, in §9.3.4, four possible stances towards the singularities of QFT.

These four stances are cases of four more-general classes of attitudes towards singularities in current theories. In §9.4, we outline this classification of four attitudes towards singularities, which we base primarily on a survey of the physics literature. While there seems to be a universal consensus that at least *some* singularities must or will be resolved, not all authors think that *all* singularities must or will be resolved. Also, there is no consensus about which singularities are to be resolved, nor about the stage of theory development at which this will happen (e.g., singularity resolution in the classical or in the quantum theory).

Briefly, the four attitudes are:

1   Singularities are resolved classically, or "at the level of current theories": the (particular) singularities do not point to QG.
2   Singularities are resolved in QG.
3   Peace with singularities: they are not resolved at any level, because there is reason to keep them in our theories.
4   Indifference to singularities: the singularities are of no significance.

On our classification scheme, it is possible to adopt different attitudes towards different types of singularities (and, indeed, this seems desirable). We present examples of each of the four attitudes, but we do not evaluate the arguments for adopting one attitude rather than another in regards to specific singularities. Two of the attitudes prompt singularity resolution, but only one suggests the need for a theory of QG. In §9.5 we briefly discuss some factors which tend to influence the choice between the two attitudes supporting singularity resolution.

9.2   Singularities in GR

Spacetime singularities are pathologies of a spacetime, and there are various ways in which spacetimes can be singular.[1] Here, we are concerned with the two most common types of spacetime singularity: geodesic incompleteness (§9.2.1), and curvature singularities (§9.2.2). Here we assume that the two notions are independent of each other.[2]

9.2.1   Geodesic incompleteness

The most common definition of spacetime singularity uses *geodesic incompleteness*: a spacetime is singular if and only if it contains an incomplete, inextendible timelike geodesic. Such a geodesic is the worldline of a freely falling test object, and the property that makes it singular is that the worldline ends within finite proper time and cannot be further extended.[3] This definition forms the basis of the Penrose and Hawking singularity theorems,[4] though it is not without problems (see Curiel, 2021, §1.1).

Geodesic incompleteness appears to be a genuine physical worry, because it means that "particles could pop in and out of existence right in the middle of a singular spacetime, and spacetime itself could simply come to an end, though no fundamental physical mechanism or process is known that could produce such effects" (Curiel, 1999, p. S140). Geodesic incompleteness leads to a lack of predictability and determinism, and so could indicate that the theory is incomplete (Earman, 1995, §2.6).[5] If the breakdown of determinism were visible to external observers, "then those observers would be sprayed by unpredictable influences emerging from the singularities" (Earman, 1992, p. 171). This would represent a nasty form of inconsistency—as Earman puts it, the laws would "perversely undermine themselves."

Various forms of "cosmic censorship" have been proposed in order to render these singularities harmless. One of these is *weak cosmic censorship*, which states that the offending singularities are hidden behind event horizons, and thus outside observers are shielded from being sprayed by any inexplicable influences. Yet, in this case, though the theory would be saved from perversely undermining itself, it would still be incomplete: if spacetime exists beyond the horizon, and if GR does not determine what occurs there, then the theory is incomplete.[6] Such singularities would then suggest the need for modification of GR, or for a new theory, or some other solution (§9.4). Being hidden behind an event horizon is not enough for to render a singularity unproblematic.

The breakdown of determinism inside black holes occurs beyond the *Cauchy horizon* (inside the event horizon): beyond this surface, the Einstein equations no longer give a unique solution. In response, Penrose

(1979) proposed *strong cosmic censorship* (SCC), which postulates that the appearance of the Cauchy horizon in Schwarzschild black holes is non-generic, and that the interior region of these black holes is in some way unstable (under small perturbation of initial data) in the vicinity of the Cauchy horizon. Any passing gravitational waves would prevent the formation of Cauchy horizons, meaning that instead, spacetime would terminate at a "spacelike singularity," across which the metric is inextendable.

SCC ensures that no violations of predictability are detectable even by local observers (i.e., an astronaut on a geodesically incomplete worldline would detect nothing up until, and presumably after, her disappearance), and so, the truth of this conjecture would render any singularities (incomplete geodesics) harmless in regards to determinism. As Dafermos and Luk (2017, p. 5) states, "The singular behaviour of Schwarzschild, though fatal for reckless observers entering the black hole, can be thought of as epistemologically preferable for general relativity as a theory, since this ensures that the future, however bleak, is indeed determined." Thus, SCC may be able to save GR from the charge of incompleteness.

Dafermos and Luk (2017) argues, however, that SCC is violated in the case of dynamical rotating vacuum black holes without symmetry (Kerr spacetimes). Dafermos and Luk (2017) shows that spacetime does not in fact terminate at a "spacelike singularity" in the interior of rotating Kerr black holes, and thus SCC is false in this case (see Doboszewski's chapter in this volume for a presentation of Kerr spacetimes and a discussion of their relationship with SCC). Instead, spacetime does extend beyond the Cauchy horizon, but the spacetime is not sufficiently smooth for GR to hold (there is a weak "lightlike singularity"). Although SCC fails, there is, arguably, no indeterminism introduced into GR, since the theory itself is not applicable in this case. For our interests, though, this scenario *does* represent an incompleteness of the theory—there is spacetime beyond the Cauchy horizon, but GR does not describe what occurs here—signalling the need for modification of the theory, a new theory, or another solution (§9.4). We conclude, then, that geodesic incompleteness is physically problematic.

9.2.2   Curvature singularities

A second important kind of singularity is *curvature singularities*. The curvature, especially the scalar quantities constructed by contracting powers of the Riemann tensor, grows without bound in some region of the spacetime.[7] This gives rise to various problems, such as unbounded tidal forces and the lack of consistency of the semi-classical approximation (discussed in §9.2.3).

Curiel (2021, §1.3) suggests that curvature singularities are tolerable because their manifestation depends on the path taken by an observer in their neighbourhood: curvature singularities are not a specific property of the *spacetime* itself, but of the spacetime together with the *trajectory* of an observer. But there are *prima facie* two problems with this argument. First, it is based on a narrow conception of a curvature singularity: the argument regards them as undesirable only by virtue of the tidal forces exerted on test objects. It argues that, in some cases, the unbounded tidal forces depend on the trajectory. Thus Curiel (1999, §1.3) argues that a curvature singularity, characterized by the tidal force on a test object, is "not in any physical sense a well-defined property of a region of spacetime *simpliciter*" (p. S128), because in some cases the manifestation of the pathology depends on the object's state of motion, i.e., the trajectory followed, and not just on the observer's location in the spacetime.

But surely there are cases where the tidal forces are unbounded regardless of the direction that the observer is coming from. Furthermore, it is not *points per se*, in and of themselves, i.e., regardless of geodesics, that are important in GR. It is the *point coincidences* of trajectories (or the coincidences of straight lines of an affine connection) that matter, so that in considering the properties of objects travelling along different geodesics coinciding at a point, one is simply deploying the standard interpretation of GR.[8] In other words, standard treatments of GR do not consider spacetimes as monadic objects, but consider in addition other structures (often referred to as "observables") defined on them—more on this later. Thus the fact that the manifestation of the singularity may sometimes depend on the state of motion of the object does not seem to be a good reason to accept such singularities into *GR*.

This idea can be generalized, and leads to an issue that we have not seen discussed in philosophical discussions of singularities in GR. Namely, it is not simply *points* (or point-coincidences of trajectories, or coincidences of straight lines of an affine connection) that make up the ontology of GR. For even if points are regarded, by substantivalists at least, as fundamental, this does not prevent the inclusion of quasi-local and, indeed, global quantities into one's ontology. This is because an ontology that accepts points as objects should surely accept sets, classes, or mereological fusions of points.[9] For example, the very definition of a manifold requires the definition of (open) coverings of this manifold, i.e., of neighbourhoods of points. Thus, a realist semantics should in principle include not just points, but also neighbourhoods.[10]

Here, we note that one's position in the realism vs. anti-realism debate will in general bear on one's attitude towards singularities. The semantic realist (including e.g., the constructive empiricist) demands that there be a well-defined ontology, and will naturally demand that the models of GR are free of singularities in order to ensure this.[11] The problem with singular spacetimes is precisely that they do *not* seem to have a well-defined ontology.

Curiel has stressed the global, rather than local, nature of at least some singularities. Singularities are not always localized at a *point* of spacetime, as is often assumed; when they are, then this is indicative of having a non-essential singularity, i.e., one that can be removed (in the case of incomplete geodesics) by extending the geodesics beyond that point. This is an important fact that is not always clear in discussions of singularities. But this is, by itself, not an argument against the semantic realist's demand for a clear ontology. That some singularities cannot be localized, and do not correspond to "missing points," so that there appear to be "no points missing," does not mean that the ontology of the model in question is automatically well-defined. The indication that a clear ontology is lacking comes from the various pathologies that we have discussed (i.e., unbounded curvature, lack of determinism, things popping in and out in the middle of the spacetime even if covered by a horizon, etc.), even if these pathologies were to get the label "global." For there is no expectation that ontology must be local, certainly if what is at issue is the ontology of *spacetime*.

9.2.3   *The argument from effective field theory*

Effective field theory gives an interesting argument against curvature singularities. Although the argument is semi-classical, and stated in the context of small quantum corrections to the classical equations of motion, the argument also puts into question the validity of the classical approximation itself near a singularity, and it can be stated in purely classical terms.

Although Earman (1996, p. 636) characterizes the semi-classical approximation as one "in which the quantum expectation value of the (renormalized) stress-energy tensor is inserted in the Einstein equations in order to calculate the backreaction for quantum fields on the spacetime metric," this is *not* what we mean by the semi-classical approximation: rather, it is *one* specific instantiation of it.[12]

The kind of semi-classical approximation that we have in mind is best stated in the language of the Lagrangian action (although it can also be stated directly in terms of the equations of motion), which has the advantage of simplicity. The argument is as follows. Consider the Einstein-Hilbert action for GR:

$$S_{EH} = \frac{1}{16\pi G_N} \int d^4x \, \sqrt{-g}\, R(g) \,, \tag{9.1}$$

in units where $c = 1$. Suppose that the underlying QG theory, of which this action is the classical limit, has a length scale ℓ, which could be the

Planck length (since Newton's constant G_N is proportional to the square of the Planck constant), or some other length scale of the theory. Then, in general, the quantum corrections will contribute terms to the action that (on dimensional grounds) are of the order of the curvature squared, and higher:

$$S_{\text{effective}} = \frac{1}{16\pi G_N} \int d^4x \sqrt{-g}\left(R(g) + \alpha\,\ell^2 R^2(g) + \cdots\right) , \qquad (9.2)$$

where α is a dimensionless parameter. Here, our notation is schematic, because R^2 indicates not just the square of the Ricci scalar, but any linear combination of squares of the Riemann tensor and its contractions (i.e., the Ricci tensor, Ricci scalar, and Weyl tensor).[13] The dots indicate possible cubic and higher-order corrections.

These corrections to the classical theory are general, in that we have only assumed that they can be written as (a possibly infinite series of) polynomials in the curvature.[14]

The origin and interpretation of these terms can be various. In an asymptotically safe theory, this action could be taken to represent the quantum effective action, containing all of the information about the quantum theory, and no need for a cutoff. In string theory, this is a perturbative effective action, where the length scale ℓ is the fundamental length of a string, and such corrections are proportional to the length of the string as compared to the typical length-scale of the spacetime.[15] In the point-particle limit, i.e., when the size of a string is small compared to the typical length scales in the spacetime, such corrections are negligible. However, the corrections become important when the spacetime is highly curved, so that its typical length-scale is sizeable compared to the string length. Another example is the asymptotic safety scenario approach to QG, where these terms are present in the action near a fixed point at which the theory is being renormalized, and their coefficients depend on the renormalization scale—physicists say that their couplings "run" (see §9.3.3).

One should distinguish this approach to the effective action from the semi-classical approximation to which Earman (1996, p. 636) refers, namely, inserting the quantum expectation value of the stress-energy tensor into the Einstein equations, in order to calculate the back-reaction of the fields on the metric. In this approach there need to be no matter fields at all. For example, in the asymptotic safety scenario the action (9.2) is purely gravitational (Niedermaier and Reuter, 2006, p. 16), and no matter fields are required. In string theory, these terms do not originate in the way that Earman suggests either; rather, they come from taking into account quantum effects in the scattering between superstrings, and calculating from there the effective action (Green et al., 1987, pp. 169–178).

The importance of these terms is that they correct Einstein's equations with higher-curvature terms. While the corrections are negligible when the curvature is small, these corrections dominate when the curvature is large—which is what happens near a curvature singularity. Thus, near a singularity, the higher-order terms dominate, and the classical approximation (i.e., the approximation by which we drop the higher-curvature terms from the action (9.2) and hence from Einstein's equations) becomes invalid. That is, near a singularity these terms cannot be dropped, because they are larger than the Einstein tensor.[16]

One can compare this to the Taylor series of a function near some point.[17] If one is very close to that point, one can use the leading term of the series (i.e., the Einstein-Hilbert term). But if one significantly goes away from this point, the higher-order terms become increasingly important, so that they eventually dominate and the Taylor series is no longer a good approximation. Near a singularity, the curvature grows without bound, the corrections in equation (9.2) dominate, and the classical action cannot be expected to be a good approximation.

While this argument underpins the intuition that "quantum effects become important near a singularity," it seems incorrect to treat the classical theory without taking into account the possibility that such terms exist. Namely, the appearance of higher-curvature terms in the action and in the equations of motion is not *forbidden* by any symmetry or other principle. One main reason why they are not considered in the classical theory seems to be simplicity: these terms are not needed, and they are small under usual situations. The corrections they give to the known tests of GR are indeed too small to take into account. But who says that the true theory of the world, even at the classical level, does not contain such terms? Thus, even at the classical level, it seems that one needs to take into account the possibility that such terms, even if small, are in principle present in the equations of motion. In other words, while the simplest assumption is that GR is given purely by the Einstein-Hilbert action (9.1), and the Einstein equations that are derived from it, corrections are not a priori excluded.

The use of Einstein's equations in a way that is consistent with the possible existence of higher-curvature terms means that, in general, the validity of Einstein's equations has to be restricted to regions of small curvature. For near singularities, one should not expect the usual Einstein's equations to be valid, because higher-order terms will dominate, regardless of whether their origin is quantum or classical.

To spell out the argument a bit more: note that, in the actions (9.1)-(9.2), we integrate over the entire spacetime. And so, the contributions of any regions of high curvature to equation (9.2) cannot be approximated by their corresponding contributions in equation (9.1), so that the Einstein-Hilbert action is not a good approximation to such regions. This also means that the derivation of the Einstein equations from the Einstein term alone, i.e., equation (9.1), cannot be trusted in such regions. And since the Einstein-Hilbert action cannot be expected to be a good approximation in those regions, neither are the Einstein equations that are derived from it. Thus we cannot expect that a putative model of Einstein's theory, equation (9.1), with curvature singularities in it, is also a model of the full action, (9.2).

This is one main reason behind physicists' talk about the "breakdown" of GR near a singularity. The use of GR, (9.1), as an approximation to the full action, (9.2), requires that curvatures are small: and curvatures are not small near a curvature singularity. It seems to us that, in the philosophical literature, this problem has not received the attention that it deserves.

The argument does not cover all curvature singularities, since it does not cover those singularities that, although giving unbounded components of the physical components of the Riemann tensor in a parallel-propagated frame,[18] do not have unbounded values for the scalar quantities appearing in the action, (9.2) (or for the corrections to Einstein's equations).[19] We do not claim that *all* curvature singularities are undesirable, only that *some* curvature singularities are undesirable, and that physicists are justified in claiming that there is a "breakdown" of GR for those types of singularities, in the sense explained earlier.

9.3   Singularities in QFT

The infinities we consider here are the UV-divergences due to the perturbative formulation of QFT, and Landau poles.[20]

9.3.1   *UV divergences*

The infinities of concern arise in the theoretical framework in which the standard model of particle physics is formulated. This has been referred to in the literature as "conventional QFT" (CQFT), "Lagrangian QFT," or "QFT with cutoffs" (Wallace, 2006, 2011).[21] Famously, calculations in this framework were historically "plagued by infinities." Mainly, and particularly for the UV-divergences, these owe to its reliance upon perturbation theory. A perturbative calculation of any particular physical process involves a summation over all possible intermediate states, and this is done at all orders of perturbation theory (though, in practice, often only the first few terms are taken and the higher-orders, it is hoped, decay rapidly). CQFT, with its local dynamics (i.e., point sources and interactions), as well as the integration over all the momentum-energy states, implies that there are an infinite number of intermediate states. Thus, if the terms are not sufficiently suppressed, perturbative calculations within the theory lead to divergent integrals.

Historically, these infinities were removed in particular theories, such as quantum electrodynamics (QED), via *renormalization*, which rendered the theory finite and (impressively) predictive. This procedure, however, was physically suspicious, and the perturbative approach to QFT itself remained intrinsically approximate and conceptually problematic. One response was the development of axiomatic QFT: instead of introducing informal renormalization techniques to treat interactions, this approach attempts to put QFT on a firm, non-perturbative footing, by specifying a mathematically precise set of axioms at the outset. Then, models of the axioms are constructed (constructive QFT). Importantly, this approach is not an attempt at QG, of physics beyond, but simply a new formulation of QFT at the level of QFT—i.e., as a combination of QM and special relativity without any singularities in the theory.[22] Unlike CQFT, it is not an intrinsically approximate theory, since it is supposed to be directly defined on Minkowski spacetime, and so remain well-defined at arbitrarily small and arbitrarily large length scales. Although there are various simplified toy models satisfying the axioms of AQFT, however, there have been no realistic models constructed. In particular, it has not been demonstrated that QED or any other successful theories in high-energy physics admit formulations that satisfy the axioms of AQFT.

On the other hand, in mainstream high-energy physics, the development of the *renormalization group* (RG) techniques led to a non-perturbative framework for studying CQFT systems at different energy scales, and ultimately to the discovery of the standard model. Instead of evaluating the integrals up to infinite momenta, the theory is only evaluated up to some finite high energy scale (short length scale) *cutoff*, and the effects of the high momenta degrees of freedom at lower energies are encoded in the dynamics of the lower energy "effective" theory. The RG analysis demonstrates that the means by which the cutoff is implemented has no bearing on the low-energy physics; the only effects that are significant at these scales are changes in the coefficients of finitely many interaction terms (the *renormalizable* interactions). As Wallace (2011, p. 119) puts it, "Renormalization theory itself tells us that if there *is* a short-distance cutoff, large-scale phenomenology will give us almost no information about its nature."

The dominant philosophical interpretation of this CQFT picture is that the UV divergences are not a real physical problem, but rather indications of the limitations of the perturbative framework of CQFT. The framework itself is taken to be inherently approximate,[23] and its models are supposed to be *effective theories*: not to be valid to arbitrarily high energy scales. "This, in essence, is how modern particle physics deals with the renormalization problem[24]: it is taken to presage an ultimate failure of quantum field theory at some short length scale, and once the bare existence of that failure is appreciated, the whole of renormalization becomes unproblematic, and indeed predictively powerful in its own right" (Wallace, 2011, p. 120). The idea is that, whatever the unknown physics of QG turns out to be, the success of the CQFT models at known energies is explained, thanks to the RG.

This interpretation of CQFT as *effective* means that the theory is not supposed to be reliable at short length scales. In particular, the need to employ a short-distance cutoff is not taken to indicate *anything* regarding the physics beyond. This is in contrast with the case in condensed matter physics, where the RG is also employed because the system is described by a theory which diverges in the UV, but in which case the divergences, and the need to employ a short length scale cutoff is consistent with the existence of something physical—we know that we cannot treat matter as continuous at arbitrarily short length scales, because matter has a discrete structure at the atomic scale. In CQFT, however, there is no empirical evidence for the existence of a real physical cutoff (e.g., a discrete structure for spacetime at extremely high energies). It is possible that the UV divergences in CQFT simply reflect limitations of the theory, rather than any new physics.

9.3.2 Landau poles

A widely held view is that if we had realistic models of AQFT which properly accounted for the dynamics without relying on approximations, and hence, did not feature the UV-divergences associated with perturbative analysis, we would not need renormalization in QFT, and would lose motivation for treating QFT as an effective theory. Those who express this view nevertheless recognise that a stronger motivation for treating QFT as effective is the existence of Landau poles—for instance, in QED. This type of infinity is thought more concerning than the UV-divergences because it is not taken to be merely due to the limitations of perturbative analysis (although, since the Landau pole in QED is normally identified through perturbative one-loop or two-loop calculations, it is possible that the pole is a sign that the perturbative approximation breaks down at strong coupling).

QED is renormalizable, so in principle it should be able to be extended to arbitrarily high energies. Yet, the renormalized coupling grows with energy scale, and becomes infinite at a finite (though extremely large) energy scale, estimated as 10^{286}eV (the original result comes from Landau et al., 1954). The existence of this "pole" could mean the theory is mathematically inconsistent. This is avoided if the renormalized charge is set to zero, i.e., if the theory has no interactions. There is indication that in QED, the renormalized charge goes to zero as the cutoff is taken to infinity (a physical interpretation of this is that the charge is completely screened by vacuum polarization). This is a case of quantum *triviality*, where quantum corrections completely suppress the interactions in the absence of a cutoff. Since the theory is supposed to represent physical interactions, the coupling constant should be non-zero, and so the Landau pole and the associated triviality might be interpreted as a symptom of the theory being effective, or incomplete (i.e., that it fails to take into account other fundamental interactions relevant at high energy scales).

QED and ϕ^4 theory are thought to be trivial in the continuum limit in this way.[25] Although the Landau pole in QED is problematic for the theory, it is usually ignored because it concerns an energy scale where QED is not thought to be valid anyway, given that electroweak unification occurs at an energy scale lower than this. It also concerns an energy scale where QFT itself is not thought to be valid, based on other motivations for QG (which we consider in §9.5).

So, the Landau pole divergences in CQFT are typically interpreted as part of the formal (mathematical) grounds—*internal* to the theory—for treating QFT as effective. These motivations are called into question by those who argue for the necessity of a "non-approximate" formulation of QFT, such as AQFT; on this view, the divergences of CQFT (including Landau poles) are not thought to be inherent to QFT, properly understood.[26] But there are also *external* grounds for treating QFT as effective, which come from the motivations for QG. These external motivations hold regardless of whether or not there are divergences inherent to QFT, but they are reasons why one might not be concerned with finding a singularity-free theory of QFT in order to describe the world at arbitrarily small length scales.

9.3.3   *Perturbative non-renormalizability of GR*

The Einstein-Hilbert action is perturbatively non-renormalizable: divergences appear in loop diagrams at first order (in the matter case, or second order in the matter-free case), and there is an expectation that infinitely more infinities appear at higher orders.[27] Treating GR as an effective theory, in the same way as the Standard Model of QFT (§9.3.1), the dimensionless parameters of (9.2) satisfy an RG equation which describes how they "run" or "flow" as the renormalization scale is changed.[28] In perturbation theory, all but a finite number of these parameters diverge at high energy scales (around the Planck scale), and we are prevented from calculating anything in this regime. According to the philosophy of effective field theory, this would be indication that the theory is not applicable at these scales, and a new theory is required in order to describe the physics here.

However, it is possible that the proliferation of infinities in the UV instead signals the limitations of the perturbative approach in this regime. The asymptotic safety scenario claims that these couplings, in the non-perturbative RG flow, do not actually diverge, but instead flow to a finite value: a "fixed point" in the UV.[29] This is similar to QCD, where the couplings flow to a fixed value of 0 in the UV, and the theory is said to be asymptotically free (the theory is non-interacting in this regime). In the case of gravity, however, the fixed point is supposed to be non-zero (the theory is interacting in at least one of the couplings), and the theory is said to be *asymptotically safe*, since the physical quantities are "safe" from divergences as the cutoff is removed (taken to infinity). If there is a fixed point, then following the RG trajectory (almost) to it, one can in principle extract unambiguous answers for physical quantities on all energy scales (Niedermaier and Reuter, 2006). At the fixed point, the dependence on the UV cutoff is lost, and the theory is *scale invariant*: it does not change as smaller length scales are probed.

As Eichhorn (2019) explains, the scale invariance protects the running couplings from running into Landau poles, and thus asymptotic safety could potentially serve as a UV-completion for the Standard Model of QFT; the search for asymptotically safe extensions of the standard model with new degrees of freedom close to the electroweak scale is ongoing. If gravity and the standard model are asymptotically free, then this removes the remaining internal motivation for treating these theories as effective: the infinities in the theories are shown to simply be due to the inapplicability of perturbative methods in the extreme UV. The full theories, treated non-pertubatively in this regime, are finite. While there are new degrees of freedom expected at high energy scales, there is no need for new physics (i.e., a new theoretical framework).[30]

9.3.4 Responses to singularities in QFT

So far, we can identify four different possibilities in response to the singularities in QFT:

i *AQFT view*: Singularities in CQFT motivate a different QFT framework, one whose theories are singularity-free, but which does not include gravity (as in AQFT);

ii *New physics*: Singularities in CQFT motivate a new, more fundamental theory at high-energy, and motivate treating our current theories as effective (applicable only at low energy scales), consistent with external motivations for QG;

iii *Effective theory view*: Ignore the Landau poles in the Standard Model and perturbative non-renormalizability of gravity (we shouldn't worry about interpreting them), since we have external reasons for thinking of these as non-fundamental effective theories, to be replaced by QG at high-energy scales (i.e., we appeal only to the external motivations for new physics, and the singularities in current theories do not count as motivations for new physics);

iv   *Asymptotic Safety view*: UV divergences due to perturbative non-renormalizability and Landau poles do not motivate new physics or a new theory according to the asymptotic safety scenario for gravity and the Standard Model; these singularities do not appear in the full (non-perturbative) theory.

There is one more response to singularities in CQFT, which has been expressed prominently by Jackiw (1999, 2000) and Batterman (2011).

v   *Emergent physics view*: Singularities in CQFT are of physical significance, but not motivation for new, more fundamental physics. The singularities are important for facilitating and understanding the emergent, low-energy physics.

In §9.4, we identify four different attitudes towards singularities; on the categorization there, (i) and (iv) fall under Attitude 1; (ii) is Attitude 2a; (iii) is Attitude 4; and (v) Attitude 3.

9.4   Four attitudes towards singularities in QG

Based on a review of the physics literature, we identify four different treatments of, or expectations about, singularities in GR and QFT, in regards to the need for a theory of QG to resolve (remove) the singularities. Here, we classify these as four different "attitudes" towards QG, based on how they answer the question, "Do the singularities point to the need for QG or not?"[31] Thus we are taking for granted that GR and QFT are *not* fundamental theories (or frameworks), and we are assuming that QG *is* supposed to be a fundamental theory.[32] Here, by "non-fundamental," we mean a theory that is restricted in its domain of applicability, i.e., an *effective theory*.

We will see that, of the four attitudes, only one suggests the need for a theory of QG. We emphasize, however, that it is possible to have different attitudes towards different types of singularities.

9.4.1   *Attitude 1: "Singularities are resolved classically—they do not point to QG"*

Here, the idea is that the singularities in GR and/or QFT (understood as effective theories) are to be resolved, roughly, "at the level of these current theories," without recourse to QG. For the GR singularities, they are to be resolved classically. For the QFT singularities, this means trying to resolve them at the level of QFT, taking QFT as the combination of SR and QM—e.g., in the way the proponents of AQFT suggest. On this view, singularities do not indicate the existence of, or a need for, a theory of QG, since they are resolved independently of it (of course, one could hold this view and still look for a theory of QG for other reasons).

Examples include "the AQFT view" mentioned in §9.3; as expressed in e.g., Fraser (2009, 2011). We can think of the "Asymptotic Safety view" in §9.3.4 as another example, where the divergences are seen as indication of the failure of perturbation theory in the high-energy regime, and are taken as motivation for looking for the non-perturbative theory, but not requiring a new theoretical framework (thus, we treat this as a case of Attitude 1 even though it does aim at a theory of QG). As a third type of example are those who try to describe the singularities in GR in a mathematically rigorous way, by various regularizations or smearings of the singularity. The motivation here is to find a rigorous mathematical theory that reproduces the predictions of GR and at the same time is well-defined, analogously to AQFT (though perhaps more at the level of solutions or specific physical systems than at the level of axioms).

We mention two proposals for a classical resolution of singularities in GR: what we could call a "dynamical singularity resolution." The idea is to show that the "unphysical" singular solutions are approximations to more realistic solutions: e.g., by including extra dimensions, or by including matter fields (we will mention one type of solutions called a "gravastar").

Gibbons et al. (1995) show that there is a purely *classical* mechanism that can resolve black hole singularities: namely, the introduction of small extra dimensions. Certain four-dimensional black hole solutions[33] that look singular from their natural four-dimensional perspective, descend from classical solutions of higher-dimensional supergravity that are completely non-singular. "Descending from higher dimensions" means that the extra dimensions are compactified, i.e., they are a small, compact space. Some of these higher-dimensional solutions are stable supersymmetric solutions of string theory.[34]

This mechanism involves a purely classical resolution of a black hole singularity. The higher-dimensional smooth solutions *evade the singularity theorems* of Penrose and Hawking, because the higher-dimensional solutions do not have compact trapped surfaces.[35] Interesting as this mechanism may be from a physical perspective: it does not seem to give a general mechanism for singularity resolution. For example, as the authors admit, it does not seem possible to get the four-dimensional Schwarzschild solution in this way.

Another example is "gravastars": astronomical objects similar to black holes conjectured by Mazur and Mottola (2001). They took into account the gravitational back-reaction of the fields of an imploding star. This imploding star forms a compact object that, from the outside, is very similar to a black hole. The matter fields are such that the interior of the compact object is a de Sitter-like space, while the exterior is the Schwarzschild geometry. The two regions are separated by a shell of fluid. The solution does not have a singularity, and no horizon. Thus, the idea is that the Schwarzschild solution is an idealization, and that if one considers an imploding star, the final product can look similar to a black hole from the outside, but be very different on the inside. Gravastars have not been completely ruled out astronomically; the search is still ongoing.

Finally, Koslowski et al. (2018) show that, taking the scale factor in certain models to be unphysical, leads to a reduced dynamical system that can be integrated through what is normally interpreted as the Big Bang singularity.

9.4.2   Attitude 2: "Singularities point to QG"

According to this attitude, the singularities in GR (QFT) are to stay because they signal the limitations of these effective theories. They are to be resolved by QG, as a more fundamental theory. The singularities make vivid the motivation for the search for a new theory, that is, if there are no problems one does not know what one is looking for. But the singularity gives one a concrete problem to focus on. Thus, singularity-resolution (of at least *some* particular singularities) is taken as a *guiding principle* motivating the search for QG. The principle of singularity resolution might also serve as as a *criterion of theory-selection*, meaning that a prospective theory of QG should not be accepted if it is incompatible with the principle.[36]

It may also be that the singularities in GR and QFT are not thought to directly point to or motivate resolution in QG, but that particular approaches to QG—developed for other reasons besides singularity resolution—naturally feature singularity resolution, even though they were not motivated by this. In this case, the fact that the theory resolves given singularities might nevertheless be promoted as evidence in support of the correctness, or pursuit-worthiness of the approach—i.e., it might be promoted as a means of *non-empirical confirmation* by its proponents. For this reason, we still class these as cases of Attitude 2.

We can distinguish two sub-categories:

2a. *Singularities as physically significant*. Here, the singularities are informative—they point to new physics.

2b. *Mathematical or structural view of singularities*. The attitude of someone who says "singularities need to be resolved for reasons of mathematical consistency and (perhaps) predictive power", while they are non-committal about whether singularities will signal "new physics" in the heuristic sense. In other words, singularities must and will be resolved by QG at the "structural" level, once a tidy theory is developed, but we do not need to focus on the problem of singularities particularly, since it is not a deep physical problem. This attitude can also signal neutrality in regards to the singularities in current non-fundamental theories, but could recognise the practical necessity of resolving the singularities in a more fundamental theory.

In regards to 2a: Most authors that we are aware of assign some sort of physical salience to singularities—they are seen as a "smoking gun" for new physics. By "new physics," of course we do not mean that the singularity itself is a physical event, which, for example, Rovelli and Vidotto (2014, p. 209) explicitly deny: "the Big Bang singularity does not appear to be a physical event, but only an artefact of the classical approximation. In this, it is analogous to the possibility for an electron to fall into the nucleus of the atom, which is predicted by the classical approximation but not by the quantum theory of the electron." The broad idea is an old one: namely, a singularity, more than just indicating the breakdown of the classical theory, is a locus where new physics can be expected: "At this point [the big bang singularity] the classical theory completely breaks down, and has to be replaced by a quantum theory of gravity" (Bojowald, 2001, p. 5227).

This analogy with the resolution of the instability of the hydrogen atom in classical electrodynamics, by quantum mechanics, is widely used in motivating the necessity of QG in resolving GR singularities.[37] Electrodynamics predicts that the electron will eventually fall down into the nucleus, because it loses energy as it orbits around it: at which point the force becomes infinite. Quantum mechanics solves this by, first, rejecting the classical picture of an electron "orbiting" around the nucleus. Second, by confining the electron to discrete energy levels, and allowing it to emit energy only in discrete packets of energy, viz. photons. And finally, the lowest energy level allowed by the theory is where the electron is, on average, located at a finite distance from the nucleus (the "Bohr radius"), so that it can never fall in. A widespread idea is that, analogously, gravity should be quantized and that QG solutions have a discrete spectrum, similarly to the discrete spectrum of quantum mechanics—and, in this way, singularities can be avoided.

Some believe that the singularities in GR and QFT are to be cured by the existence of a minimal length, even without quantization of gravity (note, too, that the minimal length need not represent an actual discretization of spacetime, but may be an operational minimal length, e.g., due to an extended probe).[38] Henson (2009) expresses this view, particularly in regards to the need for QG to resolve the divergences associated with the non-renormalizability of GR, §9.3.3), and uses it as motivation for the discreteness postulated by an approach to QG known as causal set theory.

String theory is often motivated by the analogy between the non-renormalizability of GR and that of 4-Fermi theory (which was revealed to be the effective limit of the renormalizable electroweak theory): the non-renormalizability of GR is taken to indicate that there should be renormalizable theory with (quantum) GR as an *effective field theory* (EFT), and string theory is promoted as exactly this kind of theory. Thus, the (alleged) resolution of the UV-divergence of quantum GR by string theory is presented as one of its selling points.[39]

In the context of loop QG, Bojowald and collaborators have, in a series of papers, developed models in which discretized equations appear naturally.[40] The original model (Bojowald, 2001) is an isotropic minisuperspace approximation to the cosmological (Big Bang) singularity, where the cosmological scale factor is represented by a bounded operator. The quantum evolution occurs in discrete time steps, and does not break down when the volume goes to zero—so, the model can proceed through the Big Bang singularity to a pre-Big Bang era. A similar effect is found for the Schwarzschild singularity. The model is tentative, and some of the details (e.g., the idealizations used) controversial.

Interestingly, Bojowald (2001, p. 5230) draws a similarity with renormalization in QFT, based on his use of Thiemann's 1998 technique to obtain a dense matter Hamiltonian: "it is the same mechanism which regularizes ultraviolet divergences in matter field theories and which removes the classical cosmological singularity. We have also seen that nonperturbative effects are solely responsible for this behaviour and a purely perturbative analysis could not lead to these conclusions." Thus, according to Bojowald, QG makes the Hamiltonian well-defined through a QFT technique—namely, by renormalizing the algebra of operators.

Another example is AdS/CFT, where the boundary Yang-Mills theory is used to study the singularity of a five-dimensional AdS-Schwarzschild black hole. Festuccia and Liu (2006, pp. 17–18) study the Wightman functions of the dual conformal field theory, and find signals of the black hole singularity in them. However, this singularity must be resolved in Yang-Mills theory, since the Wightman functions are only singular because the spectrum of the Yang-Mills theory has been approximated to be continuous. Namely, the singularity is an artefact of taking a large N limit;[41] but the spectrum of the theory at finite N is discrete, and the Wightman functions cannot be singular. Thus, if AdS/CFT is correct, then this implies that the black hole singularity must be resolved as well.

As in the case of Bojowald, the details of this are both tentative and not settled. However, the mechanism for singularity resolution is similar: the spectrum of the theory (in this case, the analysis proceeds by arguing through the dual theory) is argued to be discrete, and from here it follows that the correlation functions (which are dual to the observables in the gravity theory) must be finite.

There is also the possibility of a *dynamical resolution* in the quantum case, similar to the classical case. For the big bang singularity, Ashtekar et al. (2006a, b) have proposed a model in loop QG in which the big bang is replaced by a quantum bounce, so that the quantum evolution remains non-singular across the Planckian regime of the bounce.

Another way in which singularities can be "physical" is the more literal sense that the singularity is a physical *place* (or time), even if classical GR does not describe it: "the classical singularity does not represent a final frontier; the *physical* spacetime does not end there. In the Planck regime, quantum fluctuations do indeed become so strong that the classical description breaks down" (Ashtekar and Bojowald, 2006, p. 409). The reasoning behind this idea is that singularities are boundaries of spacetime which can be reached by observers in finite proper time. And so, although classical GR cannot be extended to these boundaries, the fact that observers can reach them within finite proper time calls for a different description, or an incorporation of these boundaries into the theory.[42] So, this line of reasoning can either motivate QG (Attitude 2a), or resolution at the level of GR (Attitude 1).

In regards to Attitude 2b, we found one prominent physicist, Gerard 't Hooft, who thinks that GR singularities do not have much physical significance.[43] It is only a technical problem of the theory that needs to be resolved—the resolution would be analogous to, e.g., doing a contour integral in the complex plane, in order to define an integral with a pole. Singularities are important for doing calculations (for example, to define the convergence of a series) but not for physics. 't Hooft contrasts this mathematical interest of singularities in GR with the more significant physical role that they play in QFT. In QFT, virtual particles contribute poles (i.e., a specific type of singularity) to scattering amplitudes. These poles are physically significant, because they indicate the presence of a particle which cannot be measured directly. 't Hooft also views the Landau pole in QED as significant: it signals an incompleteness of the theory, and, therefore, resolving it is an important physical question (Attitude 2a).

9.4.3 Attitude 3: "Peace with singularities—they do not point to QG"

According to this view (particular) singularities are not to be resolved at any level. There may be reasons for keeping singularities in our theories, other than that they signal the limitations of effective theories. This is the only attitude that would permit singularities in a fundamental theory. So, this attitude takes into account other reasons why particular singularities might be considered good:

3a. They can be treated as predictions of the theory without needing to be removed;

3b. They are explanatory (without pointing to any new physics);

3c. They are required for stability.

As an example of 3a is the "tolerance for singularities" in GR expressed by Earman (1996), following Misner (1969), where the singularities are treated as predictions of the theory, from which we can learn about GR physics (and potentially QG physics), without needing to remove the singularities.

An example of 3b is the "emergent physics view" mentioned earlier, §(v), towards singularities in QFT and critical phenomena, held by Batterman (2002, 2011) and Jackiw (1999, 2000). Here, the singularities are seen as necessary for an adequate description and explanation of the low-energy physics.

An example of 3c is the argument from Horowitz and Myers (1995) that a modification of GR which is completely non-singular (free, in particular, from the Schwarzschild black hole singularity) could not have a stable ground state. The reason is that such regular modifications of GR would have completely regular solutions with negative Schwarzschild mass. Thus the Schwarzschild singularity is required to avoid negative masses: "if we want the theory to have any stable lowest-energy solution, it must have singularities, in order that one may discard what would otherwise be pathological solutions" (p. 917).

The argument is not specific to GR, but holds for effective field theories of QG with arbitrary quantum corrections, which manifest themselves in higher-curvature corrections to the Einstein-Hilbert action. This is a positive argument for the use of singularities in physics. Bojowald (2001) endorses it, and it is indeed not at odds with his own argument for the disappearance of singularities, which is a non-perturbative argument: while Horowitz and Myers' argument is perturbative. Thus Horowitz and Myers leave open the possibility that non-perturbative QG may still resolve the Schwarzschild singularity (see §9.4.2).

9.4.4   Attitude 4: "Indifference to singularities—they do not point to QG"

According to this view, (particular) singularities are of no significance. These views deny the importance of the question of the resolution of singularities.

One example of this attitude is the view that the singularities in GR and QFT don't tell us anything, but it doesn't matter given that these theories are non-fundamental. In the context of the UV-divergences of QFT, this is what we called the "effective theory view" in §9.3.4. This is expressed, e.g., in Wallace (2011), which takes the perspective that the

UV-divergences of the Standard Model of QFT can be ignored because we have other reasons for thinking that this is not the right framework at the energy scales where these divergences would be a problem, and nor that we will learn anything by resolving them. In the context of the perturbative non-renormalizability of GR has, e.g., been expressed by Hossenfelder (2013, p. 37), in saying, "The Einstein-Hilbert action is ... not the fundamental action that can be applied up to arbitrarily-high energy scales, but just a low-energy approximation, and its perturbative non-renormalizability need not worry us."[44]

Another example of this attitude is an argument in Curiel (1999), that singularities are not a problem for GR, because they are not part of the manifold: they are not part of the theory. Finally, one well-known cosmological scenario, by Brandenberger and Vafa (1989), uses some aspects of the physics of strings to argue that, even though a cosmological singularity is present in the metric, it is of no consequence for string theory, whose behaviour near the singularity is completely regular: the string does not "see" the cosmological singularity. Brandenberger and Vafa (1989) find that, for a gas of strings in a compact space, the temperature increases as the volume of the space decreases, until a maximum value is reached when the space is the size of the string length. If we further shrink the volume beyond this, however, the temperature drops. Thus, infinite temperature, or infinite energy, are never present in the observables of this string theory cosmology—as opposed to the usual point-particle cosmologies. Effectively, the universe contracts and then (as we keep shrinking) it effectively expands again—where, by "effectively," we mean *from the string's point of view*. This is a consequence of the *T-duality* of string theory:[45] a duality that interchanges (i) the momentum and winding quantum numbers; (ii) the radius and the inverse radius, in units of the string length, i.e., $R \leftrightarrow \ell_s^2/R$. Small volumes in string theory are equivalent to large volumes, and so there is no short-distance singularity.

9.5 Discussion and conclusion

We may distinguish between those motivations for QG that come from within current theories (GR or QFT), and those that are external to the theories. The "internal" motivations for QG may include, e.g., problematic inconsistencies or incompleteness within GR or the Standard Model of QFT (considered individually); these internal motivations can be seen not just as motivations for QG, but as reasons for treating the current theories as in need of replacement. The "external" motivations for QG may include, e.g., goals of unification, the need to describe particular physical phenomena, etc. Singularities in GR and QFT may represent internal motivations for QG, while singularities that occur in attempts to combine GR and QFT, for instance, could potentially represent external motivations for QG (since these do not arise within the theories themselves, but through tinkering with them).

If Earman's (Earman, 1995, 1996) attitude is correct, that the space-time singularities (geodesic incompleteness) are not necessarily problematic for GR (depending on whether or not cosmic censorship holds), then these would not count as internal motivations for QG. By contrast, in QFT the Landau poles are more readily interpreted as signalling incompleteness of the theories; these are, however, typically dismissed because they occur in regimes where the theory is not thought to be applicable anyway *for external reasons*. The tension between the internal and external reasons for treating a theory as effective is reflected in differing possible attitudes towards singularities in current theories: are they to be resolved by a more fundamental theory of QG, or should we instead look to resolve them "at the level of current theories" (i.e., through developing a new framework for QFT, or modifying GR, etc.). Those who put weight on the external motivations for QG may tend to disregard alternative, internal, possibilities for singularity resolution.

Singularities do also feature in several external motivations for QG, particularly in heuristic arguments. One is the argument for the "Planck scale" as being significant for QG. Briefly: both GR and QFT are necessary in order to describe a particle of mass m whose Compton wavelength, $l_C = \hbar/mc$, is equal to its Schwarzschild radius, $l_S = Gm/c^2$, which occurs when m is equal to the Planck mass, $m_p = \sqrt{\hbar c/G}$. Geometry at this scale is thought to be ill-defined, "fuzzy," or a "quantum foam" of microscopic, rapidly decaying black holes (Wheeler and Ford, 1998).[46] Two other arguments we have discussed include the argument from effective field theory, §9.3, and the perturbative non-renormalizability of GR §9.3.3. In regards to the former, we can say that even if incomplete geodesics turn out not to represent internal motivations for QG, the argument from effective field theory shows how curvature singularities can feature as external motivations for QG: it is precisely in the regions of extreme spacetime curvature where we expect a new theory (which may be QG, or a theory "at the level of GR") to be necessary. Thus, again, it seems that the external motivations for a new theory are perhaps more influential than the internal ones. The external motivations are, however, more risky than internal ones, since they typically stem from untested combinations of assumptions and heuristic arguments (even if these are motivated by current well-tested theories). It is imperative to critically investigate the external motivations for QG more thoroughly in future work.

In general, singularities in current fundamental theories do not automatically point to the need for a new theory. Yet, there are at least two examples of singularities (curvature singularities, and Landau poles) that do arguably motivate a new theory—whether a theory "at the level of current theories" (Attitude 1), or a more-fundamental theory of QG (Attitude 2). Deciding between these alternative attitudes will—amongst other things—depend on one's disposition towards the internal versus external motivations for QG, and one's position in regards to scientific realism. If we do adopt Attitude 2 towards these singularities, and treat singularity resolution as a principle of QG, then we are accepting its legitimacy or potential fruitfulness as a guiding principle. But, as explained in §9.4.2, it could also serve in stronger roles: as contributing to pursuit-worthiness, as a criterion of theory-selection, and/or as a means of non-empirical confirmation. Establishing the legitimacy of the use of the principle of singularity resolution in these stronger roles requires more work.

Notes

1. See, e.g., Curiel (1999); Earman (1996) for a discussion of different types.
2. Cf. Curiel (1999, §1.1).
3. See, e.g., Bojowald (2007).
4. See Hawking and Ellis (1973, §8.2) and Earman (1995, §2.8).
5. Recent work has emphasized the difficulty of defining determinism in precise terms; Doboszewski (2019) gives a pluralistic definition.
6. This fact is already recognized in the literature, e.g., Wald (1992, p. 182).
7. Besides the divergence of scalar (i.e., coordinate-independent) quantities, there are also curvature singularities whereby some of the physical components of the Riemann tensor do not have a limit: see the next Section, and also Earman (1995, p. 37).
8. See Einstein (1916, p. 776) and Kretschmann (1917, pp. 576–577, 579) (cf. Sauer, 2005). Elsewhere, Einstein (1915, p. 228) considers physically real events as consisting of more general spatio-temporal coincidences.
9. Cf. Lewis (1986, pp. 211–212), Armstrong (1997, pp. 12–13), Sider (2001, pp. xvi, 7, 59); also Butterfield (2011) campaign against *pointillisme*.
10. Cf. De Haro (2022, forthcoming, §6.3).
11. Though the realist and the constructive empiricist have different degrees of belief in this ontology.
12. For a discussion of this kind of semi-classical approach, especially in the context of singularities, and some of its problems, see: Birrell and Davies (1982, §6); Parker and Fulling (1973, pp. 2357–2359); Davies (1977, pp. 402–403).
13. One often-considered term is the Gauss-Bonnet term $R_{\mu\nu\alpha\beta}R^{\mu\nu\alpha\beta} - 4R_{\mu\nu}R^{\mu\nu} + R^2$, which when integrated over the spacetime gives the topological Euler number. The correction terms can also contain covariant derivatives of the Riemann tensor, where two derivatives have the same dimension as the Riemann tensor.
14. Since these are short-distance corrections that should be subleading with respect to the Einstein-Hilbert term at long distances, negative powers are excluded.
15. The coefficients of the higher curvature terms, like α, are in this case functions of a scalar field called the "dilaton," and contain much information about the microscopics of the theory: in particular, about its 11-dimensional origin. See e.g., Green (1999, pp. 29–30).

16. The same argument can be made directly at the level of the equations of motion, without using the action.

17. The analogy here is with a function with a finite radius of convergence. In principle, it is also possible that the series converges everywhere, in which case the function is entire: but this needs to be proven, and it will only happen in special cases. For example, supersymmetry sometimes leads to this situation, where non-renormalization theorems ensure that the series only has a finite number of terms.

18. See Earman (1995, p. 37).

19. One such example is plane-fronted shockwaves which are exact solutions. Physicists normally do not consider these singularities as undesirable. See, e.g., Israel (1966, pp. 1–3); Aichelburg and Sexl (1971, pp. 304–305); Penrose (1972, pp. 102–103).

20. There are also IR-divergences in QFT, which we do not consider here.

21. But as Fraser (2016, p. 12), says, these labels are somewhat misleading, given that it is not the Lagrangian formulation that defines this framework, and that most of the successes of the standard model do not come directly from cutoff QFT structures.

22. See, e.g., Fraser (2011).

23. Cf. Fraser (2020).

24. Footnote in the original suppressed.

25. In other words, RG analysis of QED and ϕ^4 does not indicate that these theories possess a stable UV fixed point (as in case (a) of the possibilities mentioned). See Lüsher and Weisz (1987); Gies and Jaeckel (2004). This means that the Standard Model of QFT suffers Landau poles both for the electron charge, and the Higgs boson.

26. See, e.g., Fraser (2011). Note, too, that proponents of this view hold that CQFT does not count as "QFT" properly understood.

27. The one-loop divergences were shown by 't Hooft and Veltman (1974); cf. Bern (2002).

28. See, e.g., Donoghue (1997).

29. This was proposed by Weinberg (1979); see also Eichhorn (2019), Niedermaier and Reuter (2006).

30. Interestingly, this UV completion may not be fundamental; it is possible that asymptotic safety provides a step forward in our understanding of microscopic physics, with more fundamental physics to be discovered beyond. As Eichhorn (2019) explains: While providing a UV completion for some RG trajectories, a fixed point can simultaneously act as an IR attractor for a more fundamental description.

31. Three of these views are also identified in Earman (1996), following Misner (1969).

32. In fact, however, it is possible that QG is not a fundamental theory, as argued in Crowther and Linnemann (2019).

33. The solutions in question are "dilaton" black holes, which appear in supergravity and string theory: also called "black p-branes."

34. For a philosophical introduction to these solutions and the corresponding black holes, see De Haro et al. (2020, §2–3).

35. They only have marginally trapped surfaces (i.e., the null generators have zero convergence), and these are non-compact. The existence of a compact trapped surface is a global assumption of the Penrose-Hawking theorems; the local assumption is about the focusing of geodesics.

36. Cf. Crowther and Linnemann (2019).

37. Malcolm Perry, interview with J. van Dongen and S. De Haro (Utrecht, 12 July 2019). But cf. Earman (1996), who argues against the applicability of this analogy.

38. E.g., Ellis (2018) states there is a widespread sentiment among QG physicists, that the singularities in GR and QFT are due to the assumption of a spacetime continuum. For more on the minimal length, see Hossenfelder (2013).
39. Cf. Crowther and Linnemann (2019).
40. Bojowald (2001); Ashtekar and Bojowald (2006); Bojowald (2007). For a recent review of loop quantum cosmology, see Ashtekar and Singh (2011); for a critical assessment Bojowald (2020, pp. 1–2).
41. More specifically, Festuccia and Liu (2006, pp. 24) argue that the analytic continuuation of the Wightman functions and the large-N limit taken do not commute.
42. See Bojowald (2001, p. 5227).
43. Interview with S. De Haro and J. van Dongen (Utrecht, 3 May 2019).
44. While we here attribute Attitude 4 to Wallace in regards to the UV-divergences and Landau poles in QFT, we note that Wallace has in conversation expressed that he believes these divergences are in fact strong (though not conclusive) reason to pursue QG physics. Similarly, while we here attribute Attitude 4 to Hossenfelder in regards to the perturbative non-renormalizability of GR, Hossenfelder has in other media expressed that she believes this to be a promising problem to work on towards QG. Thus, Wallace and Hossenfelder both actually take Attitude 2 towards these singularities.
45. For a philosophical account of T-duality, see Huggett (2017); cf. De Haro (2021, §3.2).
46. Gryb and Thébault (2018) presents a simple model of singularity resolution in quantum cosmology that illustrates the intuitive idea of spacetime "fuzziness" at the Planck scale.

Bibliography

Aichelburg, P. C. and R. U. Sexl (1971). On the gravitational field of a massless particle. *General Relativity and Gravitation* 2 (4), 303–312.

Armstrong, D. M. (1997). *A World of States of Affairs*. Cambridge University Press.

Ashtekar, A. and M. Bojowald (2006). Quantum geometry and the Schwarzschild singularity. *Classical and Quantum Gravity* 23, 391–411.

Ashtekar, A., T. Pawlowski, and P. Singh (2006a). Quantum nature of the big bang: An analytical and numerical investigation. *Physical Review D* 73, 124038.

Ashtekar, A., T. Pawlowski, and P. Singh (2006b). Quantum nature of the big bang: Improved dynamics. *Physical Review D* 74, 084003.

Ashtekar, A. and P. Singh (2011). Loop quantum cosmology: A status report. *Classical and Quantum Gravity* 28 (213001), 1–122.

Batterman, R. (2002). *The Devil in the Details: Asymptotic Reasoning in Explanation, Reduction and Emergence*. Oxford University Press.

Batterman, R. (2011). Emergence, singularities, and symmetry breaking. *Foundations of Physics* 41, 1031–1050.

Bern, Z. (2002). Perturbative quantum gravity and its relation to gauge theory. *Living Reviews in Relativity* 5 (5).

Birrell, N. D. and P. C. W. Davies (1982). *Quantum Fields in Curved Space*. Cambridge University Press.

Bojowald, M. (2001). Absence of singularity in loop quantum cosmology. *Physical Review Letters* 86 (23), 5227–5230.

Bojowald, M. (2007). Singularities and quantum gravity. *American Institute of Physics Conference Proceedings* 910 (1), 294–333.

Bojowald, M. (2020). Critical evaluation of common claims in loop quantum cosmology. *Universe* 6 (36), 1–23.

Brandenberger, R. H. and C. Vafa (1989). Superstrings in the early universe. *Nuclear Physics B* 316 (2), 391–410.

Butterfield, J. N. (2011). Against pointillisme: A call to arms. In D. Dieks, W. J. Gonzalez, S. Hartmann, T. Uebel, and M. Weber (Eds.), *Explanation, Prediction, and Confirmation*, pp. 347–365. Springer.

Crowther, K. and N. Linnemann (2019). Renormalizability, fundamentality and a final theory: The role of UV completion in the search for quantum gravity. *British Journal for the Philosophy of Science* 70 (2), 377–406.

Curiel, E. (1999). The analysis of singular spacetimes. *Philosophy of Science* 66, S119–S145.

Curiel, E. (2021). Singularities and Black Holes. In E. N. Zalta (Ed.), *The Stanford Encyclopedia of Philosophy* (Fall 2021 ed.). Metaphysics Research Lab, Stanford University. https://plato.stanford.edu/archives/fall2021/entries/spacetime-singularities/.

Dafermos, M. and J. Luk (2017). The interior of dynamical vacuum black holes I: The C^0-stability of the Kerr Cauchy horizon. *arXiv:1710.01722* [gr-qc].

Davies, P. C. W. (1977). Singularity avoidance and quantum conformal anomalies. *Physics Letters B* 68 (4), 402–404.

De Haro, S. (2021). The empirical under-determination argument against scientific realism for dual theories. *Erkenntnis*. http://doi.org/10.1007/s10670-020-00342-0.

De Haro, S. (2022, forthcoming). Noether's theorems and energy in general relativity. In J. Read, N. Teh, and B. Roberts (Eds.), *The Philosophy and Physics of Noether's Theorems*. Cambridge University Press.

De Haro, S., J. van Dongen, M. Visser, and J. Butterfield (2020). Conceptual analysis of black hole entropy in string theory. *Studies in History and Philosophy of Modern Physics* 69, 82–111.

Doboszewski, J. (2019). Relativistic spacetimes and definitions of determinism. *European Journal for Philosophy of Science* 9 (24), 1–14.

Donoghue, J. (1997). Introduction to the effective field theory description of gravity. In F. Cornet and M. Herrero (Eds.), *Advanced School on Effective Theories: Almunecar, Granada, Spain 26 June-1 July 1995*, pp. 217–240. World Scientific.

Earman, J. (1992). Cosmic censorship. *Proceedings of the Biennial Meeting of the Philosophy of Science Association* 1992 (2), 171–180.

Earman, J. (1995). *Bangs, Crunches, Whimpers, and Shrieks: Singularities and Acausalities in Relativistic Spacetimes*. Oxford University Press.

Earman, J. (1996). Tolerance for spacetime singularities. *Foundations of Physics* 26, 623–640.

Eichhorn, A. (2019). An asymptotically safe guide to quantum gravity and matter. *Frontiers in Astronomy and Space Sciences* 5, 47.

Einstein, A. (1915). Letter to Paul Ehrenfest. In *The Collected Papers of Albert Einstein*, Volume 8A, Doc. 173, pp. 228–229. Princeton University Press.

Einstein, A. (1916). Die Grundlage der allgemeinen Relativitätstheorie. In *The Collected Papers of Albert Einstein*, Volume 6, Doc. 30, pp. 284–339. Princeton University Press.

Ellis, G., K. Meissner, and H. Nicolai (2018). The physics of infinity. *Nature Physics* 14, 770–772.

Festuccia, G. and H. Liu (2006). Excursions beyond the horizon: Black hole singularities in Yang-Mills theories (I). *Journal of High Energy Physics* 2006 (4), 44.

Fraser, J. D. (2009). Quantum field theory: Underdetermination, inconsistency, and idealization. *Philosophy of Science* 76 (4), 536–567.

Fraser, J. D. (2011). How to take particle physics seriously: A further defence of axiomatic quantum field theory. *Studies in History and Philosophy of Modern Physics* 42 (2), 126–135.

Fraser, J. D. (2016). *What is Quantum Field Theory? Idealisation, Explanation and Realism in High Energy Physics.* Ph. D. thesis, University of Leeds.

Fraser, J. D. (2020). The real problem with perturbative quantum field theory. *British Journal for the Philosophy of Science* 71 (2), 391–413.

Gibbons, G. W., G. T. Horowitz, and P. K. Townsend (1995). Higher-dimensional resolution of dilatonic black-hole singularities. *Classical and Quantum Gravity* 12 (2), 297–317.

Gies, H. and J. Jaeckel (2004). Renormalization flow of QED. *Physical Review Letters* 93 (11), 110405.

Green, M. B. (1999). Interconnections between type II superstrings, M theory and N = 4 Yang-Mills. In A. Ceresole, C. Kounnas, D. Lüst, and S. Theisen (Eds.), *Quantum Aspects of Gauge Theories, Supersymmetry and Unification. Lecture Notes in Physics*, Volume 525. Springer.

Green, M. B., J. H. Schwarz, and E. Witten (1987). *Superstring Theory*, Volume 1. Cambridge University Press.

Gryb, S. and K. P. Thébault (2018). Superpositions of the cosmological constant allow for singularity resolution and unitary evolution in quantum cosmology. *Physics Letters B* 784, 324–329.

Hawking, S. W. and G. F. Ellis (1973). *The Large-Scale Structure of Space-Time.* Cambridge University Press.

Henson, J. (2009). The causal set approach to quantum gravity. In D. Oriti (Ed.), *Approaches to Quantum Gravity: Toward a New Understanding of Space, Time and Matter*, pp. 393–413. Cambridge University Press.

Horowitz, G. T. and R. C. Myers (1995). The value of singularities. *General Relativity and Gravitation* 27 (9), 915–919.

Hossenfelder, S. (2013). Minimal length scale scenarios for quantum gravity. *Living Reviews in Relativity* 16, 2–90.

Huggett, N. (2017). Target space $\neq$ space. *Studies in History and Philosophy of Modern Physics* 59, 81–88.

Israel, W. (1966). Singular hypersurfaces and thin shells in general relativity. *Il Nuovo Cimento B* XLIV (1), 1–14.

Jackiw, R. (1999). The unreasonable effectiveness of quantum field theory. In T. Cao (Ed.), *Conceptual Foundations of Quantum Field Theory*, pp. 148–159. Cambridge University Press.

Jackiw, R. (2000). What good are quantum field theory infinities? In A. Fokas, A. Grigoryan, T. Kibble, and B. Zegarlinski (Eds.), *Mathematical Physics 2000*, pp. 101–110. World Scientific.

Koslowski, T. A., F. Mercati, and D. Sloan (2018). Through the big bang: Continuing Einstein's equations beyond a cosmological singularity. *Physics Letters B* 778, 339–3434.

Kretschmann, E. (1917). Über den physikalischen Sinn der Relativitätsposulate, A. Einsteins neue und seine ursprünglische Relativitätstheorie. *Annalen der Physik* 53 (16), 575–614.

Landau, L., A. Abrikosov, and I. Khalatnikov (1954). *The Removal of Infinities in Quantum Electrodynamics*. Doklady Akademii Nauk SSSR.

Lewis, D. (1986). *On the Plurality of Worlds*. Blackwell.

Lüsher, M. and P. Weisz (1987). Scaling laws and triviality bounds in the lattice ϕ^4 theory: (I). one-component model in the symmetric phase. *Nuclear Physics B* 290, 25–60.

Mazur, P. O. and E. Mottola (2001). Gravitational condensate stars: An alternative to black holes. *arXiv:gr-qc/0109035*.

Misner, C. (1969). Absolute zero of time. *Physical Review* 186 (5), 1328–1333.

Niedermaier, M. and M. Reuter (2006). The asymptotic safety scenario in quantum gravity. *Living Reviews in Relativity* 5 (5).

Parker, L. and S. A. Fulling (1973). Quantized matter fields and the avoidance of singularities in general relativity. *Physical Review D* 7 (8), 2357–2374.

Penrose, R. (1972). The geometry of impulsive gravitational waves. In L. O'Raifeartaigh (Ed.), *General Relativity. Papers in Honour of J. L. Synge*, pp. 101–115. Clarendon Press.

Penrose, R. (1979). Singularities and time-asymmetry. In S. Hawking and W. Israel (Eds.), *General Relativity: An Einstein Centenary Survey*, pp. 581–638. Cambridge University Press.

Rovelli, C. and F. Vidotto (2014). *Covariant Loop Quantum Gravity: An Elementary Introduction to Quantum Gravity and Spinfoam Theory*. Cambridge University Press.

Sauer, T. (2005). Albert Einstein, review paper on general relativity theory (1916). In I. Grattan-Guinness (Ed.), *Landmark Writings in Western Mathematics, 1640–1940*, pp. 802–822. Elsevier.

Sider, T. (2001). *Four-dimensionalism. An Ontology of Persistence and Time*. Oxford University Press.

Thiemann, T. (1998). Quantum spin dynamics (QSD). *Classical and Quantum Gravity* 15 (4), 839–873.

't Hooft, G. and M. Veltman (1974). One-loop divergencies in the theory of gravitation. *Annales de l'IHP Physique theorique* 20 (1), 69–94.

Wald, R. (1992). 'Weak' cosmic censorship. *Proceedings of the Biennial Meeting of the Philosophy of Science Association* 1992 (2), 181–190.

Wallace, D. (2006). In defence of naivete: The conceptual status of Lagrangian quantum field theory. *Synthese* 151 (1), 33.

Wallace, D. (2011). Taking particle physics seriously: A critique of the algebraic approach to quantum field theory. *Studies in History and Philosophy of Modern Physics* 42 (2), 116–125.

Weinberg, S. (1979). Ultraviolet divergences in quantum theories of gravitation. In S. Hawking and W. Israel (Eds.), *General Relativity, an Einstein Centenary Survey*, pp. 790–831. Cambridge University Press.

Wheeler, J. and K. Ford (1998). *Geons, Black Holes and Quantum Foam*. W.W. Norton & Company.

10 TGFT condensate cosmology as an example of spacetime emergence in quantum gravity

Daniele Oriti

10.1 Introduction

The general issue that this contribution is concerned with is the nature of space and time in quantum gravity and, more specifically, the sense in which they can be said to be emergent notions in a non-spatiotemporal fundamental theory.

This is a possibility that has been raised often and in many different contexts in the (recent) past, it is increasingly discussed in the philosophy of physics literature (Huggett and Wüthrich, 2021; Crowther, 2020; Baron, 2019; Le Bihan and Linnemann, 2019; Huggett and Wüthrich, 2018; Le Bihan, 2018; Crowther, 2014) and partially realized (to a different degree) in several quantum gravity formalisms.

First of all, we summarize what we mean by emergence in general and the different types of emergence that can be recognized in physics, and then articulate our perspective on the different 'levels' or senses in which we can specifically speak of emergence of spacetime in quantum gravity. This serves as a necessary conceptual background for the actual focus of this work. We illustrate how space and time can be shown to emerge in the quantum gravity formalism of tensorial group field theories (TGFTs), specifically in the class of models with richer 'quantum geometric' content called group field theories, at least in a well-identified given approximation and for the simple case of cosmological (i.e., homogeneous and isotropic) spacetimes.

We emphasize throughout the exposition that this particular set of results may represent a concrete example of a more general template that is applicable to other quantum gravity formalisms, thus having a much more general interest. In fact, some of the technical results we illustrate are already applicable to other quantum gravity formalisms, thanks to the fact that they share several ingredients with TGFTs. Moreover, similar results, about one aspect or the other of the spacetime emergence that we illustrate in TGFTs, have been also obtained in other quantum gravity contexts.

DOI: 10.4324/9781003219019-13

10.2 Emergence of spacetime in quantum gravity

We use a notion of emergence that is simple and general enough to accommodate all known examples of emergent phenomena in physics (Butterfield, 2011b, a): a physical behaviour or phenomenon is understood as *emergent* if it is sufficiently novel and robust with respect to some comparison class, usually associated to the class of behaviours and phenomena it emerges from. This definition can be refined in specific contexts if needed, but it suffices for our purposes.

Two important points need to be noted. First, very often some sort of approximation or limit (in the mathematical model(s) describing the phenomena under consideration) is needed to realize such emergence (Butterfield and Bouatta, 2012; Batterman, 2011, 2004; Castellani, 2000). Second, in this definition, emergence is not incompatible with reduction (understood, within the same mathematical model(s), as deduction); Indeed, reduction is usually needed to specify the comparison class entering the definition, so that one could even argue that emergence is the inverse process of reduction, at the epistemological level, i.e., that we recognize some phenomenon as *emergent* from another exactly *because* it has been shown to be deducible from the other.

10.2.1 Emergence and its many kinds

This contribution deals with physics, so we phrased the notion of emergence in physical terms. Associated with this physical notion, however, come two different types of emergence: intertheoretic and ontological emergence. We speak of a set of physical phenomena as emergent from another, if the theoretical description of the former can be reduced to the one of the latter. This intertheoretic (or epistemic) emergence amounts in fact to a relation between mathematical and conceptual models of the world, from which we imply a relation between natural phenomena described by those models. Ontological emergence is instead a relation between entities themselves, thus it is both a physical and metaphysical statement. However, the philosophical perspective on which we base our discussion accepts the following points. First. all the physical entities that we assign ontological status to are to a large extent defined by the theoretical frameworks/models we use to describe them, in the straightforward sense that the concepts we use to characterize them and the properties we assign to them are those taken from such theoretical models, and not identified independently from them. I take this to follow from some basic naturalistic attitude toward metaphysics, according to which metaphysics follows from (or should be at least strongly constrained by) physics, but it is also more or less the (often naive) attitude of working physicists.[1] Second, given this ontological emergence follows from intertheoretic emergence, once we decide to add an ontological commitment about (some of) the entities featuring the two theories related by the emergence relations. The ontological commitment is not a trivial step by any means, and should be argued for convincingly on a case by case basis. However, once this is agreed, there is not much left to argue about the existence of some ontological emergence. Obviously, non-trivial metaphysical consequences can then follow from the existence of ontological emergence, and this adds to the non-triviality of any ontological commitment. In this contribution, we focus on the current and future theoretical models of space and time, and thus on intertheoretic emergence of spacetime, and on the more physical rather than metaphysical aspects. Thus, we limit ourselves to emphasize at which level the intertheoretic relations make room for ontological emergence and what sort of emergence this could be, once the ontological commitment is agreed upon. We choose not to discuss the very many additional subtleties concerning the conditions for an ontological commitment to be justified and, more generally, concerning the complex relations between theoretical models and concepts, models and reality, including the definition of the latter, the normative vs representational character of natural laws and how these are captured by theoretical models, and many more.

Note that any intertheoretic emergence, and thus any ontological emergence in our ("naturalistic") perspective, is in itself not associated to any physical process happening in time and leading from the fundamental to the emergent description (and/or entity). It corresponds rather to a change in perspective, accompanied by a change in relevant concepts (and entities inferred in correspondence to such concepts). It is, then, a *synchronic* emergence, not corresponding to a temporal process, a priori.[2] This is the case also for the peculiar example of emergent behaviour often associated to phase transitions, whose peculiarity lies in the fact that they are *symmetric* relations between theoretical descriptions (and corresponding entities) in which none can be said to be more fundamental than the other (while both could be emergent from some more fundamental "microscopic" theory).

While this is not part of the definition, however, it is not excluded either, and many examples of physical processes can be found, that instantiate emergent behaviour, then characterized in intertheoretic terms. In this sense, the same emergence that was understood in synchronic terms at the intertheoretic level can then be understood in *diachronic* terms as something happening *in time*.[3] In the case of the emergence of time itself, by definition a temporal characterization is ruled out, which makes the understanding it even more challenging, and one can at best imagine a partial temporal characterization "up to the time when time disappears" (Oriti, 2021). We will come back to this point when discussing our specific example of spacetime emergence.

10.2.2   *Spacetime emergence and its many levels*

Let us now summarize the "levels of emergence" that can be envisioned for spacetime in quantum gravity.[4] A more detailed outline can be found in Oriti (2018) and, with a focus on the emergence of time, in Oriti (2021).

General relativity is a dynamical theory of continuum (in fact, smooth) fields defined on a (differentiable) manifold (i.e., a set of points with appropriate regularity properties). Thus, at first the underlying ontology seems to be given by these elements: fields (including the metric) and manifold. However, diffeomorphism invariance of the theory implies that values of fields at different points in the manifold can be physically equivalent and thus the manifold and its points do not really carry any physical meaning in themselves and they are thus not part of the ontology of the world, except as providing global (topological) conditions on the fields (since the set of allowed field configurations depends on the manifold topology).[5] Moreover, the theory is background independent in the sense that all fields appearing in it are dynamical entities (Giulini, 2007; Gaul and Rovelli, 2000), subject to "equations of motion" constraining their allowed values and mutual relations. And generic solutions possess no feature that can be used to single out a preferred direction of time or space, that can only be associated to geometries with distinctive isometries or to special boundary conditions (or both, like AdS spacetimes).

In this sense (absence of preferred time or space) one could claim that there is in fact no space or time in GR: there are only fields and their relations. In fact, just like physical (i.e., diffeo-invariant) observables can be constructed as relations between dynamical fields, time and space can be defined by choosing appropriate (matter) fields (components) as rods and clocks and computing spatial and temporal quantities using them and the metric field (which enters all such quantities, and in this sense encodes the physical, dynamical spacetime in GR). This is the *relational strategy* (Rovelli, 1991; Marolf, 1995; Rovelli, 2002; Dittrich, 2006) to the definition of space and time in GR, and its application can be said to implement a first kind of *emergence* of space and time within a theory that does not select one a priori. We could call it a *level* −1 of spacetime emergence. Indeed, generic fields will not behave like a perfect rod or clock and the notions of space and time that they will concur to define will not match the usual notions (basically corresponding to the Newtonian ones of our common sense), which is only the case in some special approximation and for special kinds of fields.

Special kinds of fields are also introduced and used in classical GR in order to allow for the *deparametrization* of the theory, i.e., its formulation in an entirely diffeo-invariant language and with a clearly identified notion of space and time (that's the main benefit of this procedure), in the (preferred) frame defined by such material rods and clock (that's the main shortcoming). One can then proceed to quantization for example by standard canonical quantization methods, assuming that the frame degrees of freedom are not quantized, and obtain a rather standard quantum field theory for the other fields, including the metric (Husain and Pawlowski, 2012; Thiemann, 2006; Giesel and Thiemann, 2015; Giesel and Herzog, 2018). Provided one can complete the quantization procedure (at present something that is only achieved in a rather formal sense), we end up with a theory of quantum gravity which could be valid as long as the quantum dynamical nature of the matter fields chosen as preferred reference frame can be neglected (again, we see that we have the usual notions of time and space only as the result of some idealization or approximation).

Because the metric is quantized (and dynamical), however, space and time do not remain what they were even in presence of a preferred reference frame. In fact, in a quantum theory of gravity obtained from the quantization of the metric field (Kiefer, 2013, 2012; Thiemann, 2007), they *disappear* in a more radical sense in which they had already disappeared in classical GR. Accordingly, the emergence of space and time from quantum gravity presents much more radical challenges that the one in classical GR. It does not imply any ontological emergence, since the fundamental entities remain the same, i.e., dynamical fields, and relational construction remain essential to identify physical notions of time and space. The quantum description of fields, including the metric field, however, implies that these notions (and all geometric ones: distances, curvature, volumes, causal relations) will be subject to uncertainty relations, irreducible quantum fluctuations, some form of contextuality, discreteness of observable values, and, in the case of composite systems, entanglement. The challenges to our common sense notions of realism, separability, and locality are formidable, even more than in standard quantum mechanics.

The relational strategy for the definition of space and time will be affected by the quantum properties of our relational frames, in particular our clocks (Hoehn et al., 2021; Castro-Ruiz et al., 2020; Giacomini, 2021). We must abandon any value-definiteness of spatiotemporal quantities and possibly any continuous notion of space and time, if spatiotemporal observables end up having discrete values (Rovelli and Smolin, 1995). We have to learn to deal with quantum reference frames and indefinite causal structures or temporal order (Hardy, 2005; Castro-Ruiz et al., 2020; Giacomini, 2021, see also the chapter by Lam, Letertre, and Mariani in this volume), and we are forced to abandon unitary time evolution as a key aspect of quantum dynamics. All these aspects are of truly foundational nature as well as bearer of important physical consequences, but have not been studied as much as they deserve in a full quantum gravity setting, even in quantum gravity contexts like the TGFT one we will discuss in the following where quantum reference frames are indeed used to extract a relational dynamics from non-spatiotemporal entities. Starting from such quantum realm, the emergence of space and time as we know them from GR requires a number of approximations and restrictions, which together define the semiclassical limit of quantum gravity. This includes the use of special semiclassical states, the focus on a subset of observables, and more.

The very presence of this additional step justifies speaking of a new level of emergence, level 0, common to any quantum theory of gravity and spacetime obtained by quantization of the classical one. Once more, we are dealing with an intertheoretic, synchronic emergence. Whether any physical process (a "classicalization process" of spacetime and geometry) can be put in correspondence with it can only be determined within some specific quantum gravity formalism. Even if it can, speaking of the emergence of spacetime (time, in particular) as if it was a temporal process remains impossible (outside special approximations) (Oriti, 2021) as anticipated earlier.

A new level of spacetime emergence is found in quantum gravity formalisms in which the fundamental theory is not understood as the straightforward quantization of the classical gravitational theory, because it is instead based on a new set of basic entities that do not correspond simply to the quantum counterpart of the continuum fields of GR (Oriti, 2018). This is an intertheoretic spacetime emergence of *level 1*, which by definition corresponds to an ontological emergence as well, since the spatiotemporal fields of GR (i.e., those, including the metric, whose relations define space and time as we know them) are replaced by non-spatiotemporal entities. The precise nature of such new entities varies in different quantum gravity formalisms, as does the precise degree by which they differ from continuum fields. We will deal in some detail with one specific example in the following. The step away from the very ontology of fields brings us new conceptual and technical challenges with respect to the already challenging disappearance of spacetime forced upon us by quantum spatiotemporal fields in any quantized version of GR. There is a new non-spatiotemporal ontology to make sense of, first of all; and the need to make sense of the ontological status of spatiotemporal fields, of spacetime itself, now that they are deprived of fundamental status. The precise dependence relation between fundamental entities and emergent spatiotemporal fields needs to be clarified, the best options being in the sense of functionalism or grounding, indeed apt to be applied also in the analogous context of fluid dynamics emerging from atomic or molecular physics. At the technical as well as physical level, on the other hand, the big challenge is to control the collective behaviour of the new quantum entities replacing fields, since the latter, including the metric field and any associated notion of space and time, should arise from such collective behaviour, as coarse-grained, approximate notions.

This is often referred to as the problem of the "continuum limit" in quantum gravity, since the new fundamental entities are often some discrete (and quantum) counterpart of continuum fields, and it is usually tackled by renormalization group methods and other tools form statistical and quantum many-body theory (Bahr and Steinhaus, 2017; Dittrich, 2017; Eichhorn et al., 2019; Carrozza, 2016; Finocchiaro and Oriti, 2021). It is important to stress that such continuum limit is conceptually as well as technically distinct from the classical one, which would be the only one responsible for the emergence of spacetime from quantum gravity at level 0. Indeed, working at this level 1, we could expect the continuum limit to bring us to level 0, with the emergence of spacetime as we know it from GR still requiring the additional steps we outlined earlier.

The situation becomes more complicated still, once we realize that the continuum limit of a quantum many-body theory of non-spatiotemporal entities is not unique, in general. We should rather expect it to give rise to several continuum phases of the same fundamental quantum gravity system. Since we have to match (at least approximately) the well-tested general relativistic description, at least one of these continuum phases should be geometric or spatiotemporal. That is, it should allow for an effective reconstruction of continuum spacetime and geometry, after further (classical) approximation (and a relational rewriting of the theory). But other will be non-geometric and non-spatiotemporal, not allowing such reconstruction, even approximately.

This is a *level 2* emergence, since it clearly brings further issues to be solved. At the physical level, they concern the study of the phase diagram of the fundamental quantum gravity theory and the identification of the one (or more) phase(s) in which we reach a situation corresponding to a level 0 emergence of spacetime, i.e., a situation like in quantum GR (if not directly a level -1 emergence, that of classical GR, if the classical approximation is somehow part of the continuum limit). At the conceptual level, new difficulties arise because the same kind of non-spatiotemporal entities that require a new ontology (one that does not make use of spatiotemporal notions) are recognized to be "even less spatiotemporal" now, since they may not give rise to spacetime at all, even in a continuum approximation (Oriti, 2018).

This is another point where the ontological commitment one is willing to adopt concerning the new non-spatiotemporal structures appearing in quantum gravity formalisms changes drastically the conceptual context in which the same formalisms are understood.

The non-spatiotemporal entities adopted as basic structures in several quantum gravity may be considered purely mathematical tools, instrumental for defining a quantum theory of the gravitational field alongside other matter fields, assumed to be the only truly physical entities. This does not change at all the technical difficulties to be solved, with the continuum limit, the coarse-graining and renormalization techniques, the non-geometric phases, etc, but it implies that physics is found only in level 0 of spacetime emergence, and that only at the same level we should be concerned with the associated conceptual difficulties.

If one takes seriously the same entities as (potentially) physical ones, on the other hand, taking seriously also the probably rich phase diagram of the fundamental theory becomes necessary. One is also led to consider the possible physical nature of the phase transitions separating geometric and non-geometric phases. What is the physical interpretation and what are the physical signatures of *a geometrogenesis* phase transition (Konopka et al., 2006; Oriti, 2007; Mandrysz and Mielczarek, 2019; Smolin, 2003) (Delcamp and Dittrich, 2017; Kegeles et al., 2018; Bahr et al., 2021; Koslowski and Sahlmann, 2012; Surya, 2012; Pithis and Thürigen, 2020; Ambjørn et al., 2017; Percacci, 1991), leading from a non-geometric phase to a geometric one or viceversa? What are the relevant geometric order parameters? For example, can the cosmological big bang singularity of classical GR be in fact replaced by such geometrogenesis phase transition in quantum gravity? Can we locate *in time* (e.g., in our cosmological past) this emergence of time itself? And what is the physical interpretation of the other quantum gravity phase transitions (and of the other phases)?

New issues appear, at both physical and philosophical level, deepening further the challenge of making sense of spacetime emergence. Because of them, we identify this situation as a further level, i.e., *level 3* of spacetime emergence in quantum gravity.

In the end, assigning ontological status to the new non-spatiotemporal entities suggested by quantum gravity approaches calls for the development of an ontology that is not spacetime-based, which is a philosophical work yet to be tackled, and of the utmost importance.

A schematic representation of the multi-level scenario we outlined is given in Figure 10.1.

It should be clear that both physical and philosophical challenges raised by this quantum gravity scenario are of the deepest nature. We will not discuss them here. Instead, we proceed to present a concrete example of a quantum gravity formalism in which all these levels of spacetime emergence have been realized, tentatively and partially, in recent work.

10.3   The TGFT atoms of space

The example of spacetime emergence we discuss is realized in the context of tensorial group field theories (Oriti, 2011; Gurau and Ryan, 2012; Krajewski, 2011; Rivasseau, 2016; Oriti, 2017a), and more specifically of the most "quantum geometric" subclass of such models, usually

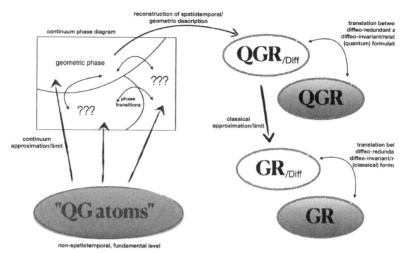

Figure 10.1 A schematic representation of the multi-level scenario for spacetime emergence.

called group field theories. We also restrict attention to 4-dimensional Lorentzian models, i.e., the most directly relevant for physics.

The fundamental entities of these quantum geometric models are quantized tetrahedra whose discrete geometry is encoded in algebraic data. Specifically, the Hilbert space associated to an individual tetrahedron is obtained starting from $\mathcal{H} = L^2(G^4; d\mu_{Haar})$ and imposing appropriate additional "geometricity" restrictions on quantum states. The Lie group G is the Lorentz group $SO(3, 1)$ (or its double cover $SL(2, \mathbb{C})$) or its rotation subgroup $SU(2)$, $d\mu_{Haar}$ is the corresponding Haar measure. The geometricity conditions can also be imposed at the dynamical level.

In this representation of the quantum tetrahedra, quantum states are then functions of group elements, each associated to one of the triangles on the boundary of each tetrahedron. One can also use different equivalent representations of the same Hilbert space, for example in terms of (unitary, irreducible) group representations associated as well to the boundary triangles, so that one can represent the same tetrahedron equivalently as a spin network vertex, i.e., a vertex with four outgoing open links, each labeled by a group representation (cf. Figure 10.2). This representation of the tetrahedral quantum geometry can be obtained by straightforward quantization of a phase space description of the (intrinsic and extrinsic) classical geometry of the same tetrahedra in terms of same kind of group-theoretic data (Baez and Barrett, 1999; Bianchi et al., 2011; Dittrich and Ryan, 2011; Baratin and Oriti, 2012; Rovelli and Speziale, 2010; Perez, 2013; Finocchiaro et al., 2020).

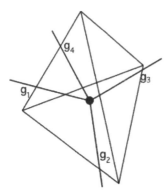

Figure 10.2 A tetrahedron with its dual graph representation, labelled by group elements.

Building on this single-tetrahedron Hilbert space, one can then define a Fock space for the quantum states of arbitrary numbers of tetrahedra: $\mathcal{F}(\mathcal{H}) = \oplus_{V=0}^{\infty} sym\{\mathcal{H}^{(1)} \otimes \mathcal{H}^{(2)} \otimes \cdots \otimes \mathcal{H}^{(V)}\}$. Field operators creating/annihilating the quanta of this Fock space, i.e., quantum tetrahedra, can then be straight forwardly introduced, to realize a field theory formulation of pre-geometric "atoms of space."

This Fock space contains quantum states associated to extended simplicial complexes formed by gluing tetrahedra to one another across shared boundary triangles (or equivalently, associated to extended 4-valent graphs obtainedby gluing open vertices to one another across their open links). The elementary gluings are obtained as maximal entanglement between the quantum degrees of freedom associated to the same triangle in the two tetrahedra to be glued (and correspondingly for the dual graphs) and the overall connectivity of such discrete structures is encoded as entanglement patterns associated to the quantum states in the TGFT Fock space (Colafranceschi and Oriti, 2021).

In terms of group irreps for $G = SU(2)$, the resulting connected states match the spin network states of canonical loop quantum gravity (Sahlmann, 2010) and the Hilbert space associated to them, for a given graph, coincide (see Figure 10.3). When considering quantum states associated to arbitrary graphs and their scalar product, i.e., the full Hilbert space of the theory, they differ, with the one in canonical LQG encoding more conditions motivated by continuum geometry, absent in the Fock space of TGFT (Oriti, 2016, 2017a). In this sense, quantum geometric TGFTs are a 2nd quantized version of LQG, with a stronger discrete flavour.

So there is a clear connection between TGFT states and discrete (piecewise-flat) geometries, at least for some of them and in a semiclassical approximation, and this discrete geometric intuition guides model building and the analysis of their quantum dynamics. It is also clear,

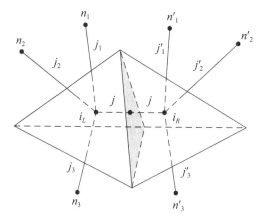

Figure 10.3 Two tetrahedra with the dual graph representation, glued along a shared triangle, labelled by group irreps.

however, that they are not at all fully spatiotemporal structures. Generic TGFT states correspond to arbitrary numbers of (partially) disconnected quantum simplices, themselves not behaving like classical geometric ones, even in the limited sense of piecewise-flat geometry. Moreover, even the very special ones with a nice simplicial geometric interpretation (thus corresponding to quantum piecewise-flat geometries) will remain distant from (quantized) continuum geometries. The list of "geometric or spatiotemporal pathologies" that generic quantum states for arbitrary collections of quantum TGFT building blocks can possess is longand it marks a huge gap with respect to the usual description of spacetime in terms of fields (including a metric field), a gap that is much wider than the already relevant gap between classical piecewise-flat geometry for extended simplicial complexes and the smooth geometry of classical GR. This justifies referring to this level of description as non-geometric and not spatiotemporal. Of course, we have to go beyond the use of the word "geometric" in both contexts, realizing that the corresponding sense in which we have "geometry" is very different in the two cases, and that our usual notions of space and time only apply to one of the two, really.

The same considerations apply at the dynamical level. A partition function for the TGFT "atoms," i.e., the quantized tetrahedra, and for the field over the group manifold of which they represent elementary quanta, will in general take the form:

$$Z = \int \mathcal{D}\varphi \mathcal{D}\varphi^* \, e^{-S_\lambda(\varphi,\varphi^*)} \quad , \tag{10.1}$$

for an action $S_\lambda(\varphi, \varphi^*) = K + U + U^*$ function of the field (and complex conjugate) and some coupling constant(s) weighting an interaction term

U given by a polynomial in the fields and encoding a pairing of their (group) arguments which is, in general, non-local (in the sense that they are not simply identified at interactions, like positions in usual local QFT; cf. Gurau and Ryan, 2012; Oriti, 2011; Krajewski, 2011; Rivasseau, 2016; Oriti, 2017a). K is instead a quadratic (in the fields) kinetic term.

It can be defined in quantum statistical mechanics setting in terms of entropy maximization for given thermodynamic constraints on the fundamental TGFT quanta (Chirco et al., 2019; Kotecha, 2019). These constraints can be motivated by discrete gravity considerations or as corresponding to a dynamical (Hamiltonian) constraint of a canonical (loop) quantum gravity framework.

The best established connection between the TGFT quantum dynamics (for quantum geometric models) and that of discrete quantum gravity and loop quantum gravity is to view the TGFT partition function as the generating function for a discrete gravity path integral, incorporating also a sum over discrete topologies. In formulae:

$$Z = \int \mathcal{D}\varphi \mathcal{D}\varphi^* \, e^{-S_\lambda(\varphi, \varphi^*)} = \sum_\Gamma \frac{\lambda^{N_\Gamma}}{sym(\Gamma)} \, \mathcal{A}_\Gamma \quad . \tag{10.2}$$

Indeed, for appropriately quantum geometric models as the ones we focus on, the TGFT partition function can be expanded perturbatively with respect to its coupling constants (weighting the interaction terms in the TGFT action), the generic result being that: a) the Feynman diagrams Γ so generated as dual to cellular (simplicial) complexes of arbitrary topology; b) the (model dependent) Feynman amplitudes $\mathcal{A}_\Gamma$ are given by lattice gravity path integrals for a (Plebanski-like) formulation of discrete gravity in terms of group-theoretic data, on the lattice dual to the Feynman diagram Γ (Baratin and Oriti, 2012; Finocchiaro and Oriti, 2020); c) the same Feynman amplitudes $\mathcal{A}_\Gamma$, when expanded in irreps of the group G, take the form of spin foam models (Reisenberger and Rovelli, 2001; Perez, 2013; Finocchiaro and Oriti, 2020), which in turn can be understood as a covariant formulation of the quantum dynamics of the spin network states of canonical loop quantum gravity.

The first point is common to all TGFT models, and a large number of results in simpler models (in particular, random tensor models) clarified combinatorial, topological and non-perturbative aspects of the sum over complexes they generate (Gurau and Ryan, 2012; Rivasseau, 2016). These simpler TGFT models also share the second point, but in general their Feynman amplitudes can be understood as discrete gravity path integrals only in the purely combinatorial sense of dynamical triangulations (Loll, 2020). The additional group-theoretic data of quantum geometric TGFT models enrich this discrete gravity description with dynamical geometric information, like in quantum Regge calculus (Hamber, 2009) and spin foam models, and are necessary for the third point to be realized.

The considerations we put forward at the kinematical level, concluding that the basic TGFT entities should be understood as pre-geometric, non-spatiotemporal entities, apply also at this dynamical level. The TGFT partition function can be seen as the generating functional for a gravitational path integral provided one can give non-perturbative and continuum meaning to the discrete structures appearing in its initial definition.

These are still far from a theory of continuum fields and spacetime and in fact generic discrete configurations appearing in the perturbative expansion would not even correspond to well-behaved simplicial geometries. Still, this connection to discrete gravity path integrals and to spin foam models is very important for TGFTs. It clarifies the interpretation and meaning of the TGFT quantum dynamics, it provides a key guide for model building, and it illustrates the strong ties to other quantum gravity approaches which in turn allow importing (and exporting) techniques and results.

The guideline provided by the connection to discrete gravity path integrals is crucial also for extending TGFT models to matter coupling. In particular, we are interested in adding degrees of freedom that can be interpreted as discretized scalar matter,[6] just like the group-theoretic variables can be interpreted as discrete geometric data. Thus, the model building strategy is to define a TGFT field and action in such a way that, in perturbative expansion, one obtains Feynman amplitudes with the form of discrete path integrals for gravity coupled to scalar fields, on the lattice dual to the Feynman diagram. Beside the obvious physical relevance of matter coupling, this extension of TGFT models is needed in order to apply the relational strategy for the reconstruction of space and time observables from the full theory.

Focusing on a single real scalar field (to be later used as a relational clock), one works with a TGFT field $\varphi(g_I, \chi)$ defined on an extended domain $G^4 \times \mathbb{R}$ and with an action of the general form (Oriti et al., 2016; Li et al., 2017):

$$S_\lambda(\varphi, \varphi^*) = K + U + U^* \quad ,$$
$$K = \int dg_I dh_I \int d\chi d\chi' \varphi^*(g_I; \chi) K(g_I, h_I; (\chi - \chi')^2) \varphi(h_I; \chi') \quad , \quad (10.3)$$
$$U = \int [dg] d\chi \, \varphi(g; \chi) ... \varphi(g'; \chi) \, \mathcal{V}(\{g\}; \chi) \quad ,$$

with a kinetic term involving some (local) dependence on quantum geometric data and that can be expanded in (infinite) powers of second derivatives of the TGFT field with respect to the scalar field variable χ, and an interaction term that is local in the scalar field data, while remaining non-local in the quantum geometric ones.

Notice that the scalar field variable (to be later used as a clock to define relational observables) enters the TGFT action in a way compatible with its use as a standard time coordinate in the analysis of the field theory itself (provided one can deal satisfactorily with the higher derivatives). This led to develop a TGFT counterpart of the "deparametrization" strategy of classical GR, and a straight forward canonical quantization of TGFT models (and consequent cosmological applications; cf. Wilson-Ewing, 2019, Gielen and Polaczek, 2020; Gielen, 2021).

TGFT models propose therefore a formulation of quantum gravity in terms of non-spatiotemporal "atoms of space," an example of the "QG atoms" in Figure 10.1, from which continuum spacetime and geometry should then be shown to emerge.

Let us now turn to how this emergence is realized in this formalism.

10.4   Continuum limit in TGFT models

In a quantum field theory formalism, albeit peculiar like the TGFT one, the full quantum theory incorporating all dynamical degrees of freedom (and arbitrary numbers of the fundamental quanta) and thus also the "continuum limit," corresponds to the evaluation of the full partition function Z (or free energy $F = \ln Z$) or, equivalently, the full quantum effective action $\Gamma[\phi] = sup_J(J \cdot \phi - F(J))$ for source J and mean field $\phi = \langle \varphi \rangle$.

This is also in line with the discrete gravity interpretation of the TGFT quanta and perturbative amplitudes. The evaluation of the TGFT partition function or effective action amounts to resumming the full perturbation series, thus the sum over triangulations weighted by a discrete gravity path integral in 10.2, including infinitely refined lattices. This is also how one would define the continuum gravitational path integral starting from the discrete one in simplicial quantum gravity approaches, as well as in spin foam models. The caveat, with respect to the discrete quantum gravity intuition, is that the TGFT picture suggest that discrete gravity and spin foam models may capture only part of the QG story, and thus to look beyond the perturbative regime, exploring the corresponding TGFT models in the full non-perturbative regime, i.e., for large values of the couplings too. In physical terms, this means being able to control the full collective quantum dynamics of the QG atoms, looking for regimes in which the discrete picture can (and should) be replaced by one in terms of continuum spatiotemporal fields.

Much recent work has been devoted to this issue, for different TGFT models and from different perspectives. A main tool is the renormalization group. Besides a number of constructive techniques proving that, at least for simpler TGFT models, a rigorous non-perturbative definition can be achieved, most results have been obtained employing functional renormalization group methods. Fully quantum geometric models have proven so far too involved to be treated at such non-perturbative level, and we are only recently gaining some understanding fo the scaling properties and basic divergence structure, from a perturbative point of view. On the other hand, simpler TGFT models have been analyzed in detail, for different choice of domain, rank, features of the dynamics, thanks also to the detailed knowledge about the underlying combinatorial structures gained in the even simpler context of random tensor models. We have many examples of perturbatively renormalizable TGFTs and a good understanding of their RG flow, at least for suitable truncations in theory space. In particular, we are gaining important knowledge about the conditions leading to interesting critical behaviour and phase transitions (Pithis and Thürigen, 2020). Reviews about this body of work can be found in Carrozza (2016); Baloitcha et al. (2020); Finocchiaro and Oriti (2021).

From the point of view of spacetime emergence in Figure 10.1, we are then trying, step by step, to complete the move from QG atoms to the (mathematical) definition of the continuum phase diagram of the fundamental theory, which would amount to be able to control the (formal properties of the) full theory, i.e., the dynamics of the QG atoms at all scales and in all regimes.

Tackling physical issues, of course, requires much more than formal mathematical control. We need to identify which continuum phase admits a rewriting of the theory in terms of spatiotemporal fields and a GR-like dynamics, at least approximately, and then do new physics starting from the fundamental theory.

We need more than full mathematical control, but also less. Of course, while computing the full quantum effective action would be ideal, it is simply not feasible, as it is not for any quantum many-body system of the kind of complexity that we expect a full quantum gravity theory to have. We need approximation schemes that allow us to get at least a partial glimpse of the continuum phase diagram, and mathematical control over the approximation schemes, so that we are able to improve them. And we need to extract physical insights from the partial, approximate picture we obtain in this way, without waiting for the full picture to become available. That's how physics works.

If the emergence of space and time takes place due to the collective dynamics of the QG atoms, we need approximation schemes that capture such collective dynamics, that correspond to some form of coarse-graining of the fundamental "atomic" dynamics, and that maintain visible the quantum nature of the same atoms (since the continuum limit is distinct from the classical one, and it could well be that quantum properties of the QG atoms are in fact responsible for key aspects of the spatiotemporal physics we want to reproduce).

The simplest approximation of the full quantum effective action, that possesses these features, is the mean field approximation. That is, the first place to look for spacetime physics, in the perspective we are advocating, is TGFT mean field hydrodynamics. It corresponds to the saddle point evaluation of the full TGFT path integral, and to approximating the full quantum effective action with the classical TGFT action $\Gamma[\phi] = S(\phi)$. It is the natural starting point, which has then to be improved.

Notice that (despite the name), from the point of view of the QG atoms, i.e., the TGFT quanta, this still amounts to working with rather highly quantum states. Indeed, it amounts to work with coherent states $|\sigma\rangle$ of the TGFT field,[7] whose expression in terms of TGFT Fock excitations is:

$$|\sigma\rangle = \exp(\hat{\sigma})|0\rangle \quad , \quad \hat{\sigma} = \int d[g]d\chi\sigma(g_I;\chi)\hat{\varphi}(g_I;\chi) \quad , \qquad (10.4)$$

where $|0\rangle$ is the Fock vacuum (no QG atoms at all), and the exponential operator acting on it can be expanded in power series to give an infinite superposition of states with increasing number o tetrahedra or spin network vertices (the QG atoms), all associated with the same wavefunction σ, i.e., a coherent state of infinite quantum gravity degrees of freedom.[8]

Still, while highly quantum, they are extremely simplified states with respect to generic quantum states of tetrahedra or spin network vertices. In particular, they do not encode quantum correlations (entanglement, connectivity information) among them. They are a first step, to be then improved, being in fact the quantum gravity counterpart of the Gross-Pitaevskii approximation in the hydrodynamics of quantum liquids. When moving to this level of description, we are then moving from the QG atoms to the full continuum description of quantum gravity, but within a specific regime of approximation, which remains quantum and focused on the collective properties of the same QG atoms, rather than their individual, pre-geometric features. Conceptually, we are solidly within a *level 1* emergence scenario for spacetime.

10.5   TGFT condensate cosmology

The task is now to obtain, from the fundamental QG theory, an effective dynamics that can then be understood in spatiotemporal terms, in the sense of quantum GR, i.e., an effective dynamics of geometric quantities as those constructed out of continuum fields including the metric, maybe still possessing quantum properties (as in a *level 0* of spacetime emergence). This is what TGFT condensate cosmology sets to do (Gielen and Sindoni, 2016; Oriti, 2017b; Pithis and Sakellariadou, 2019).

10.5.1 Emergence of spacetime in continuum approximation in specific phase

The general strategy is then to look for continuum spacetime at the level of collective dynamics of the fundamental entities. The more specific hypothesis is that the relevant phase of the QG system is a *condensate phase*, guided by the intuition of the universe as a kind of condensate, indeed, of QG entities (in the same sense in which a quantum fluid is a condensate of atoms). The second hypothesis is that the relevant dynamical regime is the hydrodynamic one, with the relevant physical information captured by the condensate wavefunction as the main dynamical variable.

Then, as we have just discussed, the first approximation we look at is mean field hydrodynamics (which of course amounts to the assumption that a Gaussian, weakly interacting regime is already good enough to unravel interesting spacetime, gravitational aspects).

Since we are working with an interacting quantum many-body system, albeit a peculiar one, it was immediately necessary to consider the rich continuum phase diagram that is expected in such systems, and thus the issue of which phase could be the relevant one to look at. The idea is that the definition of meaningful geometric and spatiotemporal quantities requires the use of a non-vanishing expectation value of the TGFT field operator (the condensate wavefunction or mean field), i.e., that we find ourselves in a broken, condensate phase. The symmetric or unbroken phase, with vanishing mean field as order parameter, would then correspond to a non-geometric and non-spatiotemporal phase, in which all (or most) such quantities are vanishing or somewhat pathological.

When considering also these aspects (conceptual and technical) of the story, we are working, in fact, within a *level 2* spacetime emergence, with the consequent more radical departure of the fundamental ontology from a spatiotemporal one, and the additional issues that we discussed earlier.

Considering now what sort of spatiotemporal physics we can extract from the TGFT hydrodynamics, for quantum geometric (GFT) models, the immediate guess is that we can only obtain cosmological dynamics, i.e., restricted to homogeneous geometries. There are two orders of reasons, both quite heuristic. One is that the conditions that identify the simplest hydrodynamic regime of a quantum many-body system, i. e., focusing on macroscopic, global variables only, corresponding to the maximal coarse-graining of microscopic details, limit oneself to close-to-equilibrium dynamics, intuitively match the ones corresponding to the homogeneous or near-homogeneous cosmological sector of gravitational physics, at least away from the cosmological singularity. Another follows from noticing that the GFT mean field has the same domain of a wavefunction of a single GFT "atom," i.e., an individual 3-simplex, encoding its (usually spacelike) quantum geometry, and that, by definition, it does not encode any notion of "local variation of geometry" at least from the point of view of discrete gravity (as a result of maximal coarse-graining). In models extending the pure (discrete) geometric domain to matter degrees of freedom, like in equation (10.3), in presence of a single scalar field the best one can do is to try to use this additional degree of freedom as a relational clock to define a notion of temporal evolution, for homogeneous geometries. This is what we focus on, in the following, while for the study of inhomogeneities the relational strategy requires the introduction of more (matter) degrees of freedom in addition to the matter clock.

The guess turns out to be supported by the following general fact, whose validity goes beyond specific models in that it applies to any TGFT model (in particular quantum geometric ones, i.e., GFTs), in which the domain $\mathcal{D}$ of the mean field $\sigma(\mathcal{D})$ (and of the fundamental field $\phi(\mathcal{D})$) is understood as the space of geometries of a single (spacelike) 3-simplex (or conjugate extrinsic geometry), plus additional data (like the scalar field of the example we are considering) that can be used to define physical reference frames. The general supporting fact is the following. It can be shown (Gielen, 2014; Jercher et al., 2022) that such domain $\mathcal{D}$ (space of geometries of a single 3-simplex, plus discrete matter values) is diffeomorphic to the space of metrics (or conjugate extrinsic curvatures) at a point in a 3d (spacelike) hypersurface, plus matter field values at the same point, which in turn is diffeomorphic to the minisuperspace of continuum homogeneous 3-geometries (or conjugate homogeneous extrinsic data), plus homogeneous matter fields. This implies that the TGFT condensate wavefunction $\sigma(\mathcal{D})$ can be understood as a wavefunction on minisuperspace, as in quantum cosmology.

However, the other general result is that this wavefunction satisfies nonlinear dynamical equations, not the linear ones of quantum cosmology (i.e., the Wheeler-DeWitt equation restricted to wavefunctions on minisuperpsace). These are the quantum equations of motion derived from the quantum effective action or, in the simpler approximation we are focusing on, the classical equations of motion of the chosen TGFT model (the direct analogue of a Gross-Pitaevskii hydrodynamic equation for a quantum fluid):

$$\int [dg']d\chi'\mathcal{K}([g],[g'];\chi,\chi')\,\sigma(g',\chi') + \lambda\frac{\delta}{\delta\varphi^*}\mathcal{V}(\varphi,\varphi^*)\big|_{\varphi=\sigma} = 0 \quad , \quad (10.5)$$

with analogous equation for the conjugate TGFT field.

They are hydrodynamic equations, thus their non-linearity is not surprising and in fact, mandatory, stemming from the presence of fundamental interactions among QG atoms. Further, these equations are in general also non-local on minisuperspace, as follows from the non-local nature of the same TGFT interactions.

We obtain, then, a non-linear and non-local extension of a quantum cosmological equation, for the condensate wavefunction, encoding an infinity of quantum gravity degrees of freedom in a coarse-grained, collective manner. This is the more precise sense in which cosmology results from quantum gravity hydrodynamics and the universe is recognized as a quantum gravity condensate.

From the point of view of the emergence scheme of Fig. 10.1, the step from QG atoms to a continuum gravitational description (yet to be fleshed out in terms of observables), needed in any *level 1 emergence*, and leading to this cosmological setting, is their hydrodynamic approximation, then looked at in a specific (condensate) phase, and thus from a *level 2 perspective*.

Let us stress however that the correspondence with quantum cosmology is limited to the kinematical aspects, since we have seen that the dynamics is necessarily non-linear in the condensate wavefunction. The distance from quantum cosmology is also important at the conceptual level. Despite the similarities, extending to the way geometric observables can be constructed, as we are going to see in the following, the fundamentally nonlinear dynamics prevents solutions to form a Hilbert space and thus any superposition principle and any quantum mechanical interpretation for the condensate wavefunction itself, as "quantum state of the universe" or the like. While the fundamental quantum gravity degrees of freedom governed by the TGFT model are treated quantum mechanically, with a HIlbert space given by the TGFT Fock space, the interpretation of the TGFT condensate wavefunction and mean field can only be of statistical and epistemic type, as appropriate for a coarse-grained quantity like the density of a quantum fluid. This picture of cosmology as quantum gravity hydrodynamics raises then a number of conceptual issues also from the contrast with the traditional framework of quantum cosmology. We leave them for future analysis.

10.5.2 *Extracting effective spacetime/geometric dynamics*

The next task is to choose a specific (class of) GFT model(s) with a good interpretation from the point of view of discrete gravity (as it appears in its perturbative expansion, and then to recast the condensate hydrodynamic equations into equations for geometric observables and to give them a "temporal" form by implementing the relational strategy. In other words, looking at our general scheme for spacetime emergence in Figure 10.1, we need to move from the effective continuum formulation of full quantum gravity to some quantum version of General Relativity, looking at specific observables.

We consider quantum geometric models with action of the general form (10.3), and specifically Lorentzian EPRL-like models (Perez,

2013; Oriti et al., 2016) or the Lorentzian Barrett-Crane model (Perez, 2013; Jercher et al., 2022). We refer to the literature for the explicit form of the GFT action (or of the spin foam amplitudes from which the GFT action can be deduced) for these models. The first class of models adopts a formulation of Lorentzian quantum geometry based on a map from data from the Lorentz group to $SU(2)$ data (Finocchiaro et al., 2020), and it is closer to canonical LQG as well, while the second uses Lorentzian data only. We adopt a notation referring to $SU(2)$ data, for simplicity, but our results apply to both types of models.

We are interested in reconstructing homogeneous cosmological dynamics and we focus on isotropic configurations only. The simplest way to do so is to restrict the condensate wavefunction to isotropic data. This restriction (Oriti et al., 2016) amounts to working with functions which depend on a single representation label, once expanded in irreps of the group, and on the scalar field values only: $\sigma^j(\chi)$, corresponding to the fact that one single metric degree of freedom, e.g., the universe volume (or the scale factor), is sufficient to determine the full geometric configuration of the universe.

The second approximation we apply is to assume that GFT interactions are subdominant compared to the free GFT dynamics, to which we then restrict attention in the first place. This is consistent (in fact, required) by the focus on lattice gravity and spin foam dynamics to guide our physical interpretation of these GFT models (since both descriptions arise in the perturbative GFT regime), and with the use of simple coherent states of the GFT field operator (since strong interactions would generate strong quantum correlations among GFT quanta, which these states do not account for).

Next, we want to recast the hydrodynamic equations (10.5) for these models, restricted to isotropic wavefunctions, in a relational evolution form. To do so, we choose the scalar field degree of freedom χ as our relational clock and we consider condensate states that are "semiclassical enough" to admit such variable as a good clock.[9] We work with "coherent peaked states" (Marchetti and Oriti, 2021b) of the form:

$$\sigma_\epsilon(j; \chi) \equiv \eta_\epsilon(j; \chi - \chi_0; \pi_0)\tilde{\sigma}(j; \chi) \quad , \tag{10.6}$$

where η is a function (e.g., Gaussian) peaked around the χ_0 value of the clock variable χ, with width given by $\epsilon \ll 1$, and depending on a second parameter π_0 governing the fluctuations in the conjugate variable to χ (related to the momentum of the scalar field, whose fluctuations are small if $\pi_0^2\epsilon \gg 1$). Notice that this semiclassicality condition only refers to the clock values, while the geometric degrees of freedom can be highly quantum. The resulting relational temporal picture will then be approximate and effective only, since it results both from some coarse-graining (it is only the collective observable corresponding to χ at the hydrodynamic level that will have the interpretation of (clock) time), and of neglecting some physical features of the system chosen as clock (the quantum properties of effective continuum scalar field).

The peaking function η allows to approximate the remaining part of the condensate wavefunction $\tilde{\sigma}$ at χ_0 and to neglect higher orders in the derivative expansion with respect to χ of the kinetic term in 10.5. The resulting equation for $\tilde{\sigma}$ is:

$$\tilde{\sigma}''_j(\chi_0) - 2i\tilde{\pi}_0\tilde{\sigma}'_j(\chi_0) - E_j^2\tilde{\sigma}_j(\chi_0) = 0, \tag{10.7}$$

where the derivatives are with respect to χ_0, $\tilde{\pi}_0 = \frac{\pi_0}{\epsilon\pi_0^2 - 1}$ and $E_j^2 = \epsilon^{-1}\frac{2}{\epsilon\pi_0^2 - 1} + \frac{B_j}{A_j}$, with A_j and B_j being the coefficients of the 0th and 2nd order terms in the expansion of the kinetic term of the GFT action in terms of derivatives with respect to the scalar field values (thus, they are functions of the quantum geometric data j and of the fundamental parameters defining the model itself). This is the effective dynamical equation describing the evolution of the condensate function with respect to the clock time χ_0.

Now we have to recast it in spatiotemporal form, and extract from it a dynamical evolution equation for some relevant geometric observable. In the scheme of Figure 10.1, it means going from a level 1 situation to a level 0 one, corresponding to (the cosmological sector of) quantum GR.

For doing so, we define the relational observables that we expect to be relevant for describing homogeneous cosmological evolution (Marchetti and Oriti, 2021b). These are expectation values of fundamental GFT operators (acting on the GFT Fock space), evaluated on the coherent peaked states (10.6), and thus well approximated by the value the condensate wavefunction takes at $\chi = \chi_0$. The geometric interpretation is guided by the discrete gravity picture of quantum states and perturbative GFT amplitudes. We have the occupation number:

$$N(\chi_0) \equiv \langle\widehat{N}\rangle_{\sigma;\chi_0,\pi_0} = \sum_j \rho_j^2(\chi_0), \tag{10.8}$$

the universe volume (constructed from the matrix elements of the 1st quantized volume operator for GFT quanta, i.e., quantized tetrahedra, with eigenvalues V_j, convoluted with field operators[10]):

$$V(\chi_0) \equiv \langle\widehat{V}\rangle_{\sigma;\chi_0,\pi_0} = \sum_j V_j\rho_j^2(\chi_0), \tag{10.9}$$

the clock (scalar field) value:

$$\frac{\langle \widehat{\chi} \rangle_{\sigma;\chi_0,\pi_0}}{N(\chi_0)} \simeq \chi_0, \tag{10.10}$$

and the scalar field momentum:

$$\langle \widehat{\Pi} \rangle_{\sigma;\chi_0,\pi_0} \simeq \pi_0 \left(\frac{1}{\epsilon \pi_0^2 - 1} + 1 \right) N(\chi_0) + \sum_j Q_j, \tag{10.11}$$

where Q_j are conserved quantities, and we have adopted the decomposition of the condensate wavefunction in terms of standard hydrodynamic variables: the density of the fluid and the phase (from which, in standard hydrodynamics, one defines the fluid velocity): $\tilde{\sigma}_j(\chi) = \rho_j(\chi)e^{i\theta_j(\chi)}$. All these observables are clearly geometric and spatiotemporal only in an approximate, coarse-grained and collective sense.

Using the hydrodynamic equation (10.7), we obtain the equations governing the relational evolution of the universe volume (Marchetti and Oriti, 2021b; Oriti et al., 2016):

$$\left(\frac{V'}{3V} \right)^2 \simeq \left(\frac{2\sum_j V_j \rho_j sgn(\rho'_j)\sqrt{\mathcal{E}_j - Q_j^2/\rho_j^2 + \mu_j^2 \rho_j^2}}{3\sum_j V_j \rho_j^2} \right)^2,$$

$$\frac{V''}{V} \simeq \frac{2\sum_j V_j[\mathcal{E}_j + 2\mu_j^2 \rho_j^2]}{\sum_j V_j \rho_j^2}, \tag{10.12}$$

where we have defined $\mu_j^2 = E_j^2 - \tilde{\pi}_0^2$ and $\mathcal{E}_j$ is another conserved quantity.

These are the *generalized (quantum-corrected) Friedmann equations in relational time* (given by the scalar field value χ_0) that our quantum gravity model gives for the emergent spacetime in the homogeneous case (as captured by the volume observable, in the presence of a real, massless, free scalar field as the only matter content of the universe).

We are now at level 0 of spacetime emergence, in our scheme (Figure 10.1), that of quantum GR. Our geometric, spacetime notions (only the volume, in this simple case) are well defined and under control, but satisfy quantum-corrected equations, are subject to (possibly) strong quantum fluctuations, *etc.*

To recover standard notions of space and time (just the usual notion volume, here), we should recover the classical equations it satisfies in GR, and check that the quantum fluctuations become negligible, in the same limit. The classical regime corresponds to the one in which the Hubble rate is small compared to the inverse Planck time, i.e., small curvature; while technically more subtle, it is reached for large universe volumes. In this regime, in which the QG fluid density is large compared to the other terms appearing in the equations, we get the approximate dynamical relations:

$$\left(\frac{V'}{3V}\right)^2 \simeq \left(\frac{2\sum_j \mu_j V_j \rho_j^2 sgn(\rho'_j)}{3\sum_j V_j \rho_j^2}\right)^2 \quad \frac{V''}{V} \simeq \frac{4\sum_j V_j[\mu_j^2 \rho_j^2]}{\sum_j V_j \rho_j^2}. \tag{10.13}$$

This means that for any GFT model (in the class we considered) in which the right-hand-side of both equations becomes approximately a constant g, we recover the usual Friedmann equations in relational time:

$$\left(\frac{V'}{V}\right)^2 = \frac{V''}{V} = 12\pi G \tag{10.14}$$

by identifying the effective constant g with Newton's constant (up to a numerical factor). This is the case, for example, if there is a single dominant mode j, in which case $\mu_j^2 = 3\pi G$ or if $\mu_j^2 = 3\pi G \; \forall j$. Thus, we have the usual solutions corresponding to an expanding universe of the standard model of cosmology.

In fact, also all the conditions that need to be verified on the quantum fluctuations of the relational clock (that has to remain semiclassical to have a good notion of temporal evolution) and of the universe volume (that have to become negligible) can be checked explicitly (Marchetti and Oriti, 2021c). In this regime, thus, we recover standard spacetime, and its emergence from full quantum gravity is complete, having realized that the proper notion of time (and space) should be defined in terms of physical frames (and not of manifold structures, which can at most play an auxiliary role). We are at level −1, in our schematic classification, the realm of classical continuum GR.

One thing we learn, looking at GR now from this emergent spacetime perspective, is that gravitational parameters, like Newton's constant, are in fact functions of the microscopic, fundamental quantum gravity parameters, those characterizing the quantum gravity atoms and their quantum dynamics. This is because μ_j^2, which becomes identified with Netwon's constant in the large-volume, late-time regime of our emergent cosmological evolution, is a function of these parameters. This is nothing surprising, since exactly the same happens in standard hydrodynamics, for example, in its relation to the atomic or molecular theory underlying it.

10.5.3 *Cosmological singularity and geometrogenesis*

This is of course not the only thing we learn, or we expect to learn, from a full quantum gravity account of the emergence of space and time from fundamentally non-spatiotemporal entities. Let us discuss a few more results obtained in this context, and the lessons we learn from them.

First, we want our theory of quantum gravity to tell us what happens to the cosmological singularity, that signals the breakdown of classical GR (or at least, whose physics cannot be accounted for by classical GR). Looking again at our quantum cosmological dynamics (10.12), we realize an important fact. Under the assumption that the "good-clock conditions" (small fluctuations in the clock value and in the scalar field momentum) remain satisfied, If at least one conserved quantity Q_j or $\mathcal{E}_j$ is non-vanishing, there exist one value j such that the fluid density $\rho_j(\chi_0)$ remains different from zero at all times χ_0. This implies that our expanding universe can be followed toward earlier times to find that its volume remains always positive and with a single turning point. That is, we find a quantum bounce instead of the big bang (thus solving the classical cosmological singularity).[11] Moreover, one can compute what happens to quantum fluctuations in the early universe (Marchetti and Oriti, 2021c), to find that they remain small at least for a specific but rather large class of solutions to the quantum dynamics.

Both results become particularly transparent for the simplest quantum gravity condensates corresponding to condensate wavefunctions which are non-vanishing only for a single mode j_0. For them, identifying $\mu_{j_0}^2 = 3\pi G$, and with all Q_j vanishing, we find the cosmological equation:

$$\left[\frac{V'}{3V}\right]^2 = \frac{4\pi G}{3} + \frac{4V_{j_0}\mathcal{E}_{j_0}}{9V} \quad , \quad V_{j_0} \approx V_{Planck}j_0^{3/2} \tag{10.15}$$

such that all solutions corresponding to $\mathcal{E}_{j_0} < 0$ describe a quantum bounce at $V_{min} = V_{j_0}N_{min} = \frac{V_{j_0}|\mathcal{E}_{j_0}|}{6\pi G}$. The relative volume fluctuations, on the other hand, behave as $\frac{\Delta V}{V}(\chi_0) \approx \frac{1}{N(\chi_0)}$, so they remain small provided the (average) occupation number does not become of order one.

Before discussing in more detail the fate of the cosmological singularity and the beginning of our universe in an emergent spacetime scenario, let us mention a couple of recent research directions showing that the consequences of such scenario do not have to be confined to the extreme conditions of the very early universe.

Cosmological perturbations, thus inhomogeneities in both geometry and matter, are important for making stronger contact with cosmological observations and because it is at the level of their field theory description that the physical aspects of spacetime dynamics can truly be probed, in their quantum gravity origin. Moreover, it is only when reproducing local spacetime features that any quantum gravity model can be said to provide an example of an emergent space and time. Cosmological perturbations in a GFT cosmology context are therefore receiving increasing attention (Gielen and Oriti, 2018; Gielen, 2019; Marchetti and Oriti, 2021a).

From the point of view of spacetime emergence, two points need to be noted, of this recent work. First, the notion of spacetime point, and thus local physics, needs to be defined relationally, in terms of appropriate physical degrees of freedom, the easiest choice in the GFT context being four massless real scalar fields, used as rods and clock, extending the relational framework developed to have a notion of temporal evolution in the homogeneous case. Second, cosmological perturbations can be identified with (relationally local) excitations over a homogeneous GFT condensate, in direct analogy with quasi-particles in a quantum fluid, and can be studied (at first) in the same mean field approximation (Marchetti and Oriti, 2021a).

The lesson is that the framework of effective (quantum) field theory on a given (quantum) background can be reproduced from (perturbative) QG hydrodynamics.[12]

Late-time cosmology is also a test bed for new physics, since the observed acceleration of cosmological expansion remains puzzling from many reasons (Brax, 2018; Burgess, 2015). Being a large-scale phenomenon, this is difficult to understand from the point of view of microphysics and not usually understood as originating from quantum gravity, although a more fundamental theory is certainly called to contribute to its understanding.

From an emergent spacetime perspective, however, the whole spacetime dynamics at both small and large scales is in fact of direct quantum gravity origin and we are encouraged to think outside the usual effective field theory mindset[13] (see also supporting results in the analogue gravity context: Finazzi et al., 2012). In the GFT cosmology context, recent work shows that the macroscopic effect of the fundamental interactions among QG atoms (that we neglected in our analysis) can be relevant and producing interesting consequences (Pithis et al., 2016; Pithis and Sakellariadou, 2017; Gielen and Polaczek, 2020).

In particular, and quite strikingly, they can produce quite naturally an accelerated dynamics at late times of a phantom dark energy type (Oriti and Pang, 2021). It is too early to consider this result as a full, compelling explanation of dark energy, and more work is needed to do so. However, it does already constitute a proof-of-principle from which we can draw one main lesson: in an emergent spacetime scenario, effective field theory intuition is bound to fail and large scale physics can be of direct quantum gravity origin.

Let us now go back to the fate of the cosmological singularity in GFT cosmology (and more generally, in an emergent spacetime scenario). What happens to the cosmological singularity in quantum gravity, in light of GFT cosmology?

We have seen that the volume evolution governed by the equations (10.12) (and, in the simplest case, (10.15)), contains quantum gravity corrections akin to a sort of 'quantum pressure'preventing it from reaching vanishing configurations, so that the classical cosmological singularity is replaced by a quantum bounce.[14] More precisely, we found that the big bang singularity is replaced by a big bounce scenario in a mean field restriction of the hydrodynamic approximation to the full quantum dynamics, within a condensate phase.

The mean field approximation should obviously be improved, but it could well be that the bouncing scenario is stable under such improvements. If that turns out to be the case, then we could really say that quantum gravity (better, GFT cosmology) predicts a cosmic quantum bounce.

Is that it? Not quite. This would be the conclusion only if the hydrodynamic approximation still holds and remains reliable at the would-be bounce, and if the whole quantum gravity system stays within the condensate phase. But even before considering these extreme possibilities, even within the hydrodynamic approximation we need a well-behaved clock to speak of evolution towards and then across the bounce. If the clock becomes subject to too strong quantum fluctuations, for example, we could have no reliable notion of evolution, and thus we could only follow the dynamics up to the point where the very notion of time and evolution ceases to make sense.

Suppose however that this does not happen. It could still be the case that fluctuations become too strong, at the would-be bounce, to invalidate the hydrodynamic approximation within which we have been able to extract a geometric, spatiotemporal description for our system of QG atoms. Then the best we could say would be, again, that the universe history can be followed backward from present day up to a point in which space and time simply disappear, we do not have a reliable spatiotemporal or geometric description, and we have to resort to the more fundamental, non-spatiotemporal description to obtain needed input for understanding the new physics of the origin of the universe. What replaces the big bang singularity, then, is the disappearance of spacetime (when read "backward in time") or its emergence (when read "forward in time").

It could still be the case that the fundamental microscopic quantum gravity theory gives us a complement to the hydrodynamic description that can still be somehow translated in those terms. One could try, for example, to add terms coming from kinetic theory to standard hydrodynamics, in order to improve the physical description, while maintaining the same intuition about the physical entities one is dealing with. Then we would know that the spatiotemporal description is not entirely valid in that regime, and that there is no bounce, actually, but we could still describe the effects of the non-spatiotemporal dynamics in the corresponding regime in spatiotemporal terms.

Not so if the quantum fluctuations affect so drastically the fundamental dynamics and the behaviour of (geometric) observables to drive the system out of the geometric, spatiotemporal phase (i.e., the condensate phase in GFT cosmology) to the phase transition separating it from a non-geometric one. Then we would face an even more radical disappearance of spacetime, and the non-spatiotemporal description in terms of the fundamental QG atoms would be necessary to capture the relevant physics, being also not translatable into spacetime language. The phase transition between non-geometric and geometric phases of the quantum gravity system would be then what truly replaces the big bang of continuum and classical GR, in this quantum gravity scenario. Geometrogenesis replaces the big bang in quantum gravity.

From the point of view of our scheme 10.1 we are then led to exploring the details of the continuum phase diagram and of the phase transitions it contains. From the conceptual point of view, we are then confronting a level 3 emergence of spacetime, in which the philosophical issues concerning the geometrogenesis phase transition should be tackled, alongside the physical ones. The two sets of issues are intertwined, and intertwined as well with the physics of cosmological evolution and the early universe. This means that, for example, in the TGFT context, the analysis of the RG flow of (ideally) fully quantum geometric models should be performed in parallel with the analysis of their emergent cosmological dynamics, with the two research lines informing each other.

We can illustrate this relation in a simple case, albeit in rather sketchy manner. This example shows how one could give a temporal characterization to a geometrogenesis phase transition, at least in the sense of localizing it in time with respect to the geometric phase, in particular making concrete the identification with the regime corresponding to the big bang singularity in the classical continuum case, and to the quantum bounce in the GFT hydrodynamic approximation.

Consider the cosmological evolution for the simplest single-mode condensates of equation (10.15). Given the expression for the volume at the bounce and of the quantum fluctuations, it is clear that these quantum fluctuations are indeed maximal at the bounce and given by:

$$\left[\frac{\Delta V}{V}(\chi_{crit})\right]_{max} \approx \frac{1}{N_{min}(\chi_{crit})} \simeq \frac{G}{\mathcal{E}_{j_0}(G)} \quad , \tag{10.16}$$

where we have indicated that the conserved quantity $\mathcal{E}$ is in fact a function of the parameters of the model and thus of the effective Newton constant G.

The point is that we expect the hydrodynamic approximation to break down when the average occupation number becomes smaller than one, and indeed this is also when the relative fluctuations in the universe volume become larger than one, i.e., for $\frac{G}{\mathcal{E}_{j_0}(G)} \geq 1$. In turn we should recall that the effective Newton constant is, really, a function of the fundamental couplings of the underlying GFT model, i.e., $G = G(\{\lambda_i\})$.

One way in which a geometrogenesis phase transition can then be associated to the physical origin fo the universe evolution is if the critical regime fo the GFT couplings corresponding to the phase transition are shown to give $\frac{G}{\mathcal{E}_{j_0}(G)} \geq 1$, signalling the breakdown of the cosmological dynamics extracted in the hydrodynamic approximation. This should also happen in the regime that, in the same hydrodynamic approximation, would correspond to the quantum bounce.

This last identification requires, on the other hand, a clear relation between the RG flow parameter used in the RG analysis of the GFT model and the geometric data used to given spatiotemporal meaning to the GFT hydrodynamic equations. In general, for GFT models based on both quantum geometric and scalar matter data (i.e., like the ones we focused on here) the RG scale would be a combination of both. To simplify the picture, let us assume that the RG scale, i.e., the "IR cut-off" in a full functional RG analysis (Carrozza, 2016; Baloitcha et al., 2020; Finocchiaro and Oriti, 2021; Pithis and Thürigen, 2020) or a mean field one (Marchetti et al., 2021) is just given by a spin label k.

The quantum bounce found in the hydrodynamic approximation has to take place, given observational constraints, not too distant from Planckian volumes, thus the minimal spin label giving $V_{min} = V_{j_0} N_{min}$ should be close to lowest end of the volume spectrum (and the lowest possible, to try to avoid N_{min} having to be necessarily too small).[15]

This implies that the hydrodynamic equation would have to be used close to the 'IR end'of the RG flow of the same GFT model, where the GFT couplings (at least some of them) will reach their critical values. The would-be quantum bounce, then, will happen exactly close to the critical regime of the underlying quantum gravity model. If it happens that, going to such critical values of the couplings, we have

$$\frac{G(\lambda)}{\mathcal{E}_{j_0}(G(\lambda))} \to \frac{G(\lambda_{crit})}{\mathcal{E}_{j_0}(G(\lambda_{crit}))} \geq 1 \quad , \qquad (10.17)$$

then the fluctuations grow too much, preventing the quantum bounce to happen, or, better, making it physically irrelevant, since the hydrodynamic approximation in which it is formulated would not be reliable. The underlying physics would then be instead characterized by the properties of the GFT phase transition, the geometrogenesis.

Let us recapitulate the logic of this tentative scenario. The geometrogenesis phase transition is a transition between two phases (not a phase in itself), a geometric phase and a non-geometric one, by definition. The question I am posing in the text is whether this transition can be somehow localized in time and I answer that this can in principle be done but only from the standpoint of the geometric phase we live in, as one could intuitively expect, since there is no notion of time that could be applied to the whole set of continuum phases, thus encompassing both geometric and non-geometric phases, or to the level of description in terms of TGFT quanta. For example, it could be localized as having happened in our past and identified with the very early stages of the evolution of the universe, if one finds that the fundamental theory locates at that time (within an effective cosmological dynamics) the regime of very strong fluctuations in geometric observables that characterizes the phase transition. In this case, we have to conclude that a geometrogenesis scenario is more appropriate than a bouncing scenario, from the perspective of the fundamental quantum gravity theory, as the appropriate account of the very early universe and as a replacement of the classical big bang singularity.

All this reasoning is tentative as much as it is sketchy, and obviously much remains to be done to put it on solid grounds, both concerning the RG analysis of TGFT models and their effective cosmological (more generally, gravitational) continuum dynamics. It shows however that even issues associated to *level 3 spacetime emergence*, exotic as they may be, can be in principle tackled in very concrete terms, within the TGFT formalism.

10.6   Conclusion: many more questions, but one concrete framework to investigate them

What we have just concluded about the geometrogenesis scenario is in fact the main message of this contribution also for spacetime emergence in general.

There are many open issues, both technical and physical, concerning the TGFT cosmology framework. We still need to go convincingly beyond homogeneity and isotropy, and towards extracting the full continuum gravitational dynamics from the fundamental quantum gravity theory (based on a number of approximations and assumptions, for sure). We need to investigate in more detail several aspects of the cosmological evolution we extracted so far (both in the very early universe and at late times), improving the approximations on which it is based (mean field hydrodynamics), and to enrich it with more physical ingredients (e.g., realistic matter content), to see if quantum gravity can truly solve current cosmological puzzles. More associated physical issues could be named, for example the precise relation between the phenomena described in different physical (relational) frames. The important point is that all of them can be tackled in detail (at least in principle) and very concretely in this template for spacetime emergence in quantum gravity.

The same is true for the many open issues at the philosophical level, concerning foundations of spacetime and quantum gravity, and of physics more generally. The framework of TGFT cosmology offers a concrete test bed for philosophical analyses concerning the nature of space and time and their emergence, of course, i.e., the focus of this contribution, at both metaphysical and epistemological levels. Indeed, a partial list of interesting research directions include: the development of a new ontology that is not grounded on space and time notions (a tantalizing as much as a daunting challenge for metaphysics); the analysis of the role of observers and agents in emergent spacetime scenarios and, more generally, in quantum gravity, where the absence of preferred or classical notions of time evolution and causal structure complicates the debate on the nature of probabilities, and where the remote character of the relevant physical phenomena makes any operational approach to physical theories questionable, while at the same time putting under strain any naive picture of physical theories as purely representational and objective (since only very indirectly grounded on empirical data); the analysis of the possible perspectival nature of cosmological evolution (and, for example, of a thermodynamic time arrow for such evolution), since this is can only be understood in relational terms and thus referring to a specific internal (to the universe) physical frame, and in a context in which the elements that are left invariant by the switch to another internal, physical frame are not fully under control; more generally, the issue of how all these elements should affect our understanding of laws of nature, since it is not obvious that any of the traditional accounts (Humeanism, primitivism, universals-based, best-systems, *etc.*) can be applied in absence of space and time as background conceptual structures, and what would be left, more generally, of any traditional form of realism, in the same context.

We hope that this contribution, if it has not provided answers to those issues, has at least made clear that these answers can be found, in principle, within the quantum gravity context we presented.

Acknowledgements

We acknowledge funding from DFG research grants OR432/3-1 and OR432/4-1. We also express our gratitude to the editor of this volume for the invitation to contribute to it and for the remarkable patience demonstrated in dealing with us.

Notes

1. Notice that this does not imply in any way that "all the entities playing a role in physical theories should be assigned ontological status", but only the weaker (almost converse) statement that all entities to which we assign ontological status play a role in physical theories that, in fact, define them.

2. A much better term would probably be "a-chronic", though, since "synchronic" presupposes the existence of some time.
3. Phase transitions are a case at hand, but even the transition from molecular dynamics to hydrodynamics (in itself not a physical process) could be associated to specific dynamical processes of molecules whose collective behaviour then requires the switch to an hydrodynamic description, so that one could say that the emergence of fluids and fluid dynamics takes place *in time* along these dynamical processes.
4. Specific quantum gravity formalisms of course will exemplify the framework, which is meant to apply in full generality, in different ways and with their own peculiarities.
5. This conclusion, although the most reasonable and well-supported one in our opinion, is not uncontroversial and the debate about the nature of spacetime in classical GR, but also about the role of the manifold, for better or worse, still goes on (see Pooley, 2013; Hoefer, 1998, as well as the contributions to the first part of this volume).
6. The use of scalar fields can be seen as a simplifying choice to actually mimic more realistic matter content (e.g., an actual set of rods and clock, if one wants to use them as relational frames).
7. They satisfy $\hat{\phi}|\sigma\rangle = \sigma|\sigma\rangle$.
8. This is exactly how the classical electromagnetic field is obtained from coherent states of photons.
9. Other constructions can be found in the literature, based on distributional relational operators and generic coherent states (Oriti et al., 2016; Pithis and Sakellariadou, 2017), or on a classical deparametrization of the GFT system with respect to the chosen scalar field clock (Wilson-Ewing, 2019; Gielen, 2021; Gielen and Polaczek, 2020, 2021). They give similar results. Another strategy for the extraction of effective relational dynamics (Bojowald et al., 2011), not relying on specific choices of quantum states, can also be applied to the GFT system (Gielen et al., 2021) and lead to a different choice of clock, making the comparison less straightforward.
10. In other words, the operator adds the individual volume contributions from the GFT quanta populating the state.
11. This behaviour can be recast in the language of an effective mimetic gravity theory (de Cesare, 2019).
12. In fact, once appropriate matter degrees of freedom are coupled to quantum geometry in the fundamental QG model, the hydrodynamic approximation encodes both the data needed to reconstruct the metric at a point and the ones allowing to define points in a diffeo-invariant manner. Therefore, there is no obstacle to attempt a reconstruction of all gravitational dynamics and full GR from it, without confining ourselves to a perturbative regime. Whether a mean field approximation is enough to recover the correct physics is, of course, a different story.
13. The very notion of scales and separation of scales can be said to be of very dubious meaning in a background independent context, and even more so in an emergent spacetime scenario.
14. It must be noted that this quantum pressure, while very similar in its effects to the one found in loop quantum cosmology, can be traced back to a never-vanishing number density, in this GFT hydrodynamics context, rather than to the discreteness of volume spectrum or absence of zero eigenvalues from it (in fact, it is present also for GFT models where the volume spectrum is continuous;see Jercher et al., 2022).
15. Obviously, this is very heuristic and we are neglecting a number of aspects of a more realistic treatment, not last the fact that we are using the effective

dynamics obtained considering condensate wavefunctions with only one spin excitation $j0$, while a non-trivial RG flow requires of course a non-trivial field dependence on the same modes j.

Bibliography

Ambjørn, J., J. Gizbert-Studnicki, A. Görlich, J. Jurkiewicz, N. Klitgaard, and R. Loll (2017). Characteristics of the new phase in CDT. *European Physics Journal C* 77 (3), 152.

Baez, J. C. and J. W. Barrett (1999). The quantum tetrahedron in three-dimensions and four-dimensions. *Advances in Theoretical and Mathematical Physics* 3, 815–850.

Bahr, B., B. Dittrich, and M. Geiller (2021). A new realization of quantum geometry. *Classical and Quantum Gravity* 38 (14), 145021.

Bahr, B. and S. Steinhaus (2017). Hypercuboidal renormalization in spin foam quantum gravity. *Physical Review D* 95 (12), 126006.

Baloitcha, E., V. Lahoche, and D. Ousmane Samary (2020). Flowing in discrete gravity models and Ward identities: A review. *The European Physical Journal Plus* 136, 982.

Baratin, A. and D. Oriti (2012). Group field theory and simplicial gravity path integrals: A model for Holst-Plebanski gravity. *Physical Review D* 85, 044003.

Baron, S. (2019). The curious case of spacetime emergence. *Philosophical Studies* 177 (8), 2207–2226.

Batterman, R. W. (2004). Critical phenomena and breaking drops: Infinite idealizations in physics. *Studies in History and Philosophy of Modern Physics* 36 (2), 225–244.

Batterman, R. W. (2011). Emergence, singularities, and symmetry breaking. *Foundations of Physics* 41 (6), 1031–1050.

Bianchi, E., P. Dona, and S. Speziale (2011). Polyhedra in loop quantum gravity. *Physical Review D* 83, 044035.

Bojowald, M., P. A. Hoehn, and A. Tsobanjan (2011). An effective approach to the problem of time. *Classical and Quantum Gravity* 28, 035006.

Brax, P. (2018). What makes the universe accelerate? A review on what dark energy could be and how to test it. *Reports on Progress in Physics* 81 (1), 016902.

Burgess, C. P. (2015). The cosmological constant problem: Why it's hard to get dark energy from micro-physics. In C. Deffayet, P. Peter, B. Wandelt, M. Zaldarriaga, and L. F. Cugliandolo. (Eds.), *Post-Planck Cosmology: Lecture Notes of the Les Houches Summer School:* Volume 100, July 2013, pp. 149–197. Oxford University Press.

Butterfield, J. (2011a). Emergence, reduction and supervenience: A varied landscape. *Foundations of Physics* 41 (6), 920–959.

Butterfield, J. (2011b). Less is different: Emergence and reduction reconciled. *Foundations of Physics* 41 (6), 1065–1135.

Butterfield, J. and N. Bouatta (2012). Emergence and reduction combined in phase transitions. In J. Kouneiher, C. Barbachoux, T. Masson and D. Vey

(Eds.), *FRONTIERS OF FUNDAMENTAL PHYSICS:* The Eleventh International Symposium, AIP Conference Proceedings 1446, 383. AIP Publishing.

Carrozza, S. (2016). Flowing in group field theory space: A review. *SIGMA* 12, 070.

Castellani, E. (2000). Reductionism, emergence, and effective field theories. *Studies in History and Philosophy of Modern Physics* 33 (2), 251–267.

Castro-Ruiz, E., F. Giacomini, A. Belenchia, and C. Brukner (2020). Quantum clocks and the temporal localisability of events in the presence of gravitating quantum systems. *Nature Communications* 11 (1), 2672.

Chirco, G., I. Kotecha, and D. Oriti (2019). Statistical equilibrium of tetrahedra from maximum entropy principle. *Physical Review D* 99 (8), 086011.

Colafranceschi, E. and D. Oriti (2021). Quantum gravity states, entanglement graphs and second-quantized tensor networks. *Journal of High Energy Physics* 7, 052.

Crowther, K. (2014). *Appearing Out of Nowhere: The Emergence of Spacetime in Quantum Gravity.* Ph. D. thesis, University of Sydney.

Crowther, K. (2020). As below, so before: 'Synchronic' and 'diachronic' conceptions of spacetime emergence. *Synthese* 198 (8), 7279–7307.

de Cesare, M. (2019). Limiting curvature mimetic gravity for group field theory condensates. *Physical Review D* 99 (6), 063505.

Delcamp, C. and B. Dittrich (2017). Towards a phase diagram for spin foams. *Classical and Quantum Gravity* 34 (22), 225006.

Dittrich, B. (2006). Partial and complete observables for canonical general relativity. *Classical and Quantum Gravity* 23, 6155–6184.

Dittrich, B. (2017). The continuum limit of loop quantum gravity – a framework for solving the theory. In A. Ashtekar and J. Pullin (Eds.), *Loop Quantum Gravity: The First 30 Years*, pp. 153–179. World Scientific.

Dittrich, B. and J. P. Ryan (2011). Phase space descriptions for simplicial 4d geometries. *Classical and Quantum Gravity* 28, 065006.

Eichhorn, A., T. Koslowski, and A. D. Pereira (2019). Status of background-independent coarse-graining in tensor models for quantum gravity. *Universe* 5 (2), 53.

Finazzi, S., S. Liberati, and L. Sindoni (2012). Cosmological constant: A lesson from Bose-Einstein condensates. *Physical Review Letters* 108, 071101.

Finocchiaro, M., Y. Jeong, and D. Oriti (2020, December). Quantum geometric maps and their properties. *arXiv:2012.11536* [gr-qc].

Finocchiaro, M. and D. Oriti (2020). Spin foam models and the duflo map. *Classical and Quantum Gravity* 37 (1), 015010.

Finocchiaro, M. and D. Oriti (2021). Renormalization of group field theories for quantum gravity: New scaling results and some suggestions. *Frontiers in Physics* 8, 649.

Gaul, M. and C. Rovelli (2000). Loop quantum gravity and the meaning of diffeomorphism invariance. *Lecture Notes in Physics* 541, 277–324.

Giacomini, F. (2021). Spacetime quantum reference frames and superpositions of proper times. *Quantum* 5, 508.

Gielen, S. (2014). Quantum cosmology of (loop) quantum gravity condensates: An example. *Classical and Quantum Gravity* 31, 155009.

Gielen, S. (2019). Inhomogeneous universe from group field theory condensate. *Journal of Cosmology and Astroparticle Physics* 1902, 013.

Gielen, S. (2021). Frozen formalism and canonical quantization in (group) field theory. *Physical Review D* 104, 106011.

Gielen, S., L. Marchetti, D. Oriti, and A. Polaczek (2021, October). Effective cosmology from one-body operators in group field theory. *arXiv:2110.11176* [gr-qc].

Gielen, S. and D. Oriti (2018). Cosmological perturbations from full quantum gravity. *Physical Review D* 98 (10), 106019.

Gielen, S. and A. Polaczek (2020). Generalised effective cosmology from group field theory. *Classical and Quantum Gravity* 37 (16), 165004.

Gielen, S. and A. Polaczek (2021). Hamiltonian group field theory with multiple scalar matter fields. *Physical Review D* 103 (8), 086011.

Gielen, S. and L. Sindoni (2016). Quantum cosmology from group field theory condensates: A review. *SIGMA* 12, 082.

Giesel, K. and A. Herzog (2018). Gauge invariant canonical cosmological perturbation theory with geometrical clocks in extended phase-space—a review and applications. *International Journal of Modern Physics D* 27 (08), 1830005.

Giesel, K. and T. Thiemann (2015). Scalar material reference systems and loop quantum gravity. *Classical and Quantum Gravity* 32, 135015.

Giulini, D. (2007). Some remarks on the notions of general covariance and background independence. *Lecture Notes in Physics* 721, 105–120.

Gurau, R. and J. P. Ryan (2012). Colored tensor models – a review. *SIGMA* 8, 020.

Hamber, H. W. (2009). Quantum gravity on the lattice. *General Relativity and Gravitation* 41, 817–876.

Hardy, L. (2005). Probability theories with dynamic causal structure: A new framework for quantum gravity. *arXiv:gr-qc/0509120*.

Hoefer, C. (1998). Absolute versus relational spacetime: For better or worse, the debate goes on. *British Journal for the Philosophy of Science* 49 (3), 451–467.

Hoehn, P. A., A. R. H. Smith, and M. P. E. Lock (2021). Trinity of relational quantum dynamics. *Physical Review D* 104 (6), 066001.

Huggett, N. and C. Wüthrich (2018). The (a)temporal emergence of space-time. *Philosophy of Science* 85, 1190–1203.

Huggett, N. and C. Wüthrich (2021). Out of nowhere: Introduction: The emergence of spacetime. *arXiv:2101.06955* [physics.hist-ph].

Husain, V. and T. Pawlowski (2012). Time and a physical Hamiltonian for quantum gravity. *Physical Review Letters* 108, 141301.

Jercher, A. F., D. Oriti, and A. G. A. Pithis (2022). Emergent cosmology from quantum gravity in the Lorentzian Barrett-Crane tensorial group field theory model. *Journal of Cosmology and Astroparticle Physics* 2022, 050.

Kegeles, A., D. Oriti, and C. Tomlin (2018). Inequivalent coherent state representations in group field theory. *Classical and Quantum Gravity* 35 (12), 125011.

Kiefer, C. (2012). Quantum gravity: Whence, whither? In F. Finster, O. Müller, M. Nardmann, J. Tolksdorf, and E. Zeidler (Eds.), *Quantum Field Theory and Gravity*, pp. 1–13. Birkhäuser.

Kiefer, C. (2013). Conceptual problems in quantum gravity and quantum cosmology. *International Scholarly Research Notices* 2013, 509316.

Konopka, T., F. Markopoulou, and L. Smolin (2006). Quantum graphity. *arXiv: hep-th/0611197*.

Koslowski, T. and H. Sahlmann (2012). Loop quantum gravity vacuum with nondegenerate geometry. *SIGMA* 8, 026.

Kotecha, I. (2019). Thermal quantum spacetime. *Universe* 5 (8), 187.

Krajewski, T. (2011). Group field theories. *Proceedings of Science* QGQGS 2011, 005.

Le Bihan, B. (2018). Space emergence in contemporary physics: Why we do not need fundamentality, layers of reality and emergence. *Disputatio* 10 (49), 71–95.

Le Bihan, B. and N. S. Linnemann (2019). Have we lost spacetime on the way? Narrowing the gap between general relativity and quantum gravity. *Studies in History and Philosophy of Modern Physics* 65, 112–121.

Li, Y., D. Oriti, and M. Zhang (2017). Group field theory for quantum gravity minimally coupled to a scalar field. *Classical and Quantum Gravity* 34 (19), 195001.

Loll, R. (2020). Quantum gravity from causal dynamical triangulations: A review. *Classical and Quantum Gravity* 37 (1), 013002.

Mandrysz, M. L. and J. Mielczarek (2019). Ultralocal nature of geometrogenesis. *Classical and Quantum Gravity* 36 (1), 015004.

Marchetti, L. and D. Oriti (2021a). Effective dynamics of scalar cosmological perturbations from quantum gravity. *arXiv:2112.12677* [gr-qc].

Marchetti, L. and D. Oriti (2021b). Effective relational cosmological dynamics from quantum gravity. *Journal of High Energy Physics* 5, 025.

Marchetti, L. and D. Oriti (2021c). Quantum fluctuations in the effective relational GFT cosmology. *Frontiers in Astronomy and Space Sciences* 8, 683649.

Marchetti, L., D. Oriti, A. G. A. Pithis, and J. Thürigen (2021). Phase transitions in tensorial group field theories: Landau-Ginzburg analysis of models with both local and non-local degrees of freedom. *Journal of High Energy Physics* 2021, 201.

Marolf, D. (1995). Almost ideal clocks in quantum cosmology: A brief derivation of time. *Classical and Quantum Gravity* 12, 2469–2486.

Oriti, D. (2007). Group field theory as the microscopic description of the quantum spacetime fluid: A new perspective on the continuum in quantum gravity. *Proceedings of Science* QG-PH, 030.

Oriti, D. (2011, October). The microscopic dynamics of quantum space as a group field theory. In J. Murugan, A. Weltman, and G. F. Ellis (Eds.), *Foundations of Space and Time*, pp. 257–320. Cambridge University Press.

Oriti, D. (2016). Group field theory as the 2nd quantization of loop quantum gravity. *Classical and Quantum Gravity* 33 (8), 085005.

Oriti, D. (2017a). Group field theory and loop quantum gravity. In A. Ashtekar and J. Pullin (Eds.), *Loop Quantum Gravity: The First 30 Years*, pp. 125–151. World Scientific.

Oriti, D. (2017b). The universe as a quantum gravity condensate. *Comptes Rendus Physique* 18, 235–245.

Oriti, D. (2018). Levels of spacetime emergence in quantum gravity. In N. Huggett, B. Le Bihan, and C. Wüthrich (Eds.), *Philosophy Beyond Spacetime*, pp. 16–40. Oxford University Press.

Oriti, D. (2021). The complex timeless emergence of time in quantum gravity. *arXiv:2110.08641* [physics.hist-ph].

Oriti, D. and X. Pang (2021). Phantom-like dark energy from quantum gravity. *Journal of Cosmology and Astroparticle Physics* 2021, 040.

Oriti, D., L. Sindoni, and E. Wilson-Ewing (2016). Emergent Friedmann dynamics with a quantum bounce from quantum gravity condensates. *Classical and Quantum Gravity* 33 (22), 224001.

Percacci, R. (1991). The Higgs phenomenon in quantum gravity. *Nuclear Physics B* 353, 271–290.

Perez, A. (2013). The spin foam approach to quantum gravity. *Living Reviews in Relativity* 16, 3.

Pithis, A. G. A. and M. Sakellariadou (2017). Relational evolution of effectively interacting group field theory quantum gravity condensates. *Physical Review D* 95 (6), 064004.

Pithis, A. G. A. and M. Sakellariadou (2019). Group field theory condensate cosmology: An appetizer. *Universe* 5 (6), 147.

Pithis, A. G. A., M. Sakellariadou, and P. Tomov (2016). Impact of nonlinear effective interactions on group field theory quantum gravity condensates. *Physical Review D* 94 (6), 064056.

Pithis, A. G. A. and J. Thürigen (2020). Phase transitions in TGFT: Functional renormalization group in the cyclic-melonic potential approximation and equivalence to O(N) models. *Journal of High Energy Physics* 12, 159.

Pooley, O. (2013). Substantivalist and relationalist approaches to spacetime. In R. Batterman (Ed.), *The Oxford Handbook of Philosophy of Physics*. Oxford University Press.

Reisenberger, M. P. and C. Rovelli (2001). Space-time as a feynman diagram: The connection formulation. *Classical and Quantum Gravity* 18, 121–140.

Rivasseau, V. (2016). The tensor track IV. *Proceedings of Science CORFU* 2015, 106.

Rovelli, C. (1991). What is observable in classical and quantum gravity? *Classical and Quantum Gravity* 8, 297–316.

Rovelli, C. (2002). Partial observables. *Physical Review D* 65, 124013.

Rovelli, C. and L. Smolin (1995). Discreteness of area and volume in quantum gravity. *Nuclear Physics B* 442, 593–622. Erratum: Ibid. 456:753–754, 1995. http://arxiv.org/abs/gr-qc/9411005.

Rovelli, C. and S. Speziale (2010). On the geometry of loop quantum gravity on a graph. *Physical Review D* 82, 044018.

Sahlmann, H. (2010, January). Loop quantum gravity – a short review. In J. Murugan, A. Weltman, and G. F. Ellis (Eds.), *Foundations of Space and Time*, pp. 185–210. Cambridge University Press.

Smolin, L. (2003). The self-organization of space and time. *Philosophical Transactions of the Royal Society of London A* 361, 1081–1088.

Surya, S. (2012). Evidence for a phase transition in 2D causal set quantum gravity. *Classical and Quantum Gravity* 29, 132001.

Thiemann, T. (2006). Reduced phase space quantization and Dirac observables. *Classical and Quantum Gravity* 23, 1163–1180.

Thiemann, T. (2007). *Modern Canonical Quantum General Relativity*. Cambridge Monographs on Mathematical Physics. Cambridge University Press.

Wilson-Ewing, E. (2019). A relational Hamiltonian for group field theory. *Physical Review D* 99(8), 086017.

Index

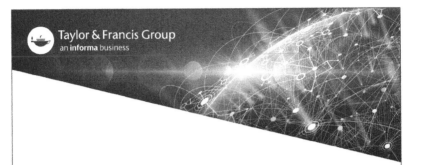

Printed in the United States
by Baker & Taylor Publisher Services